1st International Conference on Aerospace, Mechanical and Mechatronic Engineering (CAMME 2017)

IOP Conference Series: Materials Science and Engineering
Volume 211

Bangkok, Thailand
21 – 23 April 2017

ISBN: 978-1-5108-4413-1
ISSN: 1757-8981

Printed from e-media with permission by:

Curran Associates, Inc.
57 Morehouse Lane
Red Hook, NY 12571

Some format issues inherent in the e-media version may also appear in this print version.

Printed by Curran Associates, Inc. (2017)

For permission requests, please contact the Institute of Physics
at the address below.

Institute of Physics
Dirac House, Temple Back
Bristol BS1 6BE UK

Phone: 44 1 17 929 7481
Fax: 44 1 17 920 0979

techtracking@iop.org

Additional copies of this publication are available from:

Curran Associates, Inc.
57 Morehouse Lane
Red Hook, NY 12571 USA
Phone: 845-758-0400
Fax: 845-758-2633
Email: curran@proceedings.com
Web: www.proceedings.com

Table of contents

Volume 211

International Conference on Aerospace, Mechanical and Mechatronic Engineering
21–23 April 2017, Bangkok, Thailand

Accepted papers received: 14 June 2017
Published online: 27 June 2017

Preface

011001
OPEN ACCESS
International Conference on Aerospace, Mechanical and Mechatronic Engineering

011002
OPEN ACCESS
Peer review statement

Papers

012001
OPEN ACCESS
Measurement of six degree motion error of linear stage

Ryoshu Furutani.....1

012002
OPEN ACCESS
Calculation of phonon dispersion relation using new correlation functional

Ukrit Jitropas and Chung-Hao Hsu.....8

012003
OPEN ACCESS
Simulation for noise cancellation using LMS adaptive filter

Jia-Haw Lee, Lu-Ean Ooi, Ying-Hao Ko and Choe-Yung Teoh.....12

012004
OPEN ACCESS
Pitching stability analysis of half-rotating wing air vehicle

Xiaoyi Wang, Yang Wu, Qian Li, Congmin Li and Zhizhen Qiu.....18

012005
OPEN ACCESS
Selection of the battery pack parameters for an electric vehicle based on performance requirements

M Koniak and A Czerepicki.....26

012006
OPEN ACCESS
A method of computer modelling the lithium-ion batteries aging process based on the experimental characteristics

A Czerepicki and M Koniak.....35

012007
OPEN ACCESS
Integration of internet of things to reduce various losses of jatropha seed supply chain

S P Srinivasan, J Anitha and R Vijayakumar.....42

012008
OPEN ACCESS
Defining and comparing vibration attributes of AlSi10 foam and CFRP coated AlSi10 foam materials

O Çolak and L Yünlü.....48

012009
OPEN ACCESS
Inventory transparency for agricultural produce through IOT

S.P. Srinivasan, D. Sorna Shanthi and Aashish V Anand.....54

012010
OPEN ACCESS
Optimization of dynamic soaring maneuvers to enhance endurance of a versatile UAV

Imran Mir, Adnan Maqsood and Suhail Akhtar.....61

012011
OPEN ACCESS
Incipient loose detection of hoops for pipeline based on ensemble empirical mode decomposition and multi-scale entropy and extreme learning machine

Xiaowei Li, Qin Wei, Yongzhi Qu and Lin Cai.....68

012012
OPEN ACCESS
Progressive failure of composite wind blades with a shear-web spar subjected to static testing

T Y Kam and Y H Chiu.....78

012013
OPEN ACCESS
A fast learning method for large scale and multi-class samples of SVM

Yu Fan and Huiming Guo.....84

012014
OPEN ACCESS
Modeling of ultrasonic processes utilizing a generic software framework

P Bruns, J Twiefel and J Wallaschek.....91

012015
OPEN ACCESS
Kalman filter implementation for small satellites using constraint GPS data

Elmahy M Wesam, Xiang Zhang, Zhengliang Lu and Wenhe Liao.....101

012016
OPEN ACCESS
Low voltage ride-Through strategy based on virtual synchronous generator control

L Guo, J H Su, L Yu and J D Lai.....108

012017
OPEN ACCESS
The realization of the virtual synchronous generator based on the park model

L Yu, J H Su, L Guo and J D Lai.....114

012018
OPEN ACCESS
A maximum power point prediction method for group control of photovoltaic water pumping systems based on parameter identification

B Chen, J H Su, L Guo and J Chen.....121

012019
OPEN ACCESS
Power train and emission control: allocation procedure by OBD-II system for automotive technology

Porag Kalita.....128

012020
OPEN ACCESS
Unbalanced voltage control of virtual synchronous generator in isolated micro-grid

Y Z Cao, H N Wang and B Chen.....134

012021
OPEN ACCESS
Evaluation of acoustical comfort in mosque

M S Prawirasasra and S Mubarok.....141

012022
OPEN ACCESS
Effect of air velocity and direction for indirect evaporative cooling in tropical area

Tri Ayodha Ajiwiguna, Fadhlin Nugraha Rismi and Mukhammad Ramdlan Kirom.....146

012023
OPEN ACCESS
Heterogeneous carbon nano-tube window layer with higher sheet resistance improve the solar cell performance

Khomdram Jolson Singh, Thokchom Jayenta Singh, Dhanu Chettri and Subir kumar Sarkar.....150

012024
OPEN ACCESS
A performance optimization and analysis of graphene based schottky barrier GaAs solar cell

Khomdram Jolson Singh, Dhanu Chettri, Thokchom Jayenta Singh, Terirama Thingujam and Subir kumar Sarkar.....157

012025
OPEN ACCESS
Influence of oxygen flow rate on structural, optical and electrical properties of copper oxide thin films prepared by reactive magnetron sputtering

Thitinai Gaewdang and Ngamnit Wongcharoen.....164

012026
OPEN ACCESS
Tin sulphide prepared by sulphurisation process using metallic tin film precursor obtained from dc magnetron sputtering method

Ngamnit Wongcharoen and Thitinai Gaewdang.....169

012027
OPEN ACCESS
Two-stage collaborative global optimization design model of the CHPG microgrid

Qingfen Liao, Yeyan Xu, Fei Tang, Sicheng Peng and Zheng Yang.....175

012028
OPEN ACCESS
Optimal placement of fast cut back units based on the theory of cellular automata and agent

Jun Yan and Feng Yan.....183

012029
OPEN ACCESS
An energy internet-based distribution system morphology

W Yi, P Wang, X W Hu, X N Wang and F Shi.....189

012030
OPEN ACCESS
Drying characteristics of whole Musa AA group 'Kluai Leb Mu Nang' using hot air and infrared vacuum

C Kulketwong, D Thungsotanon and N Suwanpayak.....195

IOP Conf. Series: Materials Science and Engineering **211** (2017) 011001 doi:10.1088/1757-899X/211/1/011001

PREFACE

The 1st International Conference on Aerospace, Mechanical and Mechatronic Engineering(CAMME 2017) was successfully held at Thammaset University, Bangkok, Thailand on 21st – 23rd April, 2017

This was a combined conference and offered the chance for researchers, academics, students and scientists to share in subjects that are overlapping and complementary in theory and application. There were four respected and international professors that delivered keynote lectures to an erudite and enthusiastic audience.

There papers of high quality from presenters on a global scale, ranged from USA to Europe to Asia and the Far East. A rich fabric of cultures and experiences where many made new and exciting contacts for joint research or collaboration. The quality of the papers was in no small part to the extensive and experience committee and reviewers. For a new conference, it is credit to the long list of those wanting to be affiliated to the research and hard work in reviewing and ensuring the papers are truly international in quality. There was an overriding support for the conference in 2018 and we are delighted that early signs show there will be even more support and presented papers. Exceptional papers were recommended for inclusion in respected Journals.

Lastly, I would personally like to thank the fellow professors that delivered keynote speeches that were interesting and enlightening. In addition, those that planned and supported the conferences to ensure it was a success, in particular, the volunteers, and local organizing committee that added the detail to make a full success. I hope to meet you all next year and learn from your research.

Conference Chair

Professor Ian R. McAndrew PhD FRAeS

Embry Riddle Aeronautical University

UK

IOP Conf. Series: Materials Science and Engineering **211** (2017) 011002 doi:10.1088/1757-899X/211/1/011002

Peer review statement

All papers published in this volume of *IOP Conference Series: Materials Science and Engineering* have been peer reviewed through processes administered by the proceedings Editors. Reviews were conducted by expert referees to the professional and scientific standards expected of a proceedings journal published by IOP Publishing.

IOP Conf. Series: Materials Science and Engineering **211** (2017) 012001 doi:10.1088/1757-899X/211/1/012001

Measurement of six degree motion error of linear stage

Ryoshu Furutani

Tokyo Denki University, School of Engineering, Department of Mechanical Engineering, Senju-Asahi 5, Adachi, 120-8551 Tokyo, Japan

*E-mail: ryo@cck.dendai.ac.jp

Abstract. In this paper, we propose measurement system of the six degree of motion errors which is based on distance measurement by the laser interferometer. The system has six parallel laser beams and six corner cube mirrors on the linear stage, which reflect the corresponding laser beams. The error of axial direction is measured with the ordinary distance measurement method by laser interferometer. The vertical errors to the axial direction and the roll errors are measured by tilted beams using the wedge prism. The pitch and yaw errors are measured by the difference between distance of two corner cube mirrors. The experimental layout of the corner cube mirrors and the other optical devices are shown. As a result, the resolution of 66.5 nm for error of the axial direction, the resolution of 383 nm for the vertical errors to the axial direction, the resolution of 1.23 arc-sec for the roll error and the resolution of 0.429 arc-sec for the pitch and yaw errors are obtained in this system.

1. Introduction

Six Degree-of-freedom (6DOF) motion errors of a linear stage of a machine tool, a coordinate measuring machine and so on should be measured simultaneously. Accurate position and orientation can be obtained and higher compensation can be expected. However, 6DOF motion errors cannot be measured simultaneously because the stages have no enough space to setup conventional measuring instruments such as a displacement sensor and an autocollimator. Therefore, it is expected to develop the measuring instrument that can simultaneously measure the 6DOF motion errors is expected. For these reasons, many researchers have developed the measuring method of the 6DOF motion errors [1-5]. And, we developed the measuring system of 6DOF motion errors that employs the principle of the Michelson interferometer for all direction and rotational angle because of the advantage of constructing high accurate measuring system with comparative ease [6]. In this paper, we propose the 6DOF measurement system using ball lens as the target mirror of the interferometer.

2. Measurement principle

2.1. Optical system

The schematic diagram of the system is shown in Figure 1. The system is based on Michelson interferometer. The moving plate (MP) which consisted of six target mirrors (TM) and two plates, was movable and was mounted on a stage as a measuring object in situ situation. Six ball lenses were used as the target mirrors. The refractive indices of those are nearly 2. To fulfil retro reflection of a laser beam, the ball lenses were coated with silver film in a hemispheric part as shown in Figure 2. The other optical devices were fixed on an optical table. The laser beam emitted from laser source was expanded by two plano-convex lenses. The expanded laser beam was separated into the two laser beams by a beam splitter. One of the two laser beams passed toward TM and was partly tilted by three wedge prisms (WP). The other laser beam passed toward a reference mirror (RM). Each laser beam entered and was reflected by the target mirrors and the RM. Finally, the laser beam reflected from the target mirrors interferes with the laser beam reflected by the RM. When the MP moved, optical path difference between the two laser beams was varied. The displacements of the 6DOF motion errors of MP could be measured by counting the number of the interference fringes due to the two laser beams. The number of the interference fringes could be obtained by photo diodes (PD).

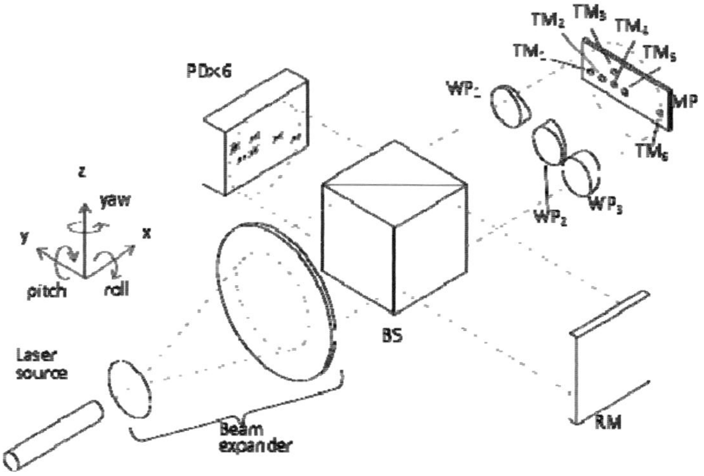

Figure 1. Configulation of optical system of 6DOF laser interferometer.

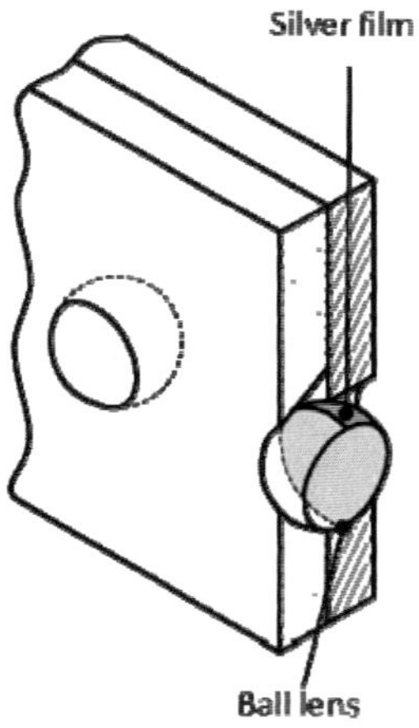

Figure 2. Moving plate(MP).

2.2. Displacement measurement in x direction

The schematic diagram of the principle of displacement measurement in x direction is shown in Figure 3. When MP moves to the x direction with a value of Δx, the relation of the displacement Δx and the number of the interference fringes which are caused by the interference of the laser beams reflected from the TM_4 and the RM is shown as following equation (1).

$$\Delta x = \frac{\lambda}{2} n_4 \tag{1}$$

Here, λ denotes a wave length of the laser beam, n_4 denotes the number of interference fringe obtained by the PD which is irradiated by the reflected laser beam from TM_4.

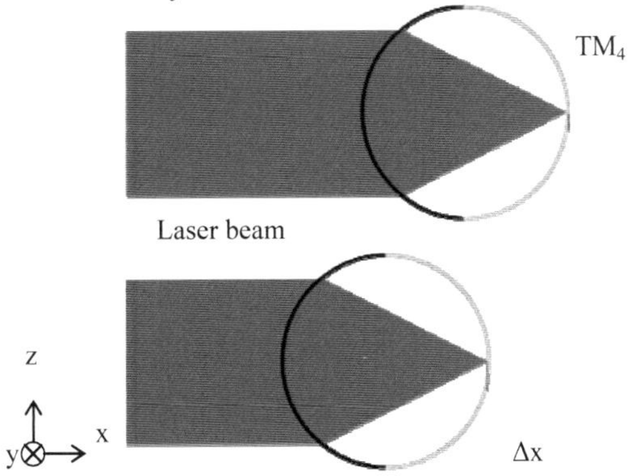

Figure 3. The displacement measurement in x-direction.

2.3. Displacement measurement in y direction

The schematic diagram of the principle of y direction displacement measurement is shown in Figure 4. A part of the laser beam is tilted with the angle θ to the x axis in the x-y coordinate system by the WP_2. When MP moves in the y direction with the value of Δy, the optical path length of the tilted laser beam changes only $\lambda n_5/2$ where n_5 is regarded as the number of interference fringes obtained by the PD which is irradiated by the laser beam reflected from TM_5. The relation of the displacement Δy and the number of the interference fringes which is caused by the interference of laser beams reflected from the TM_5 and the RM is shown as following equation (2).

$$\Delta y = \frac{\lambda}{2\sin\theta} n_5 \tag{2}$$

Here, λ denotes the wave length of the laser beam.

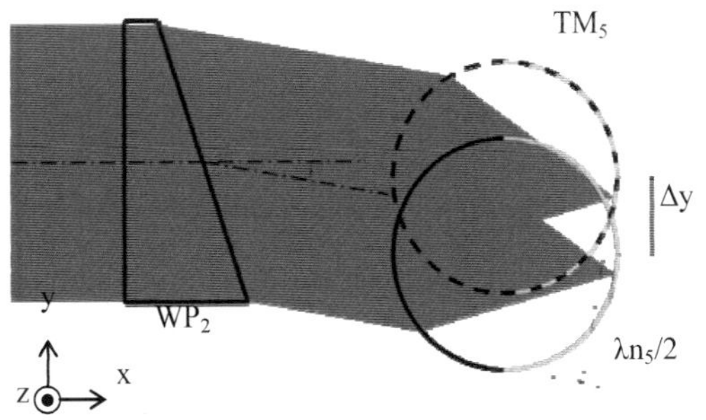

Figure 4. The displacement measurement in y direction.

2.4. Displacement measurement in z direction and roll angle measurement around x axis

The schematic diagram of the principle of z direction and roll displacement measurement is shown in Figure 5. TM_1, TM_6, WP_1 and WP_3 are used in the measurement. The displacement in z direction can be measured by the same method as that in y direction. The displacement in z direction is measured by the both laser beams tilted by the both WP_1 and WP_3 of which primary axes are rotated with 90 degree around x axis. Thus, the displacement Δz and the roll displacement $\Delta\theta_x$ can be measured as the following equation (3) and equation (4).

$$\Delta z = \frac{\lambda}{4\sin\theta}(n_1 + n_6) \tag{3}$$

$$\Delta\theta_x = \tan^{-1}\left[\frac{\lambda}{2L_{1-6}\sin\theta}(n_1 - n_6)\right] \tag{4}$$

Here, λ denotes the wave length of the laser beam, L_{1-6} denotes the distance between TM_1 and TM_6.

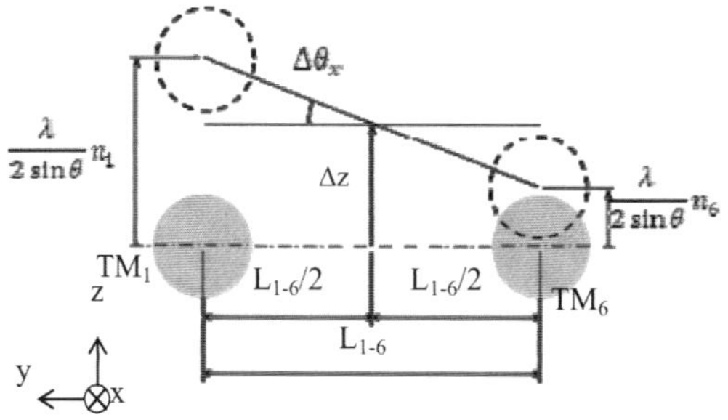

Figure 5. The displacement measurement in z direction and roll angle measurement around x axis.

2.5. Pitch and yaw angle measurement

The schematic diagram of the principle of pitch displacement measurement is shown in Figure 6. The pitch displacement can be calculated by x direction displacement of TM_3 and TM_4 and the distance between TM_3 and TM_4. The x direction displacement of TM_3 and TM_4 can be measured by the same method of Δx mentioned in 2.2. The yaw displacement can be measured as similar as the pitch displacement using TM_2 ad TM_4. Thus, the pitch and yaw displacement $\Delta\theta_y$ and $\Delta\theta_z$ are expressed as following equation (5) and equation (6).

$$\Delta\theta_y = \tan^{-1}\left[\frac{\lambda}{2L_{3-4}}(n_3 - n_4)\right] \tag{5}$$

$$\Delta\theta_z = \tan^{-1}\left[\frac{\lambda}{2L_{2-4}}(n_2 - n_4)\right] \tag{6}$$

Here, λ denotes the wave length of the laser beam, L_{3-4} denotes the distance between TM_3 and TM_4, L_{2-4} denotes the distance between TM_2 and TM_4, n_2, n_3 and n_4 denotes the number of interference fringes obtained by the PD which are irradiated by the laser beam reflected from TM_2, TM_3 and TM_4.

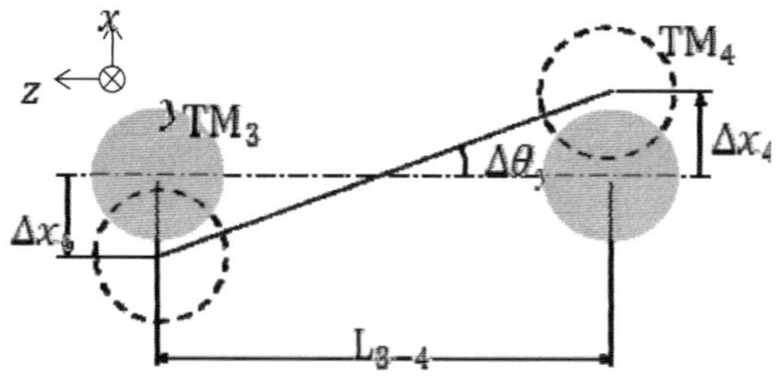

Figure 6. The pitch and yaw angle measurement.

3. Experiment

The specifications of the opthical devices are shown in Table 1. To evaluate the system, displacement of the MP was measured by the system. The displacement of the MP was also measured by 5530 Laser Calibration System manufactured by Agilent Technologies to be compared with the measured value of the system. For the purpose of giving the MP the displacement, the MP was set on positioning stages. P-733.3DD: the 3 axes piezo nanometer positioning stage manufactured by Physik Instrumente was used for the x, y and z direction movement. KGW06100, KGW06075 and KRW06360: the gonio and rotation stages manufactured by SURUGA SEIKI were used for the roll, pitch and yaw angle movements. The specifications of the used positioning stages are shown in Table 2. The number of the interference fringes were calculated by dividing measuring time by a period of a sinusoidal shaped measurement data that was fitted to the function of sine curve using the least squares method in the experiment of translational displacement measurement. In angular displacement measurement, the number of the interference fringes were counted in 0.5 times period of the sine curve because the least squares method couldn't be applied to the deformed measurement data due to low repeatability of the rotational and goniometer stages compared with the resolution of the interferometer. Regarding of translational displacement measurement, 1-10 μm displacements were given to the stage by 1 μm step. Regarding of rotational displacement measurement, the displacements from 50 to 250 arcsec were given to the stages by 50 arcsec step. Once displacement was measured, the stages were returned to the original point. Each displacement was measured five times. The results of the experiments of translational and rotational displacement measurements were shown in Figure 7 and Figure 8. The horizontal axis indicates the measured length by the Laser Interferometer (5530 Laser Calibration System). The vertical axis indicates the measured value of the proposed interferometer. The averages of five times measurements are shown. Let standard deviation be σ, the error bars indicate 3σ,. The results of the movement of 1 μm in the y direction measurement and those of 1 μm and 2 μm in the z direction measurement were not recorded in the Figure 7 because the interferometric fringes are so obscure at these locations that the number of fringes could not be counted. From Figure 7 and Figure 8, linear outputs of the interferometer were obtained. It is proved that the proposed system has the ability to measure the 6DOF motion errors. It is assumed that the errors of the y, z and roll measurements were caused by misalignment of the WP because those errors were larger than the error of x measurement whose system does not employ WP.

Table 1. The specifications of the optical devices.

λ:Wave length of the laser beam	632.57 nm
θ: Tilted angle of the laser beam	10 degree
Refractive indices of the ball lens	2.003
$L_{1\text{-}6}$: Distance between TM_1 and TM_6	48.06 mm
$L_{3\text{-}4}$: Distance between TM_3 and TM_4	7.92 mm
$L_{2\text{-}4}$: Distance between TM_2 and TM_4	8.19 mm

Table 2. The specifications of positioning stages.

	XYZ-Stage [P-733.3DD]		Gonio-Stage [KGW06100]	Gonio-Stage [KGW06075]	θ-Stage [KRW06360]
Direction	x, y	z	roll	pitch	yaw
Resolution	0.1 nm		0.002466 degree	0.0032 degree	0.004 degree
Repeatability	2nm	1 nm	less than 0.003 degree	less than 0.003 degree	less than 0.01 degree

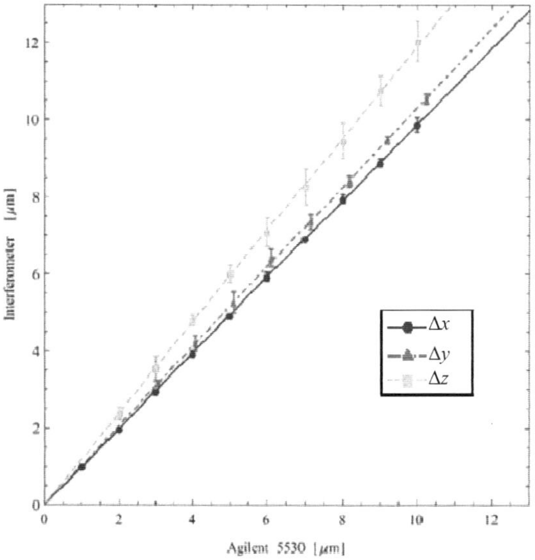

Figure 7. Results of the translation displacements.

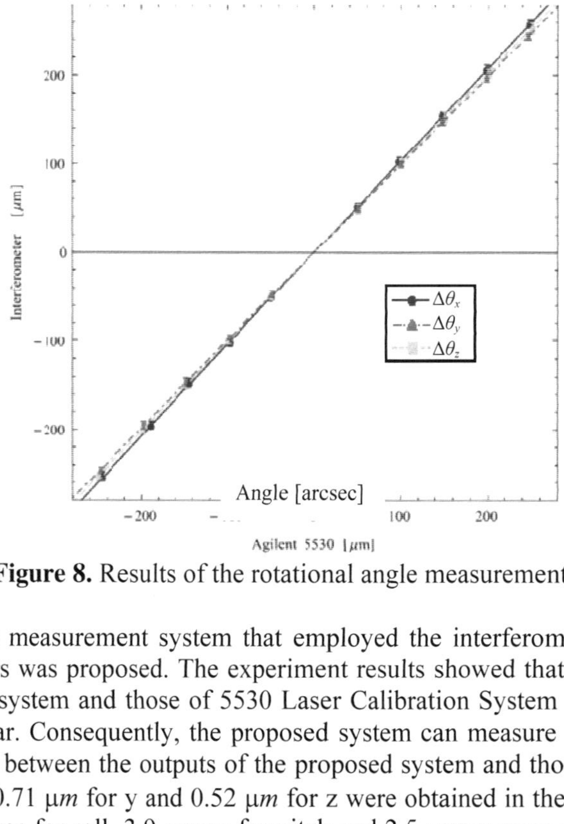

Figure 8. Results of the rotational angle measurements.

4. Conclusion

Six DOF motion errors measurement system that employed the interferometric principle using ball lens as the target mirrors was proposed. The experiment results showed that the relationship between ouputs of the proposed system and those of 5530 Laser Calibration System manufactured by Agilent Technologies were linear. Consequently, the proposed system can measure the 6DOF motion errors. As the maximum errors between the outputs of the proposed system and those of the laser calibration system, 0.09 μm for x, 0.71 μm for y and 0.52 μm for z were obtained in the range of 10 μm travel of each direction. 11.1 arcsec for roll, 3.9 arcsec for pitch and 2.5 arcsec were obtained within the range of 250 arcsec of each rotational angle. as the maximum standard deviation, 0.13 μm for x, 0.18 μm for y, 0.30 μm for z, 3.9 arcsec for roll, 3.6 arcsec for pitch and 2.2 arcsec for yaw were obtained. Due to the interferometric principle of the proposed system, higher resolution and accuracy are expected if the commercial laser interferrometer are employed as the light source of the proposed system.

References

[1] S Shimizu, H S Lee and N Imai 1994 Simultaneous measuring method of table motion errors in 6 degree of freedom, Int. J. Japan Soc. Prec. Eng. **28** p 273

[2] C H Liu, W Y Jywe and C C Hsu 2005 Development of a laser-based high-precision six-degrees-of freedom motion errors measuring system for linear stage *REVIEW OF SCIENTIFIC INSTRUMENT* **76** 055110

[3] Q Feng, B Zhang, C Cui, C Kuang, Y Zhai and F You 2008 Development of a simple system for simultaneously measuring 6DOF geometric motion errors of a linear guide Optics Express **22** 21 25805

[4] K C Fan, M J Chen and W M Huang 1998 A six-degree-of-freedom measurement system for the motion accuracy of linear stages *Int. J. Machine Tools and Manufacture* **38** p 155

[5] K C Fan and M J Chen 2000 A six-degree-of-freedom measurement system for the motion accuracy of linear stages *Precision Engineering* **24** p 15

[6] J Izawa, S Kanno, M Ozaki and R Furutani 2015 Study on detection of 6DOF motion of linear stage *Proc. ISMTII 2015*

Calculation of phonon dispersion relation using new correlation functional

Ukrit Jitropas and Chung-Hao Hsu

School of Manufacturing Systems and Mechanical Engineering, Sirindhorn, International Institute of Technology, Thammasat University, Pathum Thani 12120, Thailand

*Email: u.jitropas@gmail.com; chung-hao@siit.tu.ac.th

Abstract. To extend the use of Local Density Approximation (LDA), a new analytical correlation functional is introduced. Correlation energy is an essential ingredient within density functional theory and used to determine ground state energy and other properties including phonon dispersion relation. Except for high and low density limit, the general expression of correlation energy is unknown. The approximation approach is therefore required. The accuracy of the modelling system depends on the quality of correlation energy approximation. Typical correlation functionals used in LDA such as Vosko-Wilk-Nusair (VWN) and Perdew-Wang (PW) were obtained from parameterizing the near-exact quantum Monte Carlo data of Ceperley and Alder. These functionals are presented in complex form and inconvenient to implement. Alternatively, the latest published formula of Chachiyo correlation functional provides a comparable result for those much more complicated functionals. In addition, it provides more predictive power based on the first principle approach, not fitting functionals. Nevertheless, the performance of Chachiyo formula for calculating phonon dispersion relation (a key to the thermal properties of materials) has not been tested yet. Here, the implementation of new correlation functional to calculate phonon dispersion relation is initiated. The accuracy and its validity will be explored.

1. Introduction

In recent decades, major technology has been transformed into a very small scale, thanks to the large investment in silicon nanofabrication techniques. This trend makes great revolutions in our society. In the applications of thermoelectric devices, we seek for the materials that can conduct electricity easily, but block the motion of heat. In converse, finding better ways to dissipate heat is a key requirement as the devices become smaller and chips get faster. It is important to control heat transfer in these crystalline materials. Unlike bulk material, there are many evidences indicate the violation of Fourier's law in low dimensional lattices [1]. The aim to control heat transport grounds on the understanding of phonon behaviour.

Phonon is the physical particle representing mechanical vibration [2-3]. Above the absolute temperature, even in solid, the atoms are never in the static state. They always oscillate around their equilibrium points. Because atoms are arranged very tightly, all atoms stay very close to their neighbours and are connected to each other by chemical bonds. This can be compared to the spring-mass system, where atoms represent the mass and their chemical bonds represent the spring. When the material is heated, atoms start moving and travel far from their equilibrium points. The atom will exert a force on the neighbour atoms and push them to move. This results in the creation of "phonon", a lattice vibration wave, which is the major heat carrier and responsible for the heat transmission through the semiconductor. On the other hand, the oscillation of phonons with different frequencies can be seen as the special wave of lattice distortion that propagates through the solid. A wave has a fixed relationship between two points in space or at two different times. The wave characteristics of phonons are important for transport processes in nanostructures, where waves will experience the phenomena of reflection and refraction.

In quantum mechanics, Schrödinger equation is generally used to describe the quantum state in terms of wave functions. Phonon dispersion relation (relationship between frequencies and wave vectors) is calculated using ground-state energy obtained from Kohn-Sham equation. This approach takes an advantage of mapping a complex many electrons interaction system to an equivalence non-interaction electron system. Once the phonon dispersion is known, the essential thermal properties such as heat capacity and thermal conductivity can be further calculated. Unfortunately, the correlation energy (a crucial ingredient of Kohn-Sham equation) is an unknown term and we do not have the analytical expression for it. The quality of the correlation energy approximation will have an effect on the accuracy of the modelling system. There are many fitting correlation functionals obtained from parameterizing Ceperley-Alder near exact Monte-Carlo work such as Vosko-Wilk-Nusair, VWN (1980) and Perdew-Wang, PW92 (1992) [4-6]. Nevertheless, these functionals are in a very complex fitting-parameters form (not analytical).

Chachiyo correlation functional is introduced as a latest analytical correlation functional in a very simple form but still provides accurate results compared with the previous approaches. This correlation functional will expand the new area of phonon calculation and encourage more predictive simulation power.

2. Theory

2.1. Density functional theory

Density functional theory (DFT) is the computational technique used in quantum mechanical modelling to determine the ground state electronic structure of the system. According to the theory, properties of the system can be determined using the functional, function of another function. In this case, the energy of the system is in the functional form of the electron charge density $n(r)$.

$$E[n(r)] = F[n(r)] + \int n(r)V(r)\,\mathrm{d}r \qquad (1)$$

where $F[n(r)]$ and $V(r)$ are the universal functional and the external potential respectively.

The electron charge density that minimizes equation (1) coincides with the ground-state energy. The challenge is that the form of the universal function $F[n(r)]$ is not known. Kohn and Sham [7] equation is the useful technique for dealing with this problem by mapping the problem of interacting electrons onto the auxiliary system of non-interaction electrons with an appropriate approximation.

2.1.1. Kohn and sham equation.

In general, if the strength of electron-electron interaction disappears, the universal functional $F[n(r)]$ is the ground-state kinetic energy, as a function of its electron charge density, of the non-interaction electron system. Using Kohn-Sham approach, the ground-state energy in equation (1) is expressed in the following terms:

$$\{-\tfrac{1}{2}\nabla^2 + V_{nuc}(\mathbf{r}) + V_H[n(\mathbf{r})] + V_{XC}[n(\mathbf{r})]\}\psi_i(\mathbf{r}) = \varepsilon_i\psi_i(\mathbf{r}) \qquad (2)$$

where the first three terms are kinetic energy, electron-nucleus interaction and Hartree potential of electron charge density. $V_{XC}[n(r)] = \frac{\delta E_{xc}[n(r)]}{\delta n(r)}$ is exchange-correlation potential and $E_{xc}[n(r)]$ is exchange-correlation energy.

2.1.2. Local-density approximation.

Compared to the total energy, exchange-correlation energy may occupy a little portion but leads to the quality of calculation. This energy is unfortunately the part of energy of the system that does not have the exact expression. The more precise one can do the approximation for this energy, the more accurate the modelling system is. A general and very basic approach used to do this approximation is Local-Density Approximation (LDA) theorem. It assumes that the exchange-correlation energy of the system is similar to the exchange-correlation energy of an equal volume of the homogeneous electron gas (HEG) with the same electron charge density. Using this assumption, the exchange-correlation energy is given as the following:

$$E_{xc}[n(r)] = \int n(r)\{\varepsilon_x[n(r)] + \varepsilon_c[n(r)]\}dr \qquad (3)$$

where $\varepsilon_x[n(r)]$ and $\varepsilon_c[n(r)]$ are exchange and correlation energy per particle respectively.

LDA is exact when the electron density is uniform or slowly varies. For the system that the electron density changes rapidly, this technique may give a rough approximation. Although LDA is said to be the simplest approach to do the exchange-correlation energy approximation, it works very well and provides good results in many situations beyond its limit.

2.1.3. Exchange-correlation energy.

Exchange energy, $\varepsilon_x[n(r)]$, is the energy reduction of the Coulomb energy caused by the antisymmetric wave functions due to Pauli exclusion principle. The exchange-energy density of HEG is known analytically [8].

Correlation energy, $\varepsilon_c[n(r)]$, is the difference between the actual energy of a particular energy level of an atom and the calculated energy by the Hartree-Fock approximation. Except for high and low density limits, there is no exact expression form of correlation energy. Therefore the parameterization functional forms that fit to the near-exact Quantum Monte Carlo results of Ceperley and Alder such as VWN and PW are usually implemented.

2.2. Phonon dispersion relation

After doing various approximations, we can obtain the total energy. In addition, phonon frequencies corresponding to their wave vectors in different directions can be further calculated using dynamics matrix equation (4). Phonon dispersion relation is therefore ready to plot.

$$\det\left|\frac{1}{\sqrt{M_s M_t}}\tilde{C}_{st}^{\alpha\beta}(\boldsymbol{q}) - \omega^2(\boldsymbol{q})\right| = 0 \qquad (4)$$

where $\tilde{C}_{st}^{\alpha\beta}(\boldsymbol{q}) = \frac{1}{N_c}\frac{\partial^2 E}{\partial u_s^{*\alpha}(\boldsymbol{q})\partial u_t^{*\beta}(\boldsymbol{q})}$ are the inter-atomic force constants (IFCs) in reciprocal space.

3. Method

The typical correlation functionals are in a complex form and difficult to implement. As the latest approximation form, the Chachiyo correlation functional [9] is introduced in a very simple form. Different from VWN and PW correlation functionals that aim to fit with the near exact Monte Carlo data, Chachiyo correlation functional is derived independently from the second-order Møller-Plesset perturbation theory (MP2) [10] as shown below.

$$\epsilon_c(r_s) = a\ln\left(1 + \frac{b}{r_s} + \frac{b}{r_s^2}\right) \qquad (5)$$

where $r_s = (\frac{4\pi\rho}{3})^{-1/3}$ is the Wigner-Seitz density parameter, ρ is the electron density, a and b are known constants from high density limit.

Interestingly, this formula provides an accurate approximation of correlation energy compared to the Monte Carlo result and the complex VWN fitting form. Chachiyo correlation functional is therefore initiated here to implement in equation (3) and further calculate phonon dispersion relation. All phonon calculation is obtained from QUANTUM ESPRESSO package, which is widely used in the nanoscale simulation research area. Unfortunately, there is no open source code of pseudopotential generation that implements Chachiyo correlation functional yet because it is a new correlation functional. Chachiyo formula has to be manually inserted into LibXC [11], which is the database used in the calculation. Pseudopotential and phonon dispersion relation are then generated by APE (atomic pseudopotentials engine) and QUANTUM ESPRESSO package respectively [12-13].

4. Expected result

Good convergence and transferability of pseudopotential obtained from Chachiyo formula is anticipated. The validity of this new analytical correlation functional to extend the use of local density approximation within phonon calculation from fitting to first principle approach will be explored. It is expected to provide a good agreement with experimental data and a comparable modelling result with the typical functionals. Frist principle method will encourage more prediction power in the modelling system. The simplicity of this formula is hoped to reduce the calculation time for complex crystal.

5. Conclusion and future plan

Chachiyo correlation functional is initiated to implement the calculation of phonon dispersion relation. The key advantages of this formula are the first principle approach and the simplicity form. The phonon dispersion relation obtained can be used to further calculate and investigate important thermal properties such as heat capacity and thermal conductivity. Alternatively, besides phonon calculation, Chachiyo formula is also applicable to the calculation of electronic band structure and density of states of materials in future work.

References

[1] Yang N, Xu X, Zhang G and Li B 2012 *AIP Advances* **2** p 041410
[2] Gorishnyy T, Maldovan M, Ullal C and Thomas E 2005 *Physics World* **18** p 2
[3] Baroni S, Gironcoil De S, Corso A Dal and Giannozzi P 2001 *Reviews of Modern Physics* **73** p 515
[4] Ceperley D M and Alder B J 1980 *Phys. Rev. Lett.* **45** p 566
[5] Vosko S H, Wilk L and Nusair M 1980 *Can. J. Phys.* **58** p 1200
[6] Perdew J P and Wang Y 1992 *Phys. Rev.* B **45** pp13-244
[7] Kohn W and Sham L J 1965 *Phys. Rev.* **140** p A1133
[8] Slater J C 1951 *Phys. Rev.* **81** p 385
[9] Chachiyo T 2016 *J. Chem. Phys.* **145** p 021101
[10] Møller C and Plesset M S 1934 *Phys. Rev.* **46** p 618
[11] Marques M A, Oliveira M J and Burnus T 2012 *Computer Physics Communications* **183** p 2272
[12] Oliveira M J T and Nogueira F 2008 *Computer Physics Communications* **178** p 524
[13] http://www.quantum-espresso.org/

Acknowledgment

This work was supported by SIIT Young Researcher Grant under Contract No. SIIT2014-YRG4 and the Graduate Scholarship for Excellent Thai Students of Sirindhorn International Institute of Technology, Thammasat University, Thailand.

Simulation for noise cancellation using LMS adaptive filter

Jia-Haw Lee[1], Lu-Ean Ooi[1] *, Ying-Hao Ko[2] and Choe-Yung Teoh[2]

[1]School of Mechanical Engineering, Universiti Sains Malaysia, 14300 Nibong Tebal, Pulau Pinang, Malaysia
[2]Department of Mechanical Engineering, Faculty of Engineering and Built Environment, Tunku Abdul Rahman University College (TAR UC), Malaysia

*Email: ooiluean@usm.my

Abstract. In this paper, the fundamental algorithm of noise cancellation, Least Mean Square (LMS) algorithm is studied and enhanced with adaptive filter. The simulation of the noise cancellation using LMS adaptive filter algorithm is developed. The noise corrupted speech signal and the engine noise signal are used as inputs for LMS adaptive filter algorithm. The filtered signal is compared to the original noise-free speech signal in order to highlight the level of attenuation of the noise signal. The result shows that the noise signal is successfully canceled by the developed adaptive filter. The difference of the noise-free speech signal and filtered signal are calculated and the outcome implies that the filtered signal is approaching the noise-free speech signal upon the adaptive filtering. The frequency range of the successfully canceled noise by the LMS adaptive filter algorithm is determined by performing Fast Fourier Transform (FFT) on the signals. The LMS adaptive filter algorithm shows significant noise cancellation at lower frequency range.

1. Introduction

Random fluctuation of the sound wave is known as noise. The presence of noise in a signal alters the original signal and results in inaccurate transmission of the desired signal. It is an issue in many applications such as headphones, engines, pumps and etc. One of the examples is the noise generated by the jet turbine engine that affects the comfort level of the airlines passengers [1]. Another example is the noise presence in hearing aids due to acoustic feedback effect in which reduces the performance of the hearing aids [2]. Due to the concern on the impact of noise on human, many researches have been done in aiming to eliminate the noise in different applications.

Several methods have been proposed in noise cancellation. The traditional noise reduction technique is based on passive noise control, such as using earplugs, ear-protector, sound insulation walls and muffler [3]. Passive noise cancellation (PNC) uses physical object that installed in the system to isolate the background noise from the surrounding. PNC techniques are effective in reducing noise over a wide frequency range. Chen and Yijun (2016) reduced the tonal noise from the centrifugal fan Xu and Mao have implemented PNC in forward-curved centrifugal fan by 5dB after employing the open-cell metal foam [4]. However, the major restriction of passive noise cancellation is the isolation object is relatively big in size and unable to be fitted into small scale applications. It is also limited by the effective frequency range. Thus, active noise cancellation (ANC) is developed in order to cover a wider range of the noise frequencies especially [5]. ANC employs an electro-acoustic system to cancel the primary noise based on the principle of superposition, where an anti-noise (secondary signal) of equal amplitude but with antiphase is generated and combined with the primary noise in achieving the cancellation of noise [6]. ANC has greater advantages over PNC in terms of the capability to attenuate the low frequencies noise due to the presence of embedded control system which is able for the real time situation and well as low cost. ANC had been used in the microcontroller and achieved the cancellation of 20dB to 25dB of the 60Hz periodic noise component in the electronics signal measurement [7]. However, the implementation of ANC is complicated although it has high reliability in cancelling noise.

There are several algorithms that can be utilized in noise cancellation such as Least Mean Square Algorithm (LMS) and Filtered-x Least Mean Square Algorithm (FXLMS). LMS algorithm was introduced by Widrow and Hoff in 1960. It uses gradient-based method of steepest decent and updating coefficients of an adaptive filter [8]. The LMS algorithm is classified as adaptive filtering in which is a self-designing and time-varying system that continuously adjust its tap weight in the algorithm [9]. The LMS algorithm is implemented in aiming to minimize the noise in the input signal and producing a noise-free output. Sukhpreet and Sukhwinder used LMS algorithm in suppressing the effect of acoustic noise in the speech signals. The results show that LMS algorithm provides a greater signal-to-noise ratio value and suppressed the effect of acoustic noise [10]. In contrary, the FXLMS algorithm requires a filtered version of the reference signal as input in which the filter is having the same impulse response as the cancellation path. FXLMS had been used on cancelling the periodic noise generated by laptop fan after identification [11]. Recently, Chang and Chu (2013) developed the variable tap length and step size FXLMS in improving the convergence rate and noise reduction ratio in ANC [12]. Yang et al. (2014) reduced 8dB noise level in the cabin of high-speed elevator by implementing modified FXLMS algorithm [13]. However, the FXLMS algorithm has higher complexity compared to LMS algorithm due to the consideration of the secondary path parameter.

In this paper, the simulation of noise cancellation using LMS adaptive filter in MATLAB software is presented. The noise corrupted speech signal is used as the input for the developed LMS adaptive filter algorithm. The filtered signal of the LMS adaptive filter algorithm and the reference signal which is the noise-free speech signal are compared to examine the effectiveness of the LMS adaptive filter in cancelling the noise. Fast Fourier Transform (FFT) is then applied to both the signals in order to identify the frequency range successfully cancelled.

2. LMS adaptive filter algorithm

The LMS adaptive filter algorithm that developed in this study is shown in Figure 1. The parameters y_1 and y_2 are the inputs of the algorithm in the form of column vector. This is study y_1 is the noise corrupted signal and y_2 is the noise signal. The parameter $W(k)$ is the column weight vector of the filter at k^{th} time, which is used in the algorithm to update the subsequent column weight vector and can be represented in the equation (1) below:

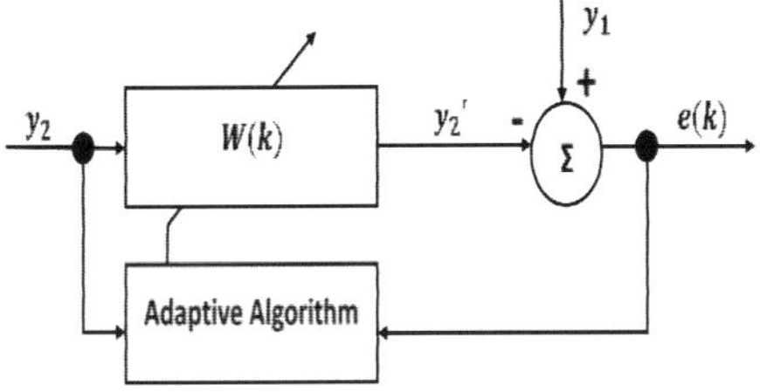

Figure 1. LMS adaptive filter algorithm.

$$W(k) = \begin{bmatrix} W_0(k) \\ W_1(k) \\ \vdots \\ W_{L-1}(k) \end{bmatrix} \tag{1}$$

The notation is the length of the filter in which is equal to the length of the input and. The weighted-noise signal y_2' is given by the multiplication of the column weight vector and the noise signal. The error *e(k)* at k^{th} time is then defined as the difference between the noises corrupted signal and the weighted-noise signal as shown the equation (2) as below. The notation is the transpose of the vector.

$$\begin{aligned} e(k) &= y_1(k) - y_2'(k) \\ &= y_1(k) - [W^T(k)y_2(k)]^T \end{aligned} \tag{2}$$

The adaptive algorithm is representing by the update of the column weight vector after each iteration and is given by equation (3). The parameter μ is the step-size that determines the convergence and stability of the filter.

$$W(k+1) = W(k) + \mu e(k) y_2(k) \tag{3}$$

At here, the LMS adaptive filter algorithm will be terminated when column weight vector iterates up to the length of filter and the final output of the algorithm is *e(k)*.

3. Methodology

Two audio files of Waveform Audio File Format (wav.) are used as input for the simulation model, in which the simulation model is developed in MATLAB software environment. These two inputs are of single channel (Monophonic sound reproduction) and with the same sampling rate (44100 Hz). Input y_1 is the audio signal of a noise corrupted speech signal (combination of speech and engine noise sound) and Input y_2 is the audio signal of the noise signal (engine noise sound). Another audio signal which is the speech sound signal (noise-free speech signal) is used as reference signal for comparison with the filtered output of the algorithm, *e(k)*. The algorithm is initiated by setting the column weight vector *W(k)* and the filtered output *e(k)* as a zero column vector as shown in equation (4) and equation (5) respectively.

$$W(k) = \begin{bmatrix} W_0(k) \\ W_1(k) \\ \vdots \\ W_{L-1}(k) \end{bmatrix} = \begin{bmatrix} 0 \\ 0 \\ \vdots \\ 0 \end{bmatrix} \tag{4}$$

$$e(k) = \begin{bmatrix} e_0(k) \\ e_1(k) \\ \vdots \\ e_{L-1}(k) \end{bmatrix} = \begin{bmatrix} 0 \\ 0 \\ \vdots \\ 0 \end{bmatrix} \tag{5}$$

The initial value for the μ is taken as any random number with the condition of $0<\mu<1$ [14]. Both the Input y_1 and Input y_2 then inserted to the developed LMS adaptive filter algorithm. Input y_2 is multiplied by the column weight vector *W(k)* to generate a weighted-noise signal (y_2'). The intermediate output y_2' is then subtracted from the Input y_1 to obtain the error *e(k)*. After each interaction, the column weight vector would adaptively updates. The filtering process would be continuing until it reaches the prescribed filter length L, which is same as the length of the inputs. The absolute difference between the filtered output *e(k)* of the LMS adaptive filter algorithm and the input y_1 (noise corrupted speech signal) are calculated. Fast Fourier Transform (FFT) is applied on both *e(k)* and y_1 to obtain the frequency function of the output. The percentage of difference for the reference signal, noise corrupted speech signal and filtered signal are calculated to compare the recovery of the speech signal from the noise corrupted speech signal upon implementing the LMS adaptive filter algorithm. Figure 2 shows the idea of the LMS adaptive filter algorithm in filtering the engine noise from the speech signal.

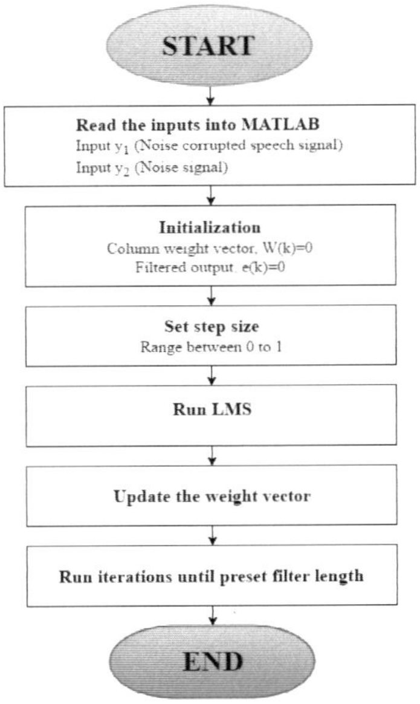

Figure 2. Flow chart of the LMS adaptive filter algorithm implementation.

4. Results and discussions

The simulation using the LMS adaptive filter algorithm attenuate the noise gradually and finally cancelled the engine noise in the noise corrupted speech signal. The results show the clean speech signal is obtained after the LMS adaptive filter is applied. The obtained clean speech signal shows low deviation with the original-free speech signal. Figure shows the signal waveform of the noise-free speech signal, engine noise signal and the noise-corrupted speech signal that had been used in this study. The signal waveform in Figure 3(a) is the noise-free speech signal in which the signal amplitude that approaches zero at intervals indicates a momentary pause in between the words during the speech. Figure 3(b) is the engine noise signal where the waveform pattern indicates that the engine noise signal can be generally classified as periodical noise due to the repetitive waveform. Figure 3(c) is the input for the simulation model which is the noise corrupted speech signal. This signal consists of the fluctuation caused by the engine noise signal and the speech signal is also altered by the noise signal. From Figure 3(c), it is difficult to identify the speech signal by analyzing the waveform.

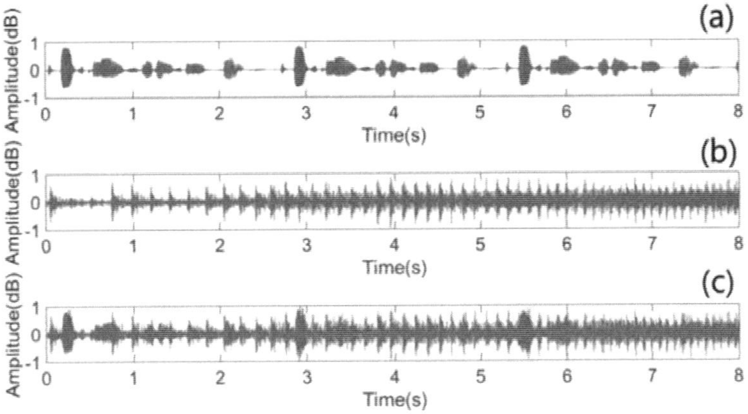

Figure 3(a). Noise-free speech signal **Figure 3(b).** Engine noise signal and **Figure 3(c).** Noise corrupted speech signal.

Figure 4 shows the comparison in between the noise corrupted speech signal and the LMS adaptive filter algorithm filtered output. It is clear that the implementation of the LMS adaptive filter algorithm on the noise corrupted speech signal reduces the engine noise signal significantly. The amplitude and the noise of the signal are reduced. This filtered output single will then compare with the original clean speech signal (noise-free speech signal). This comparison shows that the implementation of the LMS adaptive filter algorithm able to attenuate the engine noise which is classified as periodical noise effectively.

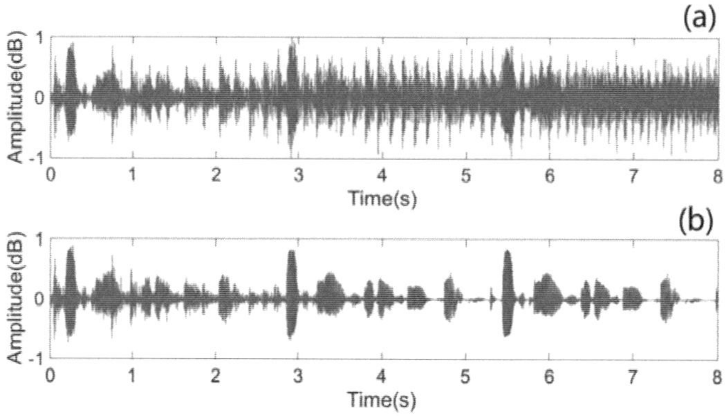

Figure 4(a). Noise corrupted speech signal and **Figure 4(b).** Filtered output.

Figure 5 below shows the comparison of the filtered signal with the original clean speech signal (noise-free speech signal). From Figure 5(c), the absolute difference between the noise-free reference signal and the filtered signal decrease with time. High difference is found at the initial filtering process. This is because at the beginning stage, the column weight vector is set to be a zero column vector. The algorithm is unable to approximate the noise signal accurately and thus results in higher error in between the original clean speech signal and the filtered signal. As the algorithm continues, the column weight vector adaptively adjusts the weight values to approximate the noise signal and thus the difference in between two signals reduced along with the iterations. The reduction implies that the LMS adaptive filter algorithm has recovered the speech signal from the noise-corrupted speech signal. The signal obtained from the simulation is folded by using Fast Fourier Transform as negative frequencies do not exist in practical. The frequency function is obtained from 0-5000Hz. The percentage difference base on the standard deviation in between the filtered signal and the original clean speech signal is 8.9% for this frequency range. The small percentage is similar with the finding of Mendiratta and Jha (2014) where the noise was suppressed with little signal distortion [15].

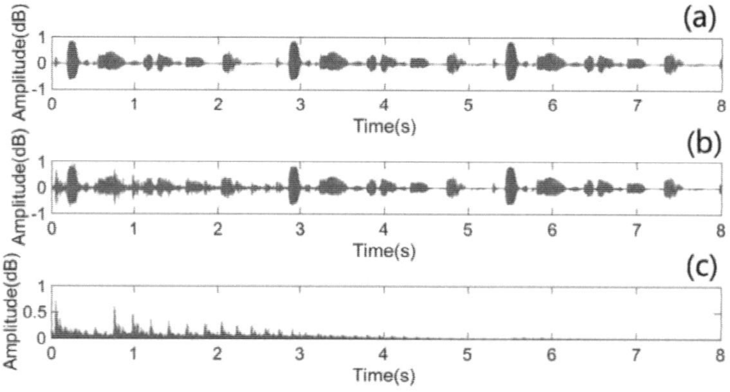

Figure 5(a). original clean speech signal **Figure 5(b).** Filtered signal and **Figure 5(c).**

Difference between original clean speech signal and filtered signal.

5. Conclusion

In this paper, the simulation model of noise cancellation using LMS adaptive filter algorithm has been presented. The simulation model shows that the LMS adaptive filter algorithm is capable to update the column weight vector adaptively in approximating the noise signal to be cancelled. The implementation of LMS adaptive filter algorithm on the noise corrupted speech signal has proven that the adaptive filter can effectively remove the noise signal at lower frequencies range (0Hz to 5000Hz). The declination in the difference between the reference signal and the filtered signal has proven that speech signal is recovered from the noise corrupted speech signal. The filtered speech signal recovered from the noise corrupted speech signal has low percentage difference which is 8.9% compared to the noise-free speech signal. This shows the performance of the LMS adaptive filter in filtering the noise signals.

References

[1] Brett RCM, Marion B and Daniel K 2013 The use of noise cancelling headphones to improve concurrent task performance in a noisy environment *Applied Acoustics* **74(1)** pp 110-15

[2] Kim N, Toon VW, Mads GC and Marc M 2013 Improved prediction error filters for adaptive feedback cancellation in hearing aids *Signal Processing* **93(11)** 3062-75

[3] Kajikawa Y, Gan WS and Sen MK 2012 Recent advances on active noise control: open issues and innovative applications *APSIPA Transactions on Signal and Information Processing* **1(3)**

[4] Chen X and Yijun M 2016 Passive control of centrifugal fan noise by employing open-cell metal foam *Applied Acoustics* **103** Part A pp 10-9

[5] Hu S, Rajamani R and Yu X 2013 Directional cancellation of acoustic noise for home window applications *Applied Acoustics* **74(3)** 467-77

[6] Wu L, Qiu X and Guo Y 2014 A simplified adaptive feedback active noise control system *Applied Acoustics* **81** pp 40-6

[7] Wesson KD, Ochshorn RM and Land BR 2009 Low-cost, high-fidelity, adaptive cancellation of periodic 60 Hz noise *J Neurosci Methods* **185(1)** pp 50-5

[8] Widrow B and Hoff ME Adaptive switching circuits *IRE WESCON Convention Record* **4** pp 96–104

[9] Hamidia M and Amrouche A 2016 Improved variable step-size NLMS adaptive filtering algorithm for acoustic echo cancellation *Digital Signal Processing* **49** pp 44-55

[10] Sukhpreet S and Sukhwinder SS 2012 Acoustic Noise Cancellation Using Block LMS FIlter In MATLAB SIMULINK *International Journal of Computer Science* **3(1)** pp 183-186

[11] Cordourier-Maruri HA and Orduna-Bustamante F 2009 Active Control of Periodic Fan Noise in Laptos: Spectral Width Requiremnts in Delayed Buffer Implementation *Journal of Applied Research and Technology* **7(2)** pp 124-35

[12] Chang DC and Chu FT 2013 Active noise cancellation with a new variable tap length and step size FXLMS algorithm *IEEE International Conference on Multimedia and Expo*

[13] Yang IH, Jeong JE, Jeong UC, Kim JS and Oh JE 2014 Improvement of noise reduction performance for a high-speed elevator using modified active noise control *Applied Acoustics* **79** pp 58-68

[14] Apolin JA, Diniz PSR, Laakso TI and Campos MLR 1998 Step-size optimization of the BNDR-LMS algorithm *9th European Signal Processing Conference*

[15] Mendiratta A and Jha D 2014 Adaptive Noise Cancelling for audio signals using Least Mean Square algorithm *International Conference Electronics Communication and Instrumentation (ICECI)*

Acknowledgment

The author would like to gratefully acknowledge Universiti Sains Malaysia (USM) Short Term Grant [account no. 60313028] for the financial funding on this project.

IOP Conf. Series: Materials Science and Engineering **211** (2017) 012004 doi:10.1088/1757-899X/211/1/012004

Pitching stability analysis of half-rotating wing air vehicle

Xiaoyi Wang*, Yang Wu, Qian Li, Congmin Li and Zhizhen Qiu

School of Mechanical Engineering, Anhui University of Technology, Ma'anshan, 243002, P.R.China

*E-mail: wangxy@ahut.edu.cn

Abstract. Half-Rotating Wing (HRW) is a new power wing which had been developed by our work team using rotating-type flapping instead of oscillating-type flapping. Half-Rotating Wing Air Vehicle (HRWAV) is similar as Bionic Flapping Wing Air Vehicle (BFWAV). It is necessary to guarantee pitching stability of HRWAV to maintain flight stability. The working principle of HRW was firstly introduced in this paper. The rule of motion indicated that the fuselage of HRWAV without empennage would overturn forward as it generated increased pitching movement. Therefore, the empennage was added on the tail of HRWAV to balance the additional moment generated by aerodynamic force during flight. The stability analysis further shows that empennage could weaken rapidly the pitching disturbance on HRWAV and a new balance of fuselage could be achieved in a short time. Case study using numerical analysis verified correctness and validity of research results mentioned above, which could provide theoretical guidance to design and control HRWAV.

1. Introduction

Flight stability must be carefully considered in the design of air vehicle to ensure safety and comfort of flight. With the development of various types of air vehicle, lots of researchers have carried out a series of studies on the stability of the aircraft. Young attempted to add tail antenna to increase the rolled stability[1]. Nakata proposed a suitable spanwise length to improve the flight stability[2]. Zhang investigated single-DOF pitching angular motion of an orbital reentry vehicle around static trim angle of attack[3].

Due to its excellent flight mobility and wide application fields, the research and development of Bionic Flapping Wing Air Vehicle (BFWAV) has become a new hot spot in recent years. The attitude of Microbat by California Institute of Technology was controlled through the flat tail and vertical tail[4]. Nano Hummingbird could achieve hovering, tumbling and other difficult movements without empennage[5]. However, there have been few studies focused on the flight stability of BFWAV. As an important factor of flight stability, pitching stability of BFWAV was also rarely mentioned in the existing literature.

Half-Rotating Mechanism(HRM) is a new type of bionic mechanism developed by our work team[6,7]. In order to meet the requirements of the bionic flying, the simplified HRM was further designed by adding a local constraint[8,9] to reduce weight. The weis-fogh effect of Half-Rotating Wing(HRW) based on HRM has been also discovered, which was helpful to generate larger lift for HRW[10]. Moreover, based on analyzing the lift of the flapping wing flight, Wang and Dong proposed an estimation method of lift of HRW[11]. Half-Rotating Wing Air Vehicle (HRWAV) is similar as

BFWAV, which used rotating-type flapping instead of oscillating-type flapping. For HRWAV, the flight stability was obviously an unavoidable problem. Based on introducing the working principle of HRW, the study on HRWAV in this paper was to focus on how to restrain the disturbance and improve the pitching stability, which was helpful to design and control HRWAV.

2. Working principle of HRW

The core component of HRWAV is HRM.

HRM as shown in figure 1 is a new kind of bionic motion mechanism as mentioned before. *AB* and *OC* respectively represented the bar of HRM and crank, while point *C* is the position of spindle. The bar of HRM would rotate the angle of $\frac{\varphi}{2}$ with angular velocity of $\frac{\omega}{2}$ counterclockwise when the crank rotated the angle of φ with angular velocity of ω counterclockwise from the initial position.

A wing panel was fixed on the bar of HRM, which was called HRW in figure 2. In the period of rotation, different acting forces acted by fluid on the wing of HRM resulted from the changing velocities of the bar at various positions. At this moment, the acting force on the wing of HRM could be simplified to the point *M*. F_R was the principal vector which was upward and vertical to the wing surface, while M_C represented the principal moment which was rotating around the spindle *C*.

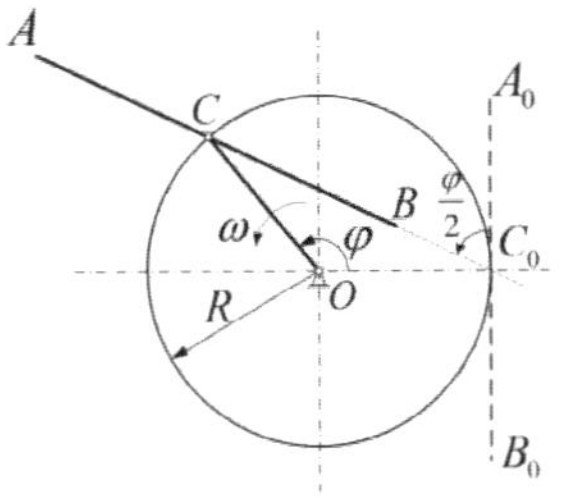

Figure 1. Principal structure of HRM.

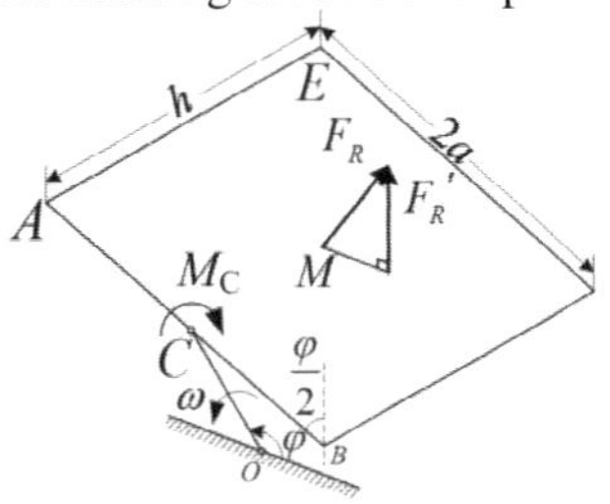

Figure 2. Working force of HRW.

Therefore, the principal vector could be represented as equation (1):

$$F_R = \frac{1}{6} C_D \rho h R \omega^2 \sin\frac{\varphi}{2}(4R^2 \sin^2\frac{\varphi}{2} + 3a^2) \tag{1}$$

where ρ was air density, C_D was the resistance coefficient of the wing panel affected by air.

F_R' was the component force of F_R in perpendicular direction.

Due to the varied direction and size of principal vector with the rotation of the bar, HRW could be symmetrically arranged in the right position. As shown in figure 3, the crank on the right rotated clockwise as well as the crank on the left rotated counterclockwise with the same angular velocity, so that the component force of both sides counteracted in horizontal direction and doubled in perpendicular direction. F_z' was twice as F_R'.

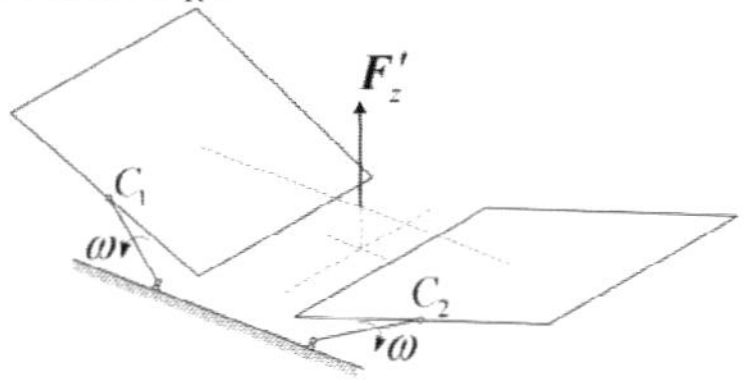

Figure 3. Symmetrical arrangement of HRW.

F_z' was acting on the middle of the line that connected the geometrical center of both wing panels, and its direction was perpendicular to the ground. This was the flight precondition for HRW.

3. Movement rule of HRWAV without empennage

3.1. HRWAV without empennage and stress simplifying

It is necessary to keep the HRWAV staying a pitching angle for generating component forces both in perpendicular and horizontal direction in order to rise and go ahead.

HRWAV is composed of empennage and fuselage that contains HRW. Figure 4 denoted the status of HRWAV without empennage in the air. Point O' was the location of the center of mass of HRWAV, point D_0 and point H respectively represented the midpoint of connecting line of the geometrical center on the left and right wing and the geometrical center of the head of fuselage. If the air resistance resulted from airflow speed being neglected, the force acting on HRWAV without empennage essentially contained gravity P_f and the principal vector of aerodynamic force F_z'. It could be seen clearly that, the fuselage of HRWAV would pitch resulted from the torque on point O' by F_z'.

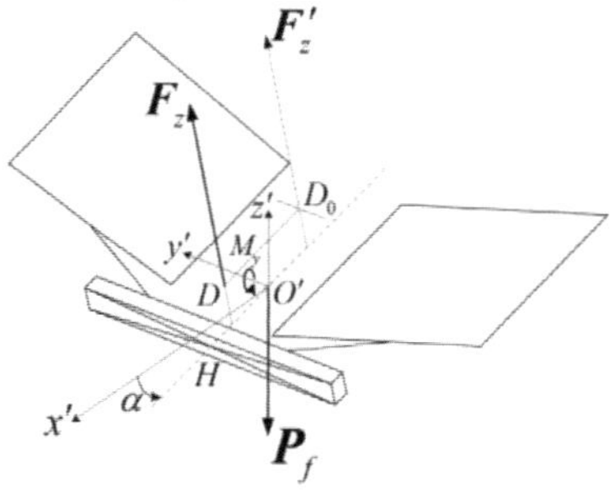

Figure 4. The body of HRWAV without empennage and its force.

F_z' was moved to point D along the longitudinal axis of fuselage and became F_z, which generated M_y simultaneously. F_z' and M_y were deduced as equation (2) and equation (3):

$$F_z = \frac{1}{3} C_D \rho h R \omega^2 \sin^2 \frac{\varphi}{2} (4R^2 \sin^2 \frac{\varphi}{2} + 3a^2) \tag{2}$$

$$M_y = \frac{1}{2} F_z h \tag{3}$$

Equation (2) can be written as equation (4):

$$F_z = \frac{1}{6} C_D \rho h R \omega^2 [R^2 \cos 2\varphi - (4R^2 + 3a^2)\cos\varphi + 3(R^2 + a^2)] \tag{4}$$

3.2. Pitching motion analysis of HRWAV without empennage

The force of HRWAV without empennage in the longitudinal symmetry plane was shown in figure 5.

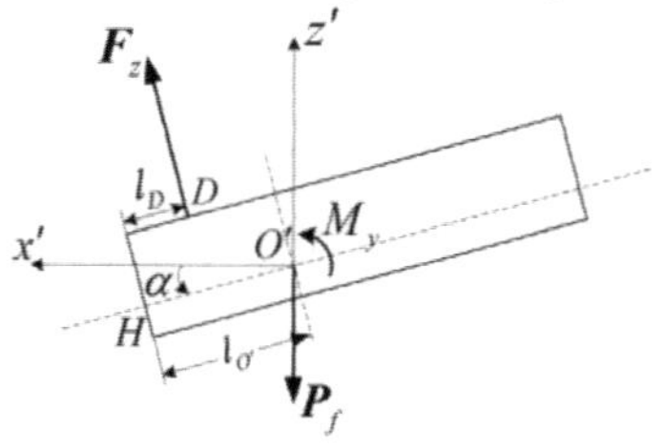

Figure 5. Force of HRWAV without empennage in the longitudinal symmetry plane.

Based on the analysis of rigid body motion, the differential equations of motion were expressed as equation (5), (6) and (7):

$$m_f \ddot{x} = F_z \sin\alpha \tag{5}$$

$$m_f \ddot{z} = F_z \cos\alpha - m_f g \tag{6}$$

$$I_y \ddot{\alpha} = M_y - F_z (l_{O'} - l_D) \tag{7}$$

where m_f represented the mass of fuselage, I_y was the moment of inertia of fuselage and $\ddot{\alpha}$ was the acceleration of pitching angle.

$\ddot{\alpha}$ was written into the function of φ and M_y in equation (7) was substituted by equation (3), then equation (7) could be got as equation (8):

$$I_y\omega^2\alpha'' = F_z\left[\frac{h}{2}-(l_{O'}-l_D)\right] \tag{8}$$

Equation (4) was put into equation (7) and α could be worked out in equation (9):

$$\alpha = -\frac{CR^2}{4I_y}\cos 2\varphi + \frac{C}{I_y}(4R^2+3a^2)\cos\varphi + \frac{3C(R^2+a^2)}{2I_y}\varphi^2 + D_1\varphi + D_2 \tag{9}$$

Equation (9) was a generalized expression about pitching angle α, which was calculated as equation (10):

$$\alpha = -\frac{CR^2}{4I_y}\cos 2\varphi + \frac{C}{I_y}(4R^2+3a^2)\cos\varphi + \frac{3C(R^2+a^2)}{2I_y}\varphi^2 - \frac{3C}{4I_y}(5R^2+4a^2) \tag{10}$$

The third item on the right of equation (9) was changed when φ was growing over time, which caused the increase of pitching angle α that make HRWAV without empennage turn over from the initial static state. Hence, it was the prerequisite that was adding other devices on HRWAV for restraining enlargement of pitching angle α.

4. Analytical solution of pitching stability of HRWAV with empennage

It was considerable to add torque whose direction was on the opposite of M_y for avoiding the enlargement of pitching angle of HRWAV without empennage with wings rotating. The torque increased automatically with the change of pitching angle of fuselage would be an optimal plan.

4.1. Effect of empennage on pitching stability

A rectangle thin plate was set as the simplified model of empennage, which was shown in figure 6. The figure on the left was the space stereogram. It illustrated the tail surface which was acted by the aerodynamic resistance force because of the increasing pitching angle of fuselage when wings of HRWAV with empennage rotate.

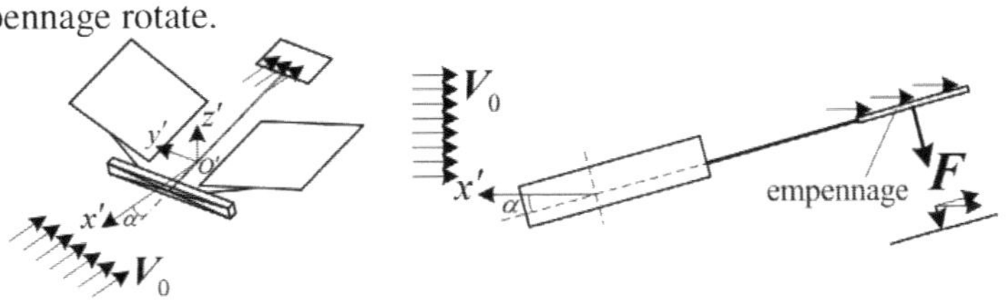

Figure 6. Resistance moment of tail surface.

The equation of pitching motion of HRWAV with empennage was given by equation (11):

$$I_y\ddot{\alpha} = M_y - F_z(l_{O'}-l_D) - M_F = F_z\left[\frac{h}{2}-(l_{O'}-l_D)\right] - M_F \tag{11}$$

According to equation (11), the moment of resistance was critical to make HRWAV with empennage balanced and move in the vertical and horizontal direction.

4.2. The motion model of HRWAV with empennage when disturbed

The pitching angle of HRWAV with empennage was stable in equilibrium state, thus the pressure element of empennage by air could be expressed by equation (12):

$$\mathrm{d}F_0 = \frac{1}{2}C_D\rho(V_0\sin\alpha)^2 b\mathrm{d}x \tag{12}$$

where α was the pitching angle of HRWAV with empennage, V_0 was the airflow speed in the horizontal direction, $V_0\sin\alpha$ was the component velocity acted vertically on the surface of empennage, $\mathrm{d}A=b\mathrm{d}x$ was the rectangular element of empennage.

Therefore, the moment of resistance of empennage in equilibrium state was in equation (13):

$$M_{F_0} = \int_A l_F\mathrm{d}F_0 = \frac{1}{2}C_D\rho(V_0\sin\alpha)^2\int_0^c l_F b\mathrm{d}x \tag{13}$$

It could be seen in figure 7 that $\ddot{\alpha}$ and $\dot{\alpha}$ were resulted in when HRWAV got instantaneous disturbance in the longitudinal symmetry plane(it was assumed that pitching angle and airflow speed were constant). $P_t=(m_f+m_e)g$ was the gravity of HRWAV with empennage, M_F was the moment of resistance of empennage, l_F was the tail arm and c was the length of empennage, $V_{rn}=V_0\sin\alpha+l_F\dot{\alpha}$ was the component velocity acted vertically on the surface of empennage, thus the pressure element of empennage by air could be got as equation (14):

$$\mathrm{d}F=\frac{1}{2}C_D\rho\left(V_0\sin\alpha+l_F\dot{\alpha}\right)^2 b\mathrm{d}x \tag{14}$$

The item with $\dot{\alpha}^2$ could be rounded off when disturbance was small, thus dF could be linearly simplified as equation (15):

$$\mathrm{d}F_1=\frac{1}{2}C_D\rho\left[\left(V_0\sin\alpha\right)^2+2V_0\sin\alpha l_F\dot{\alpha}\right]b\mathrm{d}x \tag{15}$$

The moment of resistance of empennage by instantaneous disturbance was in equation (16):

$$M_{F_1}=M_{F_0}+C_D\rho V_0\sin\alpha\dot{\alpha}\int_0^c {l_F}^2 b\mathrm{d}x \tag{16}$$

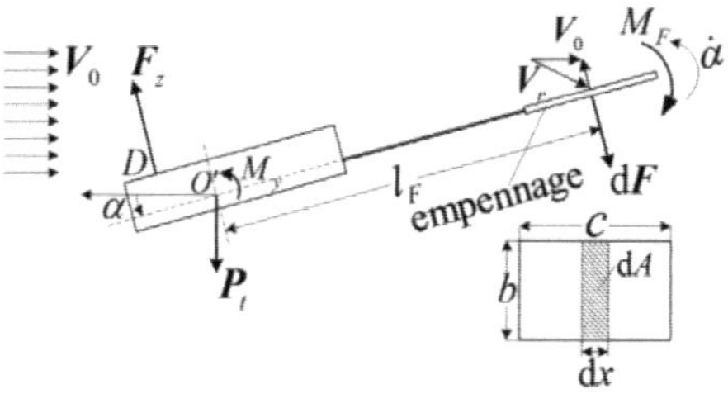

Figure 7. Instantaneous disturbance on HRWAV with empennage.

4.3. Stability analysis of HRWAV with empennage when disturbed

$\ddot{\alpha}=0$, when HRWAV with empennage was in equilibrium state, then equation (11) could be calculated as equation (17):

$$M_{F_0}=F_z\left[\frac{h}{2}-\left(l_{O'}-l_D\right)\right] \tag{17}$$

$\ddot{\alpha}\neq 0$, when HRWAV with empennage was disturbed, then equation (11) could be expressed as equation (18):

$$I_y\ddot{\alpha}=F_z\left[\frac{h}{2}-\left(l_{O'}-l_D\right)\right]-M_{F_1} \tag{18}$$

Therefore, the perturbation equation of pitching angle for HRWAV with empennage was in equation (19):

$$I_y\ddot{\alpha}+C_D\rho V_0\sin\alpha\dot{\alpha}\int_0^c {l_F}^2 b\mathrm{d}x=0 \tag{19}$$

In reference to section 3.2, the pitching angle and airflow speed were constant as soon as HRWAV got instantaneous disturbance. Thus equation (19) could be expressed as equation (20):

$$I_y\ddot{\alpha}+C_1\dot{\alpha}=0 \tag{20}$$

Under the variable substitution from equation (10) to equation (11), it could be got as equation (21):

$$\dot{\alpha}=D_3\exp\left(-\frac{C_1}{I_y\omega}\varphi\right) \tag{21}$$

Equation (21) could be solved when C_1 was substituted in, which could be equation (22):

$$\dot{\alpha}=\dot{\alpha}\big|_{\varphi=0}\exp\left(-\frac{C_D\rho V_0\sin\alpha\int_0^c {l_F}^2 b\mathrm{d}x}{I_y\omega}\varphi\right) \tag{22}$$

From equation (22), $\dot{\alpha}$ was declined as φ grew, which meant that the variation of pitching angle of HRWAV with empennage would go down and resume to equilibrium state in the end. Therefore, the pitching stability of HRWAV with empennage after being disturbed was reliable.

The equation $\varphi = \omega t$ was substituted into equation (21), which could be solved as $\alpha = -\frac{D_3 I_y}{C_1}\exp\left(-\frac{C_1}{I_y}t\right) + D_4$. The initial condition was involved, and that was equation (23):

$$\alpha = \frac{I_y \dot{\alpha}\big|_{\varphi=0}}{C_1}\left[1 - \exp\left(-\frac{C_1}{I_y}t\right)\right] + \alpha\big|_{t=0} \tag{23}$$

Eventually, it could be got as equation (24):

$$\alpha = \frac{I_y \dot{\alpha}\big|_{\varphi=0}}{C_D \rho V_0 \sin\alpha \int_0^c l_F{}^2 b\mathrm{d}x}\left[1 - \exp\left(-\frac{C_D \rho V_0 \sin\alpha \int_0^c l_F{}^2 b\mathrm{d}x}{I_y}t\right)\right] + \alpha\big|_{t=0} \tag{24}$$

It was obvious that α was close to $\alpha\big|_{t=0} + \frac{I_y \dot{\alpha}\big|_{\varphi=0}}{C_1}$, which meant that HRWAV with empennage would get a new equilibrium position compared with the initial equilibrium state.

Stability could be arrived when got instantaneous disturbance by adding empennage on HRWAV. However, theoretical derivation is mainly used for qualitative analysis, which adopted the assumption that airflow speed and pitching angle of aircraft were instantaneously invariable when being disturbed. In actual flight, the influence of change of pitching angle and airflow on the pitching stability could not be ignored. Therefore, the next analysis would be more detailed by numerical method.

5. Numerical analysis of pitching stability of HRWAV with empennage

Considering that the variation of pitching angle and airflow speed when disturbed had effect on the pitching stability of HRWAV, the influence could be calculated by combining equation (19) and equation (5) on the basis of S-Function in MATLAB/SIMULINK, among which m_f was replaced by $m_t(m_t=m_f+m_e)$. Figure 8 showed the model established by S-Function.

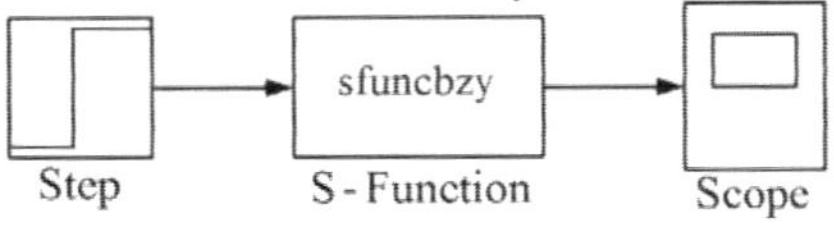

Figure 8. Simulation model of HRWAV with empennage when disturbed.

There were two initial conditions in function sfuncbzy set as:

(1) The initial pitching angle of HRWAV with empennage in equilibrium position was $\alpha = 10$, and then pitching angle was increased by $\dot{\alpha} = 0.0175$ rad/s acted by instantaneous disturbance. Figure 9 denoted the change law of the pitching angle and pitching angular velocity.

(2) The pitching angle was decreased by $\dot{\alpha} = 0.0175$ rad/s acted by instantaneous disturbance while other initial conditions were constant. Simulation was shown in figure 10.

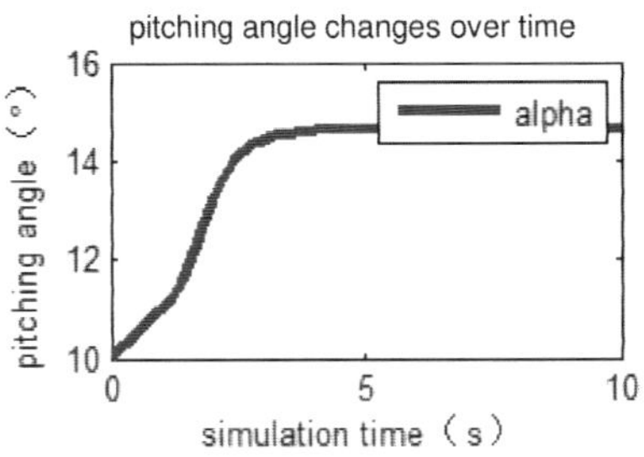

(a). Curves of pitching angle.

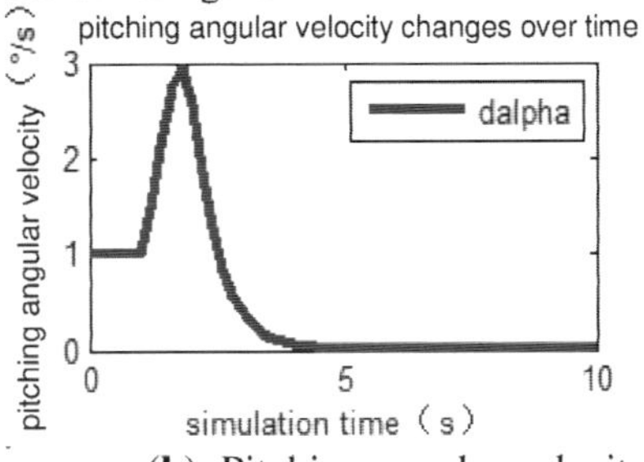

(b). Pitching angular velocity.

Figure 9. Curves of pitching angle (a) and pitching angular velocity (b) in No.1 initial condition simulation.

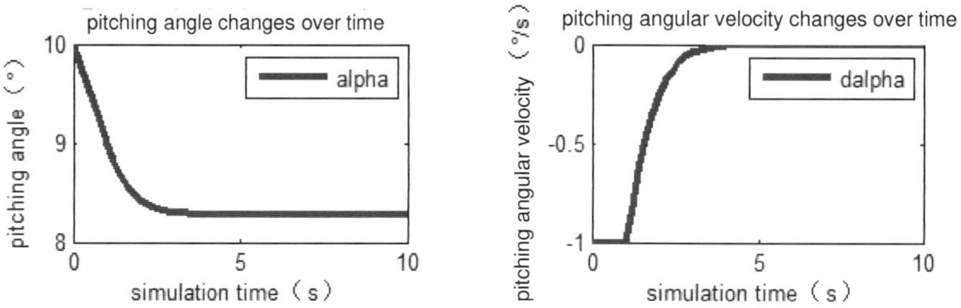

(a). Curves of pitching angle. **(b).** Pitching angular velocity.

Figure 10. Curves of pitching angle (a) and pitching angular velocity (b) in No.2 initial condition simulation.

It was apparent that the pitching angle and pitching angular velocity would both drop to zero as the instantaneous disturbance came to HRWAV with empennage from equilibrium position in a short time. The simulation results were basically consistent with the analysis of section 3.3 in the condition that the instantaneous disturbance was not large. Consequently, the pitching stability of HRWAV with empennage after the instantaneous disturbance was proved credible by numerical analysis.

6. Conclusion

HRWAV is a new flapping-imitating wing air vehicle. The pitching stability is an important part of flight stability for HRWAV. The research results in this paper are listed as follows.

(1) The pitching angle of HRWAV without empennage would increase with the rotation of wings, by which HRWAV would eventually lose its stability, even overturn and fall. Therefore, an empennage on HRWAV is essential to guarantee pitching stability.

(2) When the vertical disturbance acted on the HRWAV flying in balance, the resistance moment of empennage would rapidly weaken the pitching disturbance. HRWAV could take a new equilibrious pitching posture in a short time.

References

[1] Young L A, Briggs G A, Derby M R and Aiken E W 2002 Use of Vertical Lift Planetary Aerial Vehicles for the Exploration of Mars *Engineering Studies into Vertical Lift Planetary Aerial Vehicles*

[2] Nakata T, Liu H, Tanaka Y, Nishihashi N, Wang X and Sato A 2011 Aerodynamics of A Bio-Inspired Flexible Flapping-Wing Micro Air Vehicle *Bioinspiration & biomimetics* **6(4)** 045002

[3] Zhang H X, Yuan X X, Ye Y D and Xie Y F 2002 Research on the Dynamic Stability of an Orbital Reentry Vehicle in Pitching *Acta Aerodynamica Sinica* **20(3)** pp 247-59 (in Chinese)

[4] Keennon M and Grasmeyer J 2003 Development of Two MAVs and Vision of the Future of MAV Design *AIAA International Air and Space Symposium and Exposition: The Next 100 Years* p 2901

[5] Keennon M, Klingebiel K and Won H 2012 Development of the Nano Hummingbird: A Tailless Flapping Wing Micro Air Vehicle *50th AIAA Aerospace Sciences Meeting including the New Horizons Forum and Aerospace Exposition* p 588

[6] Qiu Z Z 2011 *Half-rotating Mechanism: Structure, Characteristics and Application* (Hefei: Press of University of Science and Technology of China) (in Chinese)

[7] Qiu H and Wang X Y 2011 Half-rotating Mechanism: A Biomimetic Mechanism of Animal Motion and Its Basic Motion Characteristics, *Mechanical Science and Technology for Aerospace Engineering* **30(2)** pp 600-4 (in Chinese)

[8] Wang X Y, Zhang Y Q, Wang H X, Dong Y P, Qiu Z Z and Chen F Q 2016 Research of Local Constraint for Simplified Half-rotating Mechanism *Applied Mechanics and Materials* **826** pp 35-9

[9] Wang H X, Wang X Y, Qiu H, Chen F Q, Qiu Z Z and Shan J H 2016 Effect of Local Constraint on Driving Torque of Driving Mechanism for Half-rotating Wing *International Conference on Design Engineering and Science (ICDES 2016)* DOI: 10.1051/matecconf/20165202005.
[10] Wang X Y, Shao C Y, Chen F Q and Qiu Z Z 2014 Numerical Analysis of Weis-Fogh Effect for Half-Rotating Wing Based on Fluent *Proceedings of the 5th International Conference on Mechanical Engineering and Mechanics* pp 614-7
[11] Wang X Y, Dong Y P, Qiu Z Z, Zhang Y Q and Shan J H 2016 Lift Estimation of Half-Rotating Wing in Hovering Flight *2nd International Conference on Mechanical Engineering and Automation Science (ICMEAS 2016)* DOI:10.1088/1757-899X/157/1/012001

Acknowledgments
This work is supported by the National Natural Science Foundation of China (Grant No. 51375014).

IOP Conf. Series: Materials Science and Engineering **211** (2017) 012005 doi:10.1088/1757-899X/211/1/012005

Selection of the battery pack parameters for an electric vehicle based on performance requirements

M Koniak*and A Czerepicki

Warsaw University of Technology Faculty of Transport, Koszykowa str. 75 Warsaw, Poland

*Email: koniakm@wt.pw.edu.pl

Abstract. Each type of vehicle has specific power requirements. Some require a rapid charging, other make long distances between charges, but a common feature is the longest battery life time. Additionally, the battery is influenced by factors such as temperature, depth of discharge and the operation current. The article contain the parameters of chemical cells that should be taken into account during the design of the battery for a specific application. This is particularly important because the batteries are not properly matched and can wear prematurely and cause an additional costs. The method of selecting the correct cell type should take previously discussed features and operating characteristics of the vehicle into account. The authors present methods of obtaining such characteristics along with their assessment and examples. Also there has been described an example of the battery parameters selection based on design assumptions of the vehicle and the expected performance characteristics. Selecting proper battery operating parameters is important due to its impact on the economic result of investments in electric vehicles. For example, for some Li- Ion technologies, the earlier worn out of batteries in a fleet of cruise boats or buses having estimated lifetime of 10 years is not acceptable, because this will cause substantial financial losses for the owner of the rolling stock. The presented method of choosing the right cell technology in the selected application, can be the basis for making the decision on future battery technical parameters.

1. Introduction

Currently use of electrically powered vehicles is the leading direction in development of public transport. Reasons include a few important factors. The main advantages are quiet operation and shifting creation of pollutants from densely inhabited areas beyond such areas. It is worth noticing that the energy may originate from renewable resources or coal fired power plants equipped with very efficient systems for exhaust flue gases treatment, which results in small negative impact on the environment. Also the exploitation cost is three times lower for electrically powered vehicle compared with combustion drive. Systems for braking energy recovery support operation wherever it is possible. Unfortunately, transport with electrically powered vehicles has certain disadvantages. Actually all of them are related to the use of chemical batteries to store energy. With high cost of batteries and substantial weight and volume, they impose significant range restrictions compared to equivalents of combustion technology. Additionally, they require infrastructure to charge at designated stations. That infrastructure may have negative impact on urban distribution grids, which also requires analyses on the investment stage. Traction batteries, being components that constitute a substantial vehicle worth, are a subject of many analyses on their efficient operation. This paper focuses on the battery selection based on predicted operation conditions.

Batteries with lithium-ion cells are of the most often used type for electrical vehicle power storage. Principle of operation of such cells is based on reversible process of lithium ion (Li+) transfer between cathode and anode, called intercalation for charging, and deintercalation for discharging [1].

Batteries of lithium-ion group share a common feature, which is the design based on positive electrode of metal oxides with lithium content, negative electrode made of porous carbon, and electrolyte with lithium salts dissolved in a mix of organic solvents. There are research projects running that aim to improve properties of cells, that focus on changing composition of specific components. This allows to build different versions and it is the differences in composition of changed components that decide about performance and operating properties of a given technology.

Currently the batteries most frequently used for transport applications are based on the following compounds [2]:

- $LiFePO_4$ (LFP);
- $Li_4Ti_5O_{12}$ (LTO);
- $LiNiMnCoO_2$ (NMC);
- $LiNiCoAIO_2$ (NCA).

2. Factors influencing the operation of a battery

2.1. How weight and volume of a battery affect suitability for transport applications

The Lithium-ion batteries have earned their position in transport applications due to some advantages. One of them is the very advantageous ratio of energy to volume or weight. This parameter is referred to as energy density and it informs, how much energy can be stored in 1 kilogram or litre. Cell manufacturers specify this value, which is important information to evaluate suitability of a given technology for a solution to design. It is important to remember, that the energy density is different for the cells alone, and different for finished battery pack that includes the cells. This is caused by presence of additional electrical, mechanical and measuring systems in the battery pack. However, this information is a good reference point to evaluate how required energy affect weight and volume of a vehicle.

Table 1. Comparison of energy densities for three types of lithium-ion cells.

Cell type	Energy density per weight [Wh/kg]	Energy density per volume [Wh/l]
LTO	90	200
LFP	130	247
NMC	150	300
NCA	240	670

Table 1 presents energy densities for four lithium-ion cell types. LTO cells are the worst in this comparison, LFP and NMC cells are on the similar level, while the NCA technology features definitely the greatest energy density.

Energy density is an important factor when selecting a battery type for a given type of vehicle. For example, passenger cars require small and high capacity batteries, which is caused by small available space and restrictions on the vehicle weight. Buses have much more space and the battery volume is not that important, but weight still can be an important factor after reaching certain level, due to limitation of gross vehicle weight allowed by regulations. Heavier battery may put limit on allowed number of passengers that can travel by the vehicle.

2.2. How climate conditions can affect operation of traction batteries

All lithium-ion battery manufacturers define how performance depends on the temperature. The restrictions are safety related and affect charging and discharging current. Exceeding such limitations may damage the battery or cause much faster aging.

Table 2 presents exemplary temperature ranges for battery operation at discharging and charging. Data in table 2 for the temperatures of cell types LTO, LFP, NMC were set at a safe level suggested by the manufacturers of the battery packs while from data sheet for NCA.

Table 2. Exemplary temperature ranges for battery operation at discharging and charging.

Cell type	Operating temperature range (discharging)	Operating temperature range (charging)
LTO	-30 to 55 °C	-20 to 55 °C
LFP	-30 to 55 °C	-20 to 55 °C
NMC	-20 to 55 °C	-0 to 55 °C
NCA	-20 to 60 °C [3]	-0 to 45 °C [3]

The table 2 data shows that the temperature ranges for charging currents are narrower than for discharge currents. This means that charging process is more electrochemically demanding, hence higher requirements for temperature parameters to meet for the process.

It is necessary to remember that besides the limit on the temperature range of cell operation, there is also relation between the temperature and allowed charging and discharging current. Figure 1 presents an example for LFP batteries charging.

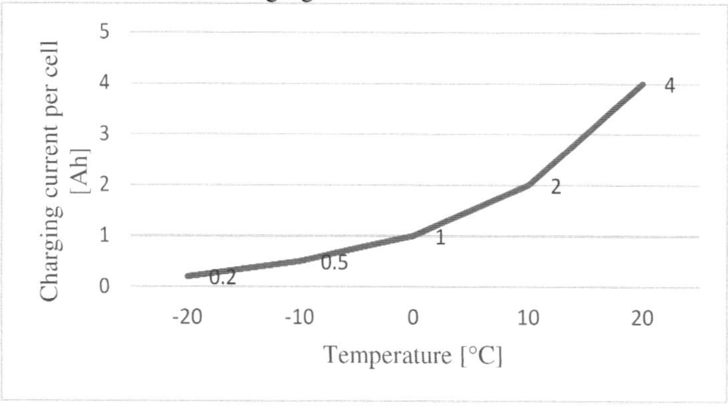

Figure 1. Exemplary relation between charging current and the cell temperature. [4]

Figure 1 reveals that charging currents for the lowest temperatures are the lowest, and increase towards temperature values close to 20°C. This means longer charging times. It is easy to imagine a situation. When a vehicle is equipped with fast-charging batteries but the temperature does not allow to take advantage of that feature, this in turn can cause delays in carrying out transport activities.

Such situation may also happen when the battery pack is too warm, then charging current is reduced. This allows to limit further heating of the battery.

Another relationship to consider is how the temperature affects available battery capacity. Unfortunately this phenomenon can be observed even in temperature range allowed by the manufacturer, and it is caused by increase of internal cell resistance with cooling of the cell. Figure 2 presents such exemplary relationship, where decrease of available capacity as a function of the cell temperature reached over 35%. There are technologies that such reduction may exceed even 50% within temperature range allowed by the manufacturer.

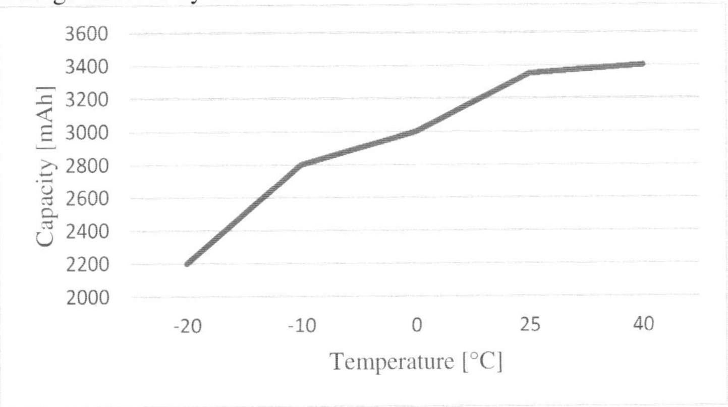

Figure 2. Relationship between available cell capacity and the cell temperature. [5]

Battery operation in too low or too high temperature contributes to faster wear. The most recommended operating temperature is as close to 20–25°C as possible.

To counteract against the above mentioned situations, it is possible to integrate passive or active systems for thermal conditioning with the battery packs. This can be a system for heating, cooling or both, according to the needs. Depending on specification of such system, it uses heated or cooled air, ventilation with ambient air, liquid heating and cooling.

Apart from obvious functionality that protects the battery pack from over-heating and over-cooling, the thermal conditioning systems aim to provide uniform temperature inside. This is important, because cells operating at different temperatures have different operating specifications, which has already been discussed in this paper. Such differences in one battery pack may cause faster wear and limit operating functionality.

2.3. *Use of heating and cooling systems for a traction battery*

Average temperature in Poland falls in between 7 and 9°C. In Warsaw extreme temperatures can be observed, from very high up to approximately 35°C, to very low, less than -15°C. Selection of battery for an electrically powered vehicle requires consideration of these conditions.

Considering very high charging currents for LTO batteries, and great amounts of heat generated in the process, they always have cooling systems with liquid coolant medium. They also allow to heat the battery in case of significant temperature fall. This guarantees that LTO battery packs are fully protected from ambient temperature variations.

This is different for LFP batteries. Due to allowable working currents and the battery size, there is no need to include liquid cooling systems, and heating is not available either. The exception are battery packs for cold climate countries, like Sweden and Norway, which have additional heating mats installed. Such mats can be included with the battery packs for Polish market on client request, but this is not a common practice. In accordance with the data from cell manufacturers, operating temperature range is between -30 to +55°C for discharging and -20 to +55°C for charging. It is clear, that fault free operation is possible down to -20°C, which will result in reduced charging current, but will not put the vehicle to a halt caused by no possibility to recharge. Adjustment of charging current to the temperature of the battery pack is governed by the BMS, Battery Management System.

It should be noted, that the remaining cell types installed in electrically powered vehicles, namely NMC and NCA, require heating systems. This is due to the operating temperatures allowed by manufacturers, and they are -20 to +55°C for discharge and 0 to +55°C for charging in case of NMC, and -20 to +60°C for discharge and only 0 to +45°C for charging in case of NCA.

2.4. Battery working currents in electric vehicles

This parameter defines allowed currents for charging and discharging, specified for temperature ~20°C. The working current is specified as a multiplier of C number, for example 5C. C current (1C) is one-hour current, which would allow to charge the battery from 0 to 100% of its capacity in one hour, or discharge the battery from 100 to 0% of capacity in one hour. In the given example, 5C is the current that exceeds one-hour current five-fold, i.e. will discharge the battery completely in one fifth of an hour, which is 12 minutes.

Table 3. Exemplary working current ranges for discharging and charging batteries.

Cell type	operating current (discharging)	operating current (charging)
LTO	5 to 10 C	5 to 10 C
LFP	3 C	1 C
NMC	2 to 3 C	1 C
NCA	2 C	0.5 C

It is shown that LTO battery offers the greatest allowed working currents. LFP and NMC technologies are comparable by this parameter, and NCA technology have greater requirements on charging.

Depending on the technology, some batteries require charging stations with sufficient power available for quick charging, and other batteries must be used in vehicles, where few hours charging does not hinder vehicle operation.

When considering working current, we must also look at the relationship between charging current intensity and the battery charge level as compared to the total capacity. This greatly affects possible selection of battery capacity, because a vehicle leaving a charging station does not have the full battery capacity. The only way to obtain full working capacity is shifting from the CC charging (Continuous Current) to CV charging (Continuous Voltage) [6].

Figure 3 shows exemplary relationship between charging current intensity and charge level for an exemplary cell. The data are collected by the authors [7].

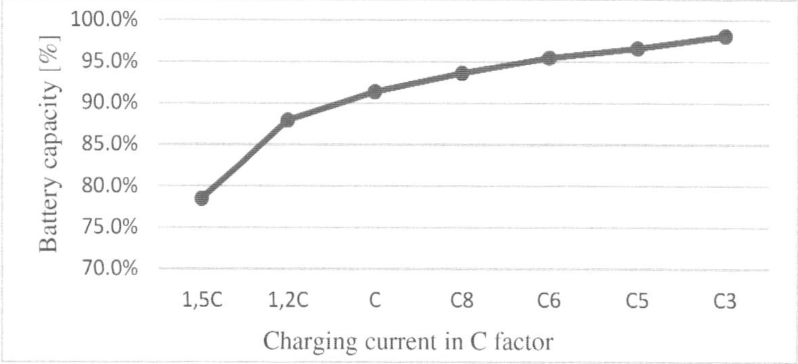

Figure 3. Sample relationship between charging current intensity and charge level of a Li-Ion cell. C factor (1C) is one-hour current, which would allow to charge the battery from 0 to 100% of its capacity in one hour, or discharge the battery from 100 to 0% of capacity in one hour.

3. Factors influencing the selection of a battery

3.1. The required number of traction battery working cycles

Manufacturer's data do not always allow to compare the numbers of work cycles. This is because the number depends on many factors: charging and discharging currents and the process depth. Therefore, comparison requires data specific for exactly the same values of the mentioned parameters.

Below is the collection of exemplary number of cycles obtained from a battery packs manufacturer. All pertain to 100% discharge depth and working currents falling in range allowed by the cell manufacturer. Getting reliable data on possible number of work cycles in a wide range of currents and discharge depths often requires performing a private study.

Table 4. Comparison of possible number of work cycles before capacity drops to 80% of rated capacity for LTO, LFP, NMC and NCA cells.

Cell type	Number of life cycles for 100% Depth of Discharge (DoD)
LTO	15000
LFP	3600
NMC	3000
NCA	500

This list shows that LTO technology has considerable advantage over other technologies, LFP and NMC are comparable again, and NCA shows much worse results. We need to notice that 15,000 full cycles with 3C current at 25°C and full discharge depth is a confirmed value that has been achieved experimentally by the LTO cells manufacturer. The tests are still in progress and carrying more such cycles is possible before the capacity drops to 80% of the rated capacity.

The questions arises, how a battery capable of going for only 500 cycles has found its use for automotive applications. Solution to this issue is the right use of the battery capacity. It is important that depth of discharge is related to the loss of capacity caused by number of work cycles. It turns out that the smaller the cycling range, for example 30–70%, the longer its service life can be. So the battery type and capacity should be selected also with consideration of the discharge depth. Cell manufacturers sometimes provide such data, and if not, they need to be obtained by making measurements. Figure 4 presents an exemplary relation for LiFePo4 battery [8].

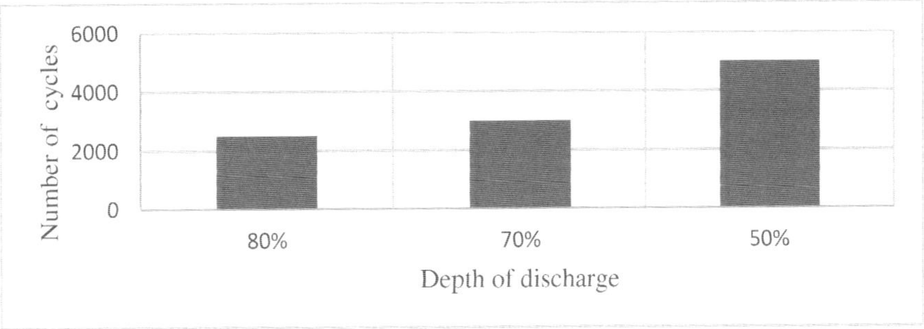

Figure 4. Relation between depth of discharge and loss of capacity caused by number of work cycles. [8]

3.2. Economic aspect of traction batteries selection

The decision making process on the traction battery selection heavily depends on economic aspect. Figure 5 shows price relations for specific lithium technologies.

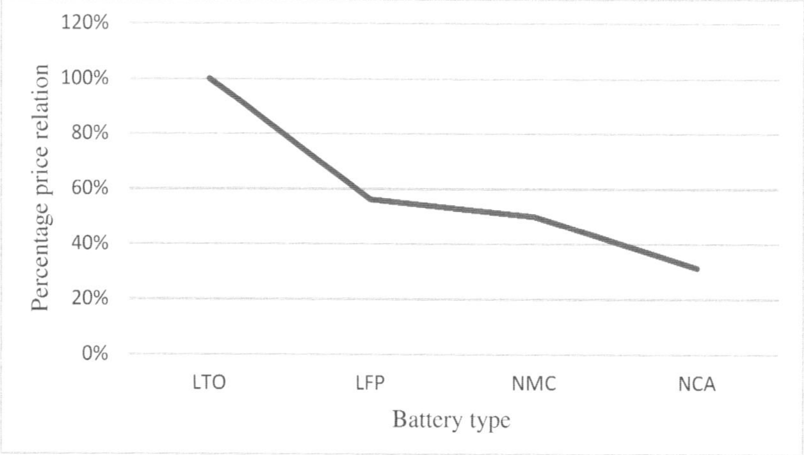

Figure 5. Price relations for specific lithium technologies for the same capacity.

Prices for lithium-ion batteries remain at high level, roughly at USD 250–900 per kWh. This is the reason for limitations in development of electrically powered communication. The abovementioned features and restrictions must be considered when deciding about investment in electrical vehicles. Appropriate operation conditions can significantly extend the lifetime. Drop of capacity by 20% of the rated capacity is assumed as the right moment for decommissioning. However, operating the batteries below 80% of initially available capacity is also considered. LFP cells aging is greatly related to change of structure (porosity) of the negative electrode, loss of material and damages in positive electrode structure, and oxidation processes. If the inner impedance increases, amount of reactive lithium and active material changes, which in turn reduces available energy and power. With loss of capacity caused by aging, the charging current reduces and amount of produced heat increases. For 60% of maximum capacity, the halt times required to replenish energy increase by over 40%. That is because the battery capacity reduction causes drop of respective charging current. Continued use of the traction batteries below recommended level of 80% capacity must be agreed and approved by the manufacturer.

When choosing the batteries it is also better to consider planned battery replacement during the vehicle life time. For example, when the planned life time of the vehicle is 12 years and whenever technical conditions allow, a LTO battery can be replaced with NMC cells battery. It is cheaper by approximately 55%, so potential replacement will not be an economical burden to the investment, and considering drop of prices and technology development it may turn out that after 6 years the savings can be even greater.

3.3. Exemplary selection of a battery with consideration of factors weights

In the "Renewable Energy Powered Hybrid Innovative Sailing Yacht" project the battery was selected by looking at the weighting sum of specific factors for each of the considered technologies. The following section describes criteria used to select parameters and weights.

Battery capacity is related to the characteristic daily specifications handed over by the project partner:

- work mode during port docking, without connected on shore power supply,
- work mode for going under sails,

- work mode for going with electric motor only.

The third specified mode was the hardest in terms of loading the battery, and required nearly 40 kWh of energy. So the selected battery must cover that need and work range must not reach 100% Depth of Discharge (DoD). Therefore, the capacity is not included in the table of weights.

Volume and weight are important, considering the planned place of batteries installation. The room will be small, and carrying the battery in will require going along steep stairs with narrow passages. The battery will be divided into battery packs maximum 50 kg each to allow carrying; the number of packs should be possibly the lowest, because the container cost increases with each additional battery pack.

Climate effect is important for operation, which has been explained in this paper. All analysed batteries can be used in the project, but they differ by allowed working temperatures in higher ranges, which may appear in Turkey, where the yacht may be operated. It is reflected with assigned points. Cooling system is not required considering the working currents, nevertheless LTO batteries are equipped with such system in accordance with the manufacturer recommendations.

The part on charging current shows that the assigned weight is 0% and scoring is skipped. This is caused by assumption that the battery will be charged during planned, few hours long docking at port, so there is no requirement for quick charging with high current.

Number of cycles is an important factor, and in this case NCA score was lifted from 1 to 3, because replacement of cells is possible after 5 years, which still does not exceed the assumed cost.

The last criterion is the price, which is a significant factor overall, and decisive in this particular case. This is because US dollar price increased during the project, and the energy required by the project partner allowed to purchase NCA battery, which allows for required 5 years of the unit operation according to simulations. Table 5 presents exemplary selection of a battery with consideration of factors weights.

Table 5. Exemplary selection of a battery with consideration of factors weights for LTO, LFP, NMC and NCA cells.

	Volume	Weight	Climate effect	No cooling system	Level of charging currents	Number of work cycles	Price	Total
percentage distribution	5%	10%	15%	10%	0%	30%	30%	100%
LTO	1	1	10	5	0	10	1	5.45
LFP	4	4	5	10	0	5	5	5.35
NMC	5	5	3	10	0	4	6	5.2
NCA	10	10	2	10	0	1	10	6.1

4. Conclusion

This paper discusses the issues of batteries selection for electric vehicle, taking into account many operation criteria. This problem is obviously a complex one. Selection of a battery for an electrically powered vehicle requires detailed analysis of many factors, such as: weight and volume, charging currents, route characteristics, depth of discharge, operating temperatures related to seasons of the year in the given area, potential integration of conditioning and heating. Selection method of the suitable technical solution is described, using comparison of the weighting sum of specific factors for each considered option. Correct selection of the battery and appropriate operation can extend the service life between replacements, thus reducing operating costs of the electrically powered vehicle.

References

[1] Linden D and Reddy T (ed.) 2002 McGraw Hill chapter 2 Electrochemical Principles and

Reactions *Handbook of Batteries Third Edition* p 229

[2] Sierszyński M, PikulaM, Fuć P, Lijewski P, Siedlecki M and Galant M 2016 Overview of solutions for lithium-ion batteries used in electric vehicles *International journal of energy and environment* pp 105-11

[3] 2012 Lithium-ion Battery Overview (PDF) *Lighting Global* **10**

[4] 2012 *A123 Systems ALM 12V7 User's Guide*

[5] Panasonic Datasheet for NCR18650b 3400 mAh [http://na.industrial.panasonic.com/sites/default/pidsa/files/ncr18650b.pdf]

[6] Al-Haj Hussein A and Batarseh I 2011 IEEE Transactions on Vehicular Technology *A Review of Charging Algorithms for Nickel and Lithium Battery Chargers* **60** 3 pp 830-8

[7] Koniak M, Czerepicki A and Tomczuk P 2016 Test bench for battery energy storage selection for use on solar powered motor yachts *Scientific Journals of the Maritime University of Szczecin* **46** pp 197-202

[8] Victron Energy Datasheet 12,8 Volt Lithium-Iron-Phosphate Batteries Victron Energy 2017 [https://www.victronenergy.com/upload/documents/Datasheet-12,8-Volt-lithium-iron-phosphate-batteries-Smart-EN.pdf]

Acknowledgment

The project "Renewable Energy Powered Hybrid Innovative Sailing Yacht" is funded by the National Research and Development Centre under the ERA – NET TRANSPORT; ERA-NET-TRANSPORT-III/5/2014.

IOP Conf. Series: Materials Science and Engineering **211** (2017) 012006 doi:10.1088/1757-899X/211/1/012006

A method of computer modelling the lithium-ion batteries aging process based on the experimental characteristics

A Czerepicki* **and M Koniak**

Warsaw University of Technology, Faculty of Transport, Koszykowa str. 75 Warsaw, Poland

*Email: a.czerepicki@wt.pw.edu.pl

Abstract. The paper presents a method of modelling the processes of aging lithium-ion batteries, its implementation as a computer application and results for battery state estimation. Authors use previously developed behavioural battery model, which was built using battery operating characteristics obtained from the experiment. This model was implemented in the form of a computer program using a database to store battery characteristics. Batteries aging process is a new extended functionality of the model. Algorithm of computer simulation uses a real measurements of battery capacity as a function of the battery charge and discharge cycles number. Simulation allows to take into account the incomplete cycles of charge or discharge battery, which are characteristic for transport powered by electricity. The developed model was used to simulate the battery state estimation for different load profiles, obtained by measuring the movement of the selected means of transport.

1. Introduction

REP SAIL research and development project aims to build an electrically powered yacht. Energy is accumulated in lithium-ion batteries, like in most other electrically powered means of transport. Selection of power source was based on analysis of yacht operating conditions, including predicted energy consumption in various sailing modes, and operation characteristics of the battery obtained with experimental measurements [1, 2].

The REP SAIL model (figure 1) was built using an experimental measurements of battery capacity dependent from currents of charging or discharging. It assumes that the battery is constructed from cells with identical parameters. Operation characteristics for a single battery cell were obtained in the form of discrete functions $\{ E(I,t) \}$ that represent relationship between collected energy E and time t at constant charge/discharge current I, for various currents $I \in \{ I_1, I_2, \dots I_k\}$ in the range between the minimal current I_1 to the maximum current allowed for specific cell type I_k. Measurement data are stored in the relational database. For other currents than measured, the spline form interpolation of the $E(I,t)$ is used. The REPSAIL behavioral model consists of measurement data and interpolation method [3, 4, 5]. The algorithm of the computer simulation of the battery exploitation needs an characteristics of charge/discharge battery current in time, and the initial battery charge level. Next the battery state in time $E(I,t)$ is calculated as iterations. The model aims to predict battery condition at any given moment, provided that the load and charging processes over time $I(t)$ are known.

Proposed model was implemented as a computer program [6] and was subject to validation tests. At the time of writing of this article there was no operational electrical yacht prototype, so the validation process used measurement data obtained with test drives of electrically powered vehicle. We need to notice that versatility is an important feature of the proposed model, which allowed for validation with the use of different input data than initially planned. This model construction method has a drawback of time consuming nature, and advantage of very precise representation of real condition.

Despite the abovementioned advantages, the battery model that authors of the paper proposed does not consider aging processes specific to lithium-ion batteries. Electrically powered vehicles working cycle is usually characterized by frequent changes in battery charge and discharge modes, so the aging process can be a significant factor in predicting the battery charging level and its general state of health.

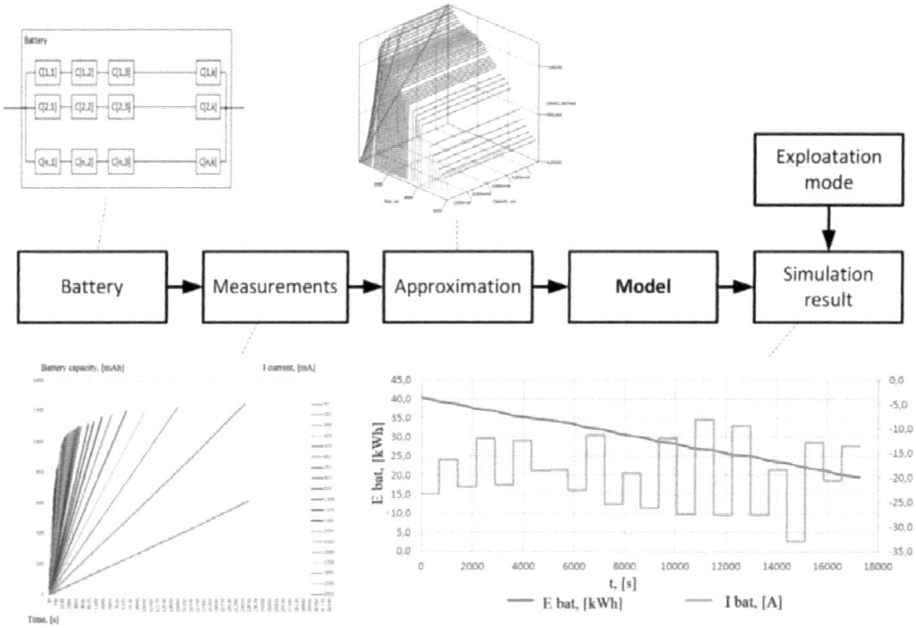

Figure 1. Proprietary model of REP SAIL battery.

This paper presents proposition to complement the battery model with aging module, aging module operation algorithm, and computer modelling results of battery behavior at long term operation at a specific profile of cyclic load.

2. Assumptions in the battery aging model

Li-Ion batteries are subject to aging process. They gradually lose capacity as the number of the charge and discharge cycles increase. The references describe many models of battery aging. Authors [7, 8] indicate that aging process simulation based on model of electrochemical processes running inside cell requires high computational power and is of little use practically. Empirical models on the other hand require large collection of experimental data, which is money and time consuming. Analytic models of cells offer poor adaptation to various types of batteries. Statistic methods are quite the opposite, they can be easily adjusted to many types of batteries, however effectiveness heavily depends on the size of collected data [9, 10]. Once strong and weak points of specific model types of battery aging have been analysed, we made a decision to construct the REP SAIL aging model based on empirical data. That is due to great scope of experimental data collected during 2 years of the project run. Besides, such approach is compliant with the concept of behavioural model of the battery discussed earlier.

At the foundation of the REP SAIL battery aging model lies an assumption, that the battery aging process is not regular at operation with various currents, which hinders derivation of general approximation formula. Therefore, range of allowed battery operation currents has been divided into intervals. It is assumed that aging process for all currents within one interval range is identical. Interval boundaries are selected based on analysis of the battery cell measurement characteristics. On one hand, it is advantageous to increase number of intervals to eliminate errors that originate from battery behaviour differences for currents near lower and upper limit of an interval. On the other hand, necessity of large number of experimental measurements significantly extends time needed to create the battery model.

A charge/discharge cycle will be defined as such change of the battery cell state, when the total change of energy exceeds maximum capacity of the cell E_{max}. We assume that the energy transferred during the cell charging process (positive value) and the energy consumed during operation (negative value) are counted separately. This allows to count the number of battery cycles in real conditions operation, when the battery cannot be fully discharged or fully charged. At the execution of subsequent cycle m the present maximum cell capacity E_{max} is reduced by the value $\Delta E(m)$ calculated with the experimental data for the number of cycles m.

3. Construction of the battery aging model

Next, we will discuss the aging model construction process for the battery built with Panasonic NCR 16850B cells. The cell type selection was based on the type of batteries in electrically powered vehicle, which was used for experimental measurements of real operating characteristics. This allowed to compare the results of computer modelling with the actual battery wear.

The battery has been built with the cells configured as presented in Table 1 to provide appropriate operating characteristics specific for design of electrically powered vehicle.

Table 1. Battery configuration.

	Capacity, [Ah]	Voltage, [V]
Single Cell	3.4	3.6
Battery	408	648

Next, a computer simulation has been run for the process of battery operation, with real input data acquired with measurements, but aging process was not included. Negative current value I corresponds to battery discharging, positive value corresponds to charging. In the given example, battery charging occurs while driving and braking (partial energy recovery at regenerative braking). Figure 2 presents simulation at a scale of a single battery cell for 200 minutes drive without long stopping periods.

Table 2 presents summary of the modelled section and compares the results of the computer simulation with the real measurements of battery condition changes.

Table 2. Results of discharge modelling.

			Simulation		Real measurement	
	E cell, [Ah]	Cells in battery	E bat, [Ah]	Discharge, %	E bat, [Ah]	Discharge, %
Drive start	1.699	120	203.88	49.97%	146.79	35.98%
Drive end	1.082	120	129.84	31.82%	88.24	21.63%
Battery level change, total			74.04	18.15%	58.55	14.35%
Battery level change, 1/h				5.44%		4.31%

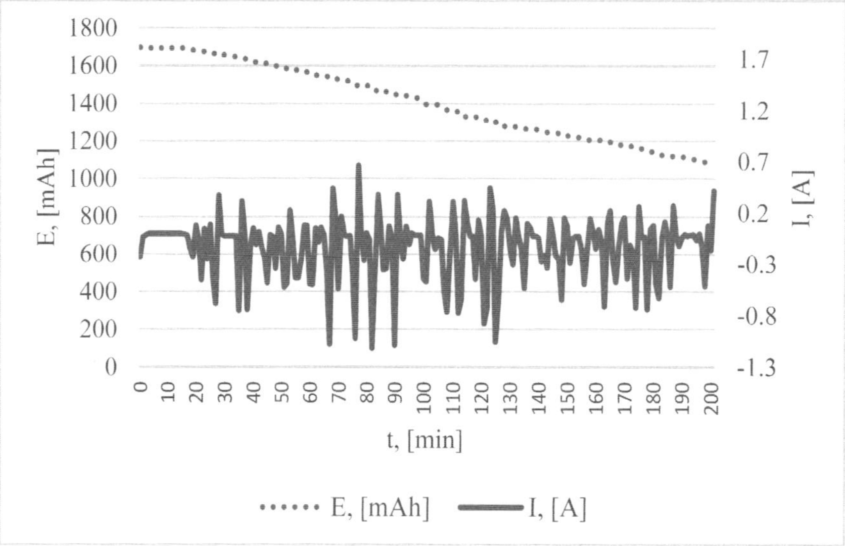

Figure 2. Battery operation process modelling with the use of real data, without aging.

The data in the Table 2 show that the computer simulation of battery use over 1 hour differs from real measurements by ~1% of maximum capacity, which indicates correct representation of battery operation process in the model.

To include the aging process in the battery model, we carried out experimental measurements that aimed to estimate the battery capacity reduction at repetitive charging and discharging with constant current at selected intervals.

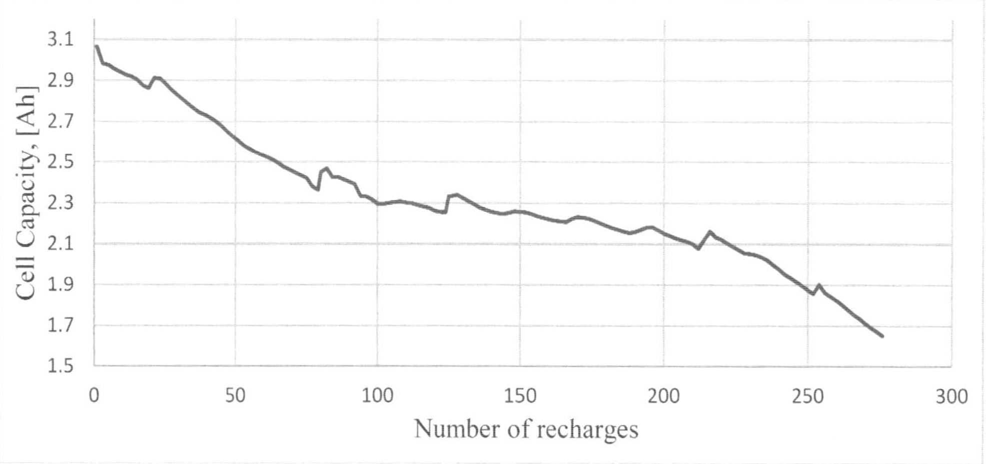

Figure 3. Change of the battery capacity in relation to the number of discharge and charge cycles.

We decide that for each discharging, constant current $I_k \in \{ I \}, k = 0..n$ there will be experimental measurements of relationship $E(I_k, t)$ with the battery fully charged at the beginning, i.e. $E_0(I_k, t = 0) = 100\%$, and fully discharged at the end. The battery is charged with the specified current after each discharge. To obtain characteristics of battery capacity changes as a function of number of cycles *n*, we will do measurements repetitively until the battery capacity is reduced to 50% compared to initial level. Experimental characteristic for battery aging for a specific current I_k=1C is presented on Figure 3. (Note: I=1C denotes charging current, which allows to charge in 1 hour; for battery under tests it is 1.6A.)

IOP Conf. Series: Materials Science and Engineering **211** (2017) 012006 doi:10.1088/1757-899X/211/1/012006

By repeating the measurement series for different values of the constant current I_k we can obtain a set of experimental characteristics, as presented in Table 3.

Table 3. Relationship between battery aging factor and number of cycles.

Current interval no.	Maximal charging current in the interval I,[mA]	Number of cycles n	Battery capacity reduction E_{max}, %	
			Total for n cycles	Averaged for 1 cycle
1	400	100	2.92%	0.0292%
2	800	100	6.08%	0.0608%
3	1200	100	14.45%	0.1445%
4	1600	100	25.14%	0.2514%
5	2000	100	34.07%	0.3407%

4. Consideration of the battery aging process in the computer model

Studies on how number of battery charging cycles reduces its capacity were carried out for constant currents. In practice, use of a battery in electrically powered transport vehicle means frequent changes of absolute current value and direction of the current. Including this factor in the battery model required to take the following assumptions, which facilitate count of battery work cycles:

1) for any time period t_i, where the battery loading current $I(t_i)$ falls into one interval (Table 3) we will assume, that the aging process is in line with characteristics for that specific interval;
2) we will sum up partial changes of energy ΔE_i within each of the current intervals, positive and negative separately (Figure 4);
3) when total energy flow ΔE_i in an interval i exceeds maximum running battery capacity E_{max}, counter of total number of cycles will be incremented $n := n + 1$, and the value $\Delta E_i := 0$;
4) after the number of cycles n is increased, the maximum battery capacity E_{max} is reduced in accordance with the factor defined in Table 3, i.e. $E_{max} := E_{max}(100 - W)$, where W is the battery aging coefficient (the last column in Table 3).

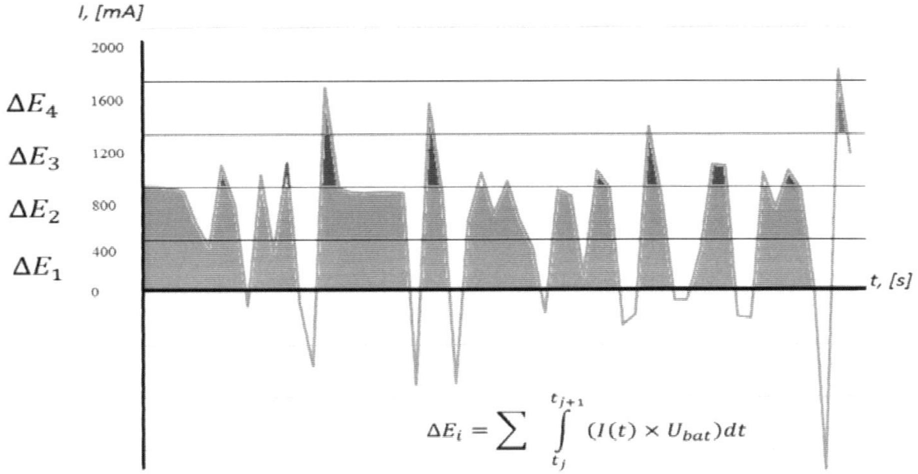

Figure 4. Concept of battery energy changes counting by the current intervals.

5. Modelling of the battery aging based on the real drives data

To show the effects in the battery model operation, a computer simulation of electrically powered vehicle drives has been run for a selected route and a long period. Source data contain real electric vehicle drives along the route, obtained by everyday drives over one month. The data has been extrapolated to longer range of time (year) to emphasize effects of the battery aging. Figure 5 shows the simulation result.

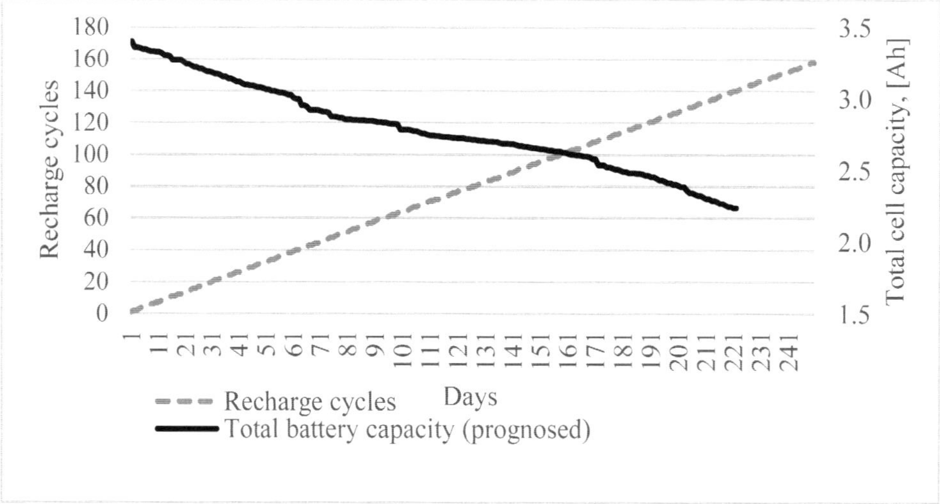

Figure 5. The result of the modelling of the battery aging.

The simulation allows for rough estimation of the battery wear at continuous operation over a specified time range. Practical verification of obtained results is complex due to lack of real data over such long time. Therefore, verification of the model is planned only after gathering necessary experimental data, which will allow to evaluate level of the battery aging. Just like for the verification of the battery model without aging, we can expect high consistency between the modelling results and the control measurement at the end of one year operation period.

6. Conclusion

The paper presents enhancement of the computer model of the Li-Ion battery used in electrically powered transport vehicles with the battery aging process caused by repetitive charge and discharge cycles. The battery model has been evaluated by the research team of the Warsaw University of Technology under the REP SAIL project. The model belongs to the class of behavioural models, because it is based on the experimental measurements. Extrapolation of the battery behaviour depending on its load distribution over time is carried out by the computer program, that uses base of measurements that describe relation between battery energy and the standardized currents. Specific character of the proposed model derives from its close relationship with the cell type used in the battery design. Each cell requires separate experimental measurements, and that significantly extends time needed to create the model and requires consideration of its imperfections. Advantages of the model include high level of consistency between modelling results and the actual results obtained with direct measurements. Error of the battery level estimation for 1 hour modelling falls within 2–3% range.

Model extension with the battery aging processes has been proposed. In order to include this feature in the computer model, the battery capacity changes has been studied for specific charging currents, with dependence of number of chargings. Obtained experimental data was the foundation to calculate the battery aging coefficients for specific loading current intervals. Battery work cycle count format has been proposed, based on integration of function that describes the battery loading power over time, separately for different current intervals.

Modified battery model was demonstrated to assess reduction of the battery capacity in regular operation over one year. Input data included measurement data of electrically powered urban transport vehicle, for which consistency had been experimentally observed between modelling of battery charge level and registered actual state. Verification of the battery aging model is planned once there is available information about actual wear after one year of service.

References

[1] Koniak M, Czerepicki A and Tomczuk P 2016 Test bench for battery energy storage selection for use on solar powered motor yachts *Scientific Journals ot the Maritime University of Szczecin* **46** pp 197-202

[2] Koniak M, Czerepicki A, Tomczuk P, et al 2016 Simulation of the battery pack exploitation based on real measurement data of applied chemical cells MATEC Web of Conferences E D P Sciences No. 77 pp. 1-5 http://dx.doi.org/10.1051/matecconf/20167712002

[3] Koniak M, Kazinski A and Czerepicki A 2016 Mathematical model to estimate the electrical parameters of the autonomous vehicle battery II-nd International correspondence Scientific and Practical Conference “Prospects of Transport Complex Development" 4th – 6th of October 2016, Minsk, Belarus

[4] Sangwan V, Sharma, A, Kumar R and Rathore A K 2016 Estimation of battery parameters of the equivalent circuit models using meta-heuristic techniques *In Power Electronics, Intelligent Control and Energy Systems (ICPEICES), IEEE International Conference on IEEE* pp 1-6

[5] Xia B, Wang H, Tian Y, Wang M, Sun W and Xu Z, 2015 Energies State of charge estimation of lithium-ion batteries using an adaptive cubature Kalman filter **8(6)** pp 5916-36

[6] Czerepicki A, Koniak M, Kazinski A 2016 Computer system for modeling of charging and discharging processes of battery of the yacht powered by electric energy *Scientific Papers of Warsaw University of Technology* (Transport Publishing House of Warsaw University of Technology) **113** pp 63-73

[7] Millner A 2010 Modeling lithium ion battery degradation in electric vehicles *IEEE Conference on Innovative Technologies for an Efficient and Reliable Electricity Supply* pp 349–356

[8] Spotnitz R 2013 Simulation of capacity fade in lithiumion batteries *Journal of Power Sources* **113(1)** pp 72–80 http://dx.doi.org/10.1016/S0378-7753(02)00490-1

[9] Berecibar M, Gandiaga I, et al 2016 Critical review of state of health estimation methods of Li-ion batteries for real applications *Renewable & Sustainable Energy Reviews* **56** pp 572–87 http://dx.doi.org/10.1016/j.rser.2015.11.042

[10] Barré A, Deguilhem B, et al 2013 A review on lithiumion battery ageing mechanisms and estimations for automotive applications *Journal of Power Sources* **241** pp 680–9 http://dx.doi.org/10.1016/j.jpowsour.2013.05.040

Acknowledgment

The project "Renewable Energy Powered Hybrid Innovative Sailing Yacht" is funded by the National Research and Development Centre under the ERA – NET TRANSPORT; ERA-NET-TRANSPORT-III/5/2014.

Integration of internet of things to reduce various losses of jatropha seed supply chain

S P Srinivasan[1*], J Anitha[2] and R Vijayakumar[3]

[1] Professor and Head, Department of Mechanical Engineering, Rajalakshmi Engineering College, India
[2] Assistant Professor, Department of Information Technology, Rajalakshmi Engineering College, India
[3] Assistant Professor, Department of Computer Science & Engineering, Rajalakshmi Engineering College, India

[*]Email: srinivasan.sp@rajalakshmi.edu.in

Abstract. The evolution of bio fuel supply chain has revolutionized the organization by restructuring the practices of the traditional management. A flexible distribution system is becoming the need of our society. The main focus of this paper is to integrate IoT technologies into a cultivation, extraction and management of Jatropha seed. It was noticed that major set-back of farmers due to poor supply chain integration. The various losses like information about the Jatropha seed availability, the location of esterification plants and distribution details are identified through this IoT. This enables the farmers to reorganize the land resources, yield estimation and distribution functions. The wastage and the scarcity of energy can be tackled by using the smart phone technologies. This paper is proposes a conceptual frame work on various losses involved in the supply chain of Jatropha seed.

1. Introduction

Jatropha in an adaptable species, its strength as a crop is its ability to grow in any part of the country with a minimum input and can be propagate easily. The uncultivable wastelands are planned to utilize for cultivating Jatropha seed which has lately emerged as major sources of bio-fuel[1] in the country. An integrated Jatropha bio-fuel project has three stages i) the first stage is production of seeds of Jatropha is the plantation[2] stage ii) extraction stages of bio-fuel iii) the final stage distribution of bio-fuel. Identifying and minimizing the losses in each stage becomes essential. The main focus of this paper is to reduce the losses of Jatropha supply chain[3]. The supply chain management success included improved relationships between warehousing and transportation within companies as a result of reduced inventory and better response time to customer requests for products and services this is done by integrating the IoT technologies in cultivation, extraction and management of Jatropha seed. Hence to assimilate the IoT technologies the foremost important factor is to forecast the various losses of Jatropha. The various losses which has been addressed in this paper are as follows, losses in the yield of the seed, barriers in Identification of factory and issues in inventory details.

The key factors that can influence the seed yield of Jatropha plant are:

- Climate
- Quality of the soil

IOP Conf. Series: Materials Science and Engineering **211** (2017) 012007 doi:10.1088/1757-899X/211/1/012007

- Irrigation

- **Climate**

Rainfall is an important factor when it comes to the yield. The optimum range is believed to be between 500mm to 600mm.Climate does not appear to be an important factor, considering the fact that this plant can survive in a wide range of temperatures. However, warmer climates appear to result in better yields.

- **Quality of the soil**

Sandy soils[4] with ample drainage are ideal for a high jatropha yield. The condition of the ground water[5] is an important factor as well. Irrigated soil seems to provide the best yield in terms of efficiency.

- **Irrigation**

Jatropha handles dryness very well and it is possible to live almost entirely of humidity in the air. - See Cape Verde where rainfall[6] is as low as 250 mm a year. Differences are expressed in what is optimum rainfall as some readings say 600 mm and some say 800 mm whilst some areas in India report good crops with rainfall of 1380 mm. Under irrigation 1 500 mm is given.500 - 600 mm of rainfall is the limit. Below it the production depends on the local water condition in the ground

It is important to rectify yield losses which in turn increase the productivity of jatropha seed.

In order to identify the location of the factory it is essential to estimate the production of the seed. The yield estimation is done using Seed_Estimation algorithm. The factory is identified for esterification of plant. Inappropriate selection of factory can lead to losses in cost of transportation and Insufficient Factory capacity can route to wastage of seed. Factory losses can be controlled by applying appropriate metrics for process capacity of the factory and neighborhood Search algorithm[7] is used to find the shortest destination from the source point. Inventory management[8] is a significant component of supply chain management. Supply chain costs need to be minimized by handling inventory levels in numerous production and distribution operations related with diverse chain stage. The system proposed also take responsibility of handling inventory issues in supply scheduling, transactions and ordering which creates the inventory losses such as cost, time, and labor. Innovative and efficient methodology to determination of the most probable excess stock level and shortage level required for inventory optimization[9] in the supply chain such that minimal total supply chain cost is ensured. Big data technology is used to save large data such as yield estimation factory details and inventory details. The wastage of scarcity of energy can be tackled by using the smart technologies. This paper propose a conceptual frame work on reducing losses involved in Jatropha seed.

The underlying figure-1 shows the losses and issues in Jatropha supply chain

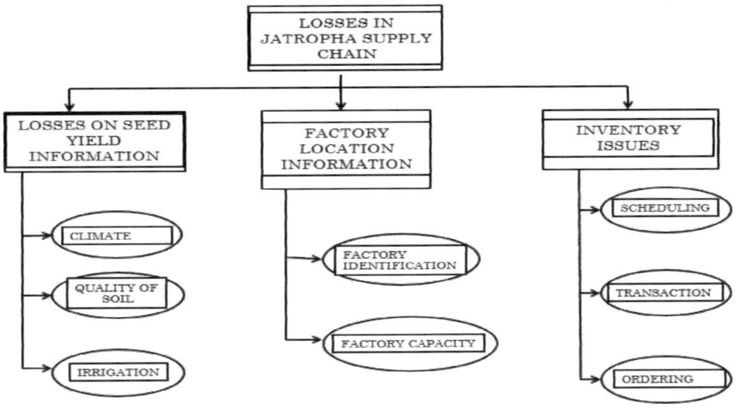

Figure 1: Losses in Jatropha Supply Chain

2. Related Works

Ji-chun Zhao et al. (2010)[10] has shown the greenhouse monitor system based on IOT technology has certain precision of monitor and control. According to the need surrounding monitor, this system has realized the automatic control on the environmental temperature, humidity factors.

Arindam Giri el al.,(2016)[11] has proposed a framework that consists of smart devices, WSN, and Internet by which agricultural processes will be automated. Using smart mobile phones the farmer can control the agricultural tools such as an automated water sprayer to be used in the field of agriculture.

S.Geetha et al., (2016)[12] have used wireless sensor network(WSNs) GSM and Android phone supporting android application and System for plant disease monitoring. With the information in the web server the farmer can automatically control irrigation using android application.

3. METHODOLOGY

The system Architecture consists of three layers. They are Application layer, Resource layer and service layer.

Application layer: This layer functionality is to retrieving and receiving the demand from external data source such as customer, retailer and factory and trigger the service layer to process the information.

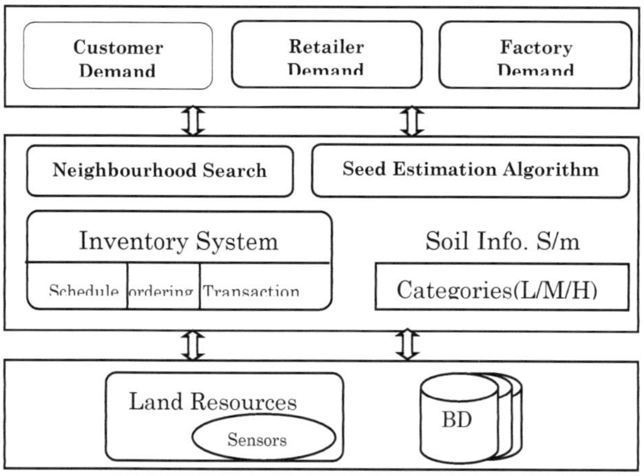

Figure 2: System Architecture Diagram

Service layer: This is the core part of the framework which has been proposed to control the losses of Jatropha seed supply chain. It is classified into four components to handle the predicted losses. Follows are the details of the four components.

Inventory System can forecast the demand and can handle the issues of scheduling, ordering and transaction

Neighborhood search algorithm is used to optimize the search of factory details.

Soil information system to predict the estimation of water content and they are categories as medium, low and high. The predicted outcome is captured by seed estimation algorithm.

Seed estimation algorithm calculate the yield estimation of jatropha seed and the extracted seed is dispatched for esterification of biofuel by identifying the optimal Factory.

Resource layer: It consist of two components as follows

M_w - Mass of Water

M_m - Mass of Mineral

V_t – Total Volume

Þ$_w$ – density of water

Land resource component consist of sensor which is used to sense the moisture content of the soil and the information is given to soil information system to predict the estimation of water content.

Big data This component provides the functionality of storage and retrieving various data such has land resource details, inventory detail, factory details, supply chain network member data and production information.

The Figure 3 shows the step by step procedure for measuring the Volumetric Water Content (Θ). The predicted Θ value is used to categories the soil into three types as High (H), Medium (m) and Low (L).

The Seed Estimation algorithm is proposed for total Jatropha seed yield estimation.

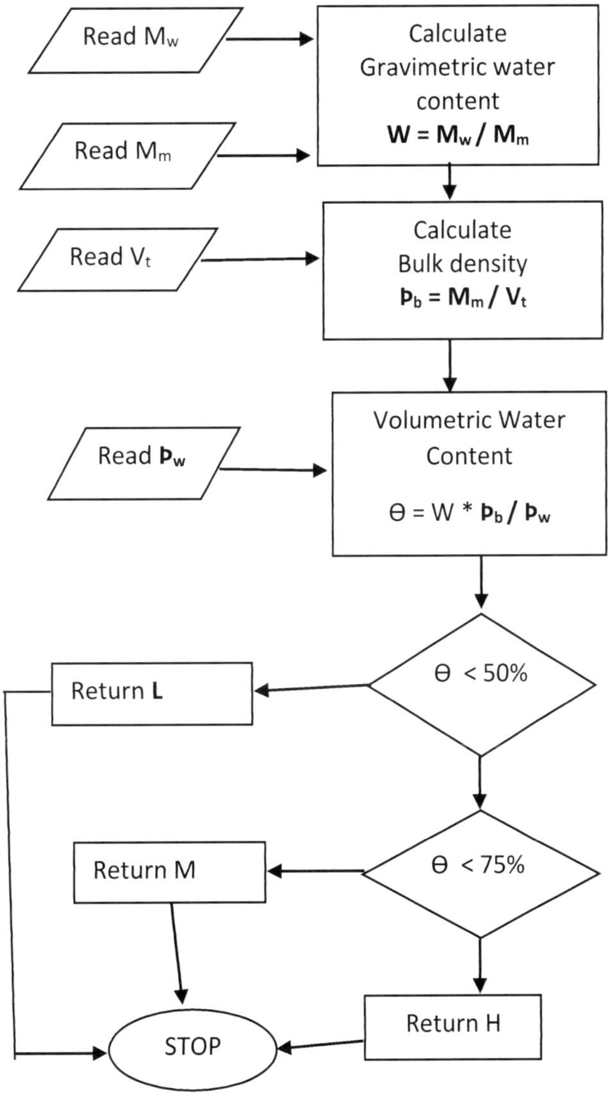

Figure 3: Flow chart Categorizing Soil

Seed Estimation Algorithm:

This algorithm is used to calculate the total seed estimation, for which soil type, collar diameter, age and height of the tree is given as input. The sensor senses the soil information and categories the type of the soil (H, M, L). Based upon the type of soil the production value (P) has been set and total Jatropha seed has been estimated.

IOP Conf. Series: Materials Science and Engineering **211** (2017) 012007 doi:10.1088/1757-899X/211/1/012007

```
Seed_Estimation(Soil_Type, Col_Dia[n], Age[n], HeighT[n]):
//Estimating Total number of Jatropha Seed
//Input:Coller diameter, Height, Age of tree, Soil Type, a,b,c
//Output: Total estimation of Jatropha Seed
If Soil_Type = H
    Then P = 1.0
Else If Soil_Type = M
    Then P = 0.75
Else
    P = 0.5
For i ← 1 to n
YSi = P * { Col_Diai + c*Agei/In(HeighTi) }[13]
```

$$\text{Total\_Estimation} = \sum_{i-1}^{n} YS_i$$

Neighborhood Search Algorithm:

This algorithm is proposed to locate optimal factory for extraction of biofuel. The input parameters are total seeds (S_n), capacity of factory (C_n) and Locations of factories (L_n). The step by step procedure to find the shortest distance from the source of the seed has been given.

```
Step 0:
Total Seed S ← {S1, S2, ... , Sn }
Factory_Details ← {(C1, L1),(C2,L2),(C3,L3), ... ,(Cm,Lm)}
Step 1:
For i ← 1 to n
For j ← 1 to m
        Find the distance by using Euclidian Distance. (i.e., one point from current location point,
                another point from Factory Location)
Sort distance of factories in non-decreasing order.
Step 1.1:
        Select the min distance factory
Step 1.2:
        If Si <= Cj
                Move Si to Cj
                Reduce Si from Cj
        Else
                Select next min distance factory
                Goto step 1.2
```

4. Conclusion

Supply chain management is focused by all the researchers in the world today, Bio fuel supply chain management is one of the important research direction. The biggest challenge faced by farmers and manufactures of bio fuel has been addressed in this proposal by applying various technologies in each stage of the supply chain process. This paper addresses the problem of supply chain losses and proposes a step-by-step procedure for predicting yield of seed estimation and conceptual frame work of IoT to reduce the losses in Jatropha supply chain. The proposed framework helps to make more optimal decision including how to control costs, improve demand forecast and enhance yield of Jatropha seed the future work will focus on constraint-based optimization, advanced forecasting and modeling of system.

References

[1] Akbar Emil, et al. 2009 "Characteristic and composition of Jatropha curcas oil seed from Malaysia and its potential as biodiesel feedstock feedstock." *European journal of scientific research* 29.3: pp 396-403.

[2] Ariza-Montobbio Pere, et al. 2010 "The political ecology of Jatropha plantations for biodiesel in Tamil Nadu, India." *The Journal of Peasant Studies* 37.4 p 875-97.

[3] Srinivasan, S.P. & Malliga, P. 2010 A new approach of adaptive neuro Fuzzy Inference System (ANFIS) modeling for yield prediction in the supply chain of Jatropha. *In Industrial Engineering and Engineering Management*, 17th International Conference, IEEE.

[4] Ogunwole, J. O., et al. 2008 "Contribution of Jatropha curcas to soil quality improvement in a degraded Indian entisol." *Acta Agriculturae Scandinavica Section B–Soil and Plant Science* 58.3 p 245-251.

[5] Valizadeh, Maryam, Syafiie Syafiie, and Intan Salwani Ahamad 2014 "Optimal planning of biodiesel supply chain using a linear programming model." Industrial Engineering and Engineering Management (IEEM), 2014 IEEE International Conference on. IEEE.

[6] Abdurrahman, Mehamed Ahmed, Gebremedhn Mehari Gebru, and Tsigabu Teame Bezabih. "Sensor Based Automatic Irrigation Management System." *International Journal of Computer and Information Technology*: 2279-0764.

[7] Liu, Bo, Keyao Wang and Ran Zhang. 2016 "Variable neighborhood based memetic algorithm for distributed assembly permutation flowshop." Evolutionary Computation (CEC), IEEE Congress on. IEEE.

[8] Palak, Gokce, Sandra D. Eksioglu, and Joseph Geunes. 2013 "Models for Cost Efficient and Environmentally Friendly Inventory Replenishment Decisions for Perishable Products." IIE Annual Conference. Proceedings. Institute of Industrial and Systems Engineers (IISE).

[9] Behera, Soumit K., et al. 2010 "Evaluation of plant performance of Jatropha curcas L. under different agro-practices for optimizing biomass–a case study." Biomass and bioenergy 34.1 p 30-41.

[10] Zhao Ji-chun, et al. 2010 "The study and application of the IOT technology in agriculture." Computer Science and Information Technology (ICCSIT), 2010 3rd IEEE International Conference on. **2** IEEE.

[11] Giri, S. Dutta and S. Neogy 2016 "Enabling agricultural automation to optimize utilization of water, fertilizer and insecticides by implementing Internet of Things (IoT)," 2016 International Conference on Information Technology (InCITe) - The Next Generation IT Summit on the Theme - Internet of Things: Connect your Worlds, Noida, p 125-131.

[12] Geetha S. and Sathya Priya R., 2016. Smart Agriculture Irrigation Control Using Wireless Sensor Networks, GSM and Android Phone. Asian Journal of Information Technology, **15** p 3780-86.

[13] Tewari, V. P., and D. K. Mishra 2013 "Modelling Seed Yield Estimation in Jatropha Curcas Plantations in Gujarat." Indian Forester 139.9 p 784-789.

IOP Conf. Series: Materials Science and Engineering **211** (2017) 012008 doi:10.1088/1757-899X/211/1/012008

Defining and comparing vibration attributes of AlSi10 foam and CFRP coated AlSi10 foam materials

O Çolak[1*] and L Yünlü[2]

[1]Anadolu University, Faculty of Engineering, Department of Mechanical Engineering 26555 Eskisehir, Turkey
[2]Mehmet Akif Ersoy University, Faculty of Engineering, Department of Mechanical Engineering, 15030 Burdur, Turkey

*E-mail: oguzcolak@anadolu.edu.tr

Abstract. Now, Aluminum materials have begun being manufactured as porous structures and being used with additive composite materials through emerging manufacturing technologies. These materials those porous structures have also begun being used in many areas such as automotive and aerospace due to light-weighted structures. In addition to examining mechanical behavior of porous metallic structures, examining vibration behavior is important for defining characteristic specifications. In this study, vibration attributes belong to %80 porous AlSi10 foam and CFRP coated %80 porous AlSi10 foam are determined with modal analysis. Modal parameters such as natural frequencies and damping coefficient from frequency response functions at the end of hammer impact tests. It is found that natural frequency of CFRP coated AlSi10 foam's is 1,14 times bigger than AlSi10 foam and damping coefficient of CFRP coated AlSi10 foam is 5 times bigger than AlSi10 foam's with tests. Dynamic response of materials in various conditions is simulated by evaluating modal parameters with FEM. According to results of the study, CFRP coating on AlSi10 foam effect vibration damping and resonance avoidance ability positively.

1. Introduction

While designing mechanical systems, Physical attributes of materials that generate parts should be known well. Modal analysis is one of the most effective methods for defining physical attributes and it is used commonly. Natural frequencies, damping coefficient and mode shapes could be found by modal analysis [1]. Frequency interval without resonance could also be found by natural frequencies. Additionally, when Effects of force on materials will wither could be determined by damping coefficient. Materials that used in the study could be produced by several methods. In this study, Powder metallurgy (PM) method is used. PM method works with metal powders such as aluminum and TiH2 and also utilizes high pressure and heat effects. After pressure implementation and heating processes, TiH2 powders are vaporized. So materials get porous structure [2]. Interest on using aluminum foam increases because of high energy absorption capacity and low density attributes with in automotive, aerospace and railways where need safety and lightness. Because of porous structures, metallic foams have also good impact absorption ability. Under low stress, metallic foams have energy absorption ability more than non-porous metals [3]. Despite all these good attributes, aluminum foams isn't suitable directly uses due to low stress resistance. They should be supported with composite materials at using areas. Those composite materials bear supportive and protective purposes. Some composite component materials such as GFRP and CFRP materials are used for coating metallic foams. In previous studies, GFRP coated aluminum foams' attributes have examined [4]. In this study, CFRP coated aluminum foams' vibration responses have defined.

2. Experimental study

2.1 Materials properties

In experimental study, two parts of AlSi10 foam and two parts of CFRP coated AlSi10 foam specimens were tested for modal measurements. Ingredients, pore structures, specimen dimensions belong to specimens is shown at Table 1.

Table 1. Properties of specimens.

Specimens No	Specimens Name	Ingredient	Coating	Dimensions	Pore
1	AlSi10 foam	%82,3Al,%17Si,%0,14Ti,%0,56O	None	50×125×10	%80
2	CFRP coated AlSi10 foam	%82,3Al,%17Si,%0,14Ti,%0,56O	CFRP	50×125×12	%80

Aluminum foam materials could be manufactured by various methods. Aluminum foam component's percentages could be regulated in those methods. These specimens are named as "Alulight GmbH" by the company that invents this material. Additionally, There are some another types of aluminum foams invented such as "Shunk GmbH AlSi7" [5].

Powder metallurgy method is used for manufacturing Alulight. In this method, two kind of powder is mixed. This mixture should be homogeneous as possible. Mixed metal powders consist of aluminum alloy powders and porous maker powders. TiH2 powders are used for getting porous structure. In this mixture, H2 elements remove from the material when heat interacts the mixture. Main materials percentages are defined depending on wanted porous amount. These materials are manufactured at five main steps. Firstly, all powders is mixed. Turbula mixer is used for mixing process. Mixed powders could be compressed in two ways as hot or cold. Then, heat treatment is implemented on compressed parts. Heat effect begins at 380°C and the optimum temperature is 450°C. Finally, porous structure is got when porous maker powders are vaporized. Whole processes are showed at Figure 1 [6,7].

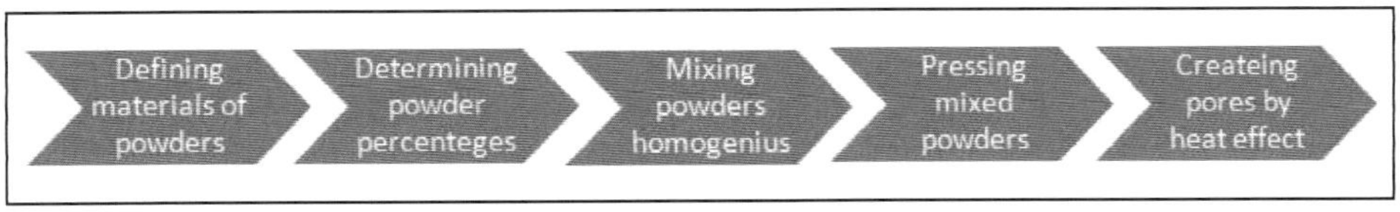

Figure 1. Alulight foam material manufacturing process.

Two numbers of CFRP coated and non-coated foams have been provided and tests were repeated on each specimen. So, determined natural frequencies damping coefficients after experiments are evaluated together and materials characteristic specifications were examined more accurate. In addition, to get more data from each specimen, data was achieved from two points on each specimen. Characteristic natural frequencies and damping factors were examined by averaging data got from tests.

In this study, either porous AlSi10 foams or CFRP coated porous AlSi10 foams attributes such as natural frequencies and damping factor were defined. Vibration damping capacities of porous structure materials are higher than non-porous structure materials. But also, the more pore amounts rise, the lower mechanical strength on porous structure materials are [8]. Besides, how natural frequencies and damping factors are effected coating foams with CFRP materials were observed by experiments. Specimens which used in experiments are showed at Figure 2.

Figure 2. AlSi10 foam (left) and CFRP coated AlSi10 foam (right) materials specimens.

2.2 The experimental setup

To obtain natural frequencies and vibration damping factors belong to specimens controlled experimental facilities that testing materials are needed. Therefore, the experimental setup which system output gets when the defined inputs are implemented was prepared. Designed experimental setup operates with converts analog data to digital data to show natural frequencies and damping factors on computer interface.

Experimental setup consists of main five components: Such as; 1) Material and fixing tools 2)Impact Hammer 3)Acceleration Sensor 4)DAQ/Amplifier system and 5)MalDAQ Cutpro™ Modal Analysis software.

3. Results evaluation

3.1 Experimental data evaluation

In experiments, achieved data from specimen had the same distances between each other. So specimens surfaces were made grid. In each specimen, two different positions of sensor were defined and then data were achieved from surface. Mean value of achieved data were calculated to define general characteristic natural frequencies and damping factors. 5 measurement points of specimens tested in experiments.

Natural frequency numbers were calculated in range of 100-5000 Hz. In this range, frequency of maximum (peak) amplitude value is the natural frequency. Obtained measurements were analyzed by using two different methods. It was found that both results of methods were the same.
Firstly, half power method was used [9].

From the FRF measurement, value of f_n was found as 417 Hz. Equation 1 is applied to measure natural frequency.

$$\omega_n = 2.pi.f_n \quad (1)$$

$$\omega_n = 2622 \; rad/s$$

In addition, damping factor is found applying half power method. In half power method, frequency numbers of 0,707*Amplitude is calculated for peak amplitude. Calculated numbers and natural frequency were evaluated together to find damping factor.

$$\xi = \frac{f_2 - f_1}{2.f_n} \quad (2)$$

$A = 2.8187.10^{-4}$ [N/m]

Frequencies of corresponding amplitude for $0,707A = 1,99.10^{-4}$ [N/m] were found : $f_1 = 413Hz$, $f_2 = 419Hz$, $\xi = 7,2.10^{-3}$

Second method operates with extreme values. In this method, the ranges between extreme points are examined on real part graph of FRF [10].

Natural frequency is found by intersection point of x axis and the curve .

$$f_n = 417Hz, \omega_n = 2622\ rad/s \qquad \xi = \frac{f_{maksreal} - f_{minreal}}{2.f_n} \tag{3}$$

Measurements achieved on each specimen's surface were examined and natural frequency and damping factor were calculated. Average values of natural frequency and damping factor was gotten to attain characteristic vibration response. Calculation results are showed on Table 2 as detailed.

Table 2. Test and the calculation results of AlSi10 foam materials and CFRP coated AlSi10 foam materials.

Material	Specimen and Exp. Number	Natural Frequency f_n	Damping Factor ξ	Average Natural Frequency f_n	Average Damping Factor ξ
AlSİ10 Foam	Spec.1, point 2	417	$7{,}2 \times 10^{-3}$	513	$1{,}1 \times 10^{-2}$
	Spec.1, point 4	503	$16{,}1 \times 10^{-3}$		
	Spec.2, point 2	407	$12{,}3 \times 10^{-3}$		
	Spec.2, point 4	475	$8{,}42 \times 10^{-3}$		
CFRP coated AlSi10 Foam	Spec.3, point 2	551	$43{,}4 \times 10^{-3}$	450	$4{,}1 \times 10^{-2}$
	Spec.3, point 4	608	$35{,}2 \times 10^{-3}$		
	Spec.4, point 2	426	$40{,}0 \times 10^{-3}$		
	Spec.4, point 4	467	$43{,}9 \times 10^{-3}$		

3.2. Finite element method analysis of experimental data

Obtained results with the study were controlled by various ways to reinforce results. Data of experiments are showed at Table 2. Making use of that data and mechanical material specifications, digital simulations were executed on computer. These simulations consist of fundamental modal analysis and FEM. Modal analysis were executed by using average test results of AlSi10 foam materials on "Ansys Workbench program".

Fastening surface of specimens were modeled according to press jaws in CAD design. In addition, because of pressing pressure, specimen's fastening surfaces were dented 1 mm inward. Specimens were modeled similar to test conditions.

Mechanical, vibration, and general material specifications were defined on "Ansys Workbench". Only enough specifications were entered for modal analysis. Entered specifications are showed at Table 3.

Table 3. AlSi10 material specifications

Temperatures °C	Young Modulus Pa	Poison Ratio	Bulk Modulus Pa	Shear Modulus Pa
22	5,333e+008	0,3325	5,3065e+008	2,0011e+008
		Constant Damping Ratio		
		7,2e-003		
		Density		
		53 kg/m³		

After study of FEM (finite element method), results that are showed on Table 4 was found as the same as result as experiments. Natural frequency value of one side fixed specimens found after FEM analysis on 450,97 Hz in Mode Shape 1.

3.3. Achieving mathematical model for first mode

Various results were found with experiment specimens. In this chapter, vibration transfer functions and system response were created as mathematically with test results. So according to results of experimental tests, how the material vibrates time dependent was found under outer effects.

By virtue of results from the tests, modal matrix, stiffness matrix, modal masses and frequency transfer functions of system responses were found. An application for specimen 1 and point 2 was explained [10].

$$k_1 = \frac{-1}{2.\xi.h_{11}} \tag{4}$$

$$m_1 = \frac{k_1}{\omega_{n1}^2} \tag{5}$$

$$u_{11} = \sqrt{-2.\xi.\omega_{n1}^2.h_{11}} \tag{6}$$

$$s = j.\omega_n \tag{7}$$

$$\frac{x_1}{F_1} = \frac{u_{11}u_{11}}{s^2 + 2.\xi.\omega_n.s + \omega_n^2} \tag{8}$$

Equation 4,5,6,7 and 8 were implemented step by step. Then results were gained.

$h_{11} = -3,76.10^{-11} m/N$ $\xi_1 = 7,2.10^{-3}$ $k_1 = 1,85.10^{12} N/m$ $m_1 = 269058kg$

$u_{11} = 1,93.10^{-3}$ $\frac{x_1}{F_1} = \frac{3,7.10^{-6}}{s^2 + 2.7,2 + 10^{-3}.\omega_n.s}$

FRF of the system was found [9].

$$c = 2.\xi.\sqrt{k.m} \left[N.\frac{s}{m} \right] \tag{9}$$

$$\varnothing = tan^{-1}\left(\frac{ImG(r=1)}{ReG(r=1)}\right) \tag{10}$$

$$r = \frac{\omega}{\omega_n} \tag{11}$$

In case of damping factor "ξ" is between 0-1 and the system is second order delayed, time-amplitude correlation (system response) was created in Equation 6.

$$x_{(t)} = \frac{v_0}{\sqrt{1-\xi^2}.\omega_n}.e^{-t.\xi.\omega_n}.sin\left(\omega_n.\sqrt{1-\xi^2}.t + \varnothing\right) \text{ [m]} \tag{12}$$

$$v_0 = \frac{F.\Delta t}{m} \text{ [N.s/m]} \tag{13}$$

$$T_0 = \frac{1}{f_n}, \Delta t = \frac{T_0}{20} \tag{14}$$

When unknown phrases put in the equation 12;

$$x_{(t)} = F.1,7.10^{-13}.e^{-17,24.t}.sin(2620.t - 90) \text{ [m]}$$

System response was found depend on impulse force magnitude (F) and time (t). The system response provides to calculate vibration amplitude which is occurred due to impulse force effect ranging between 100-1000 Hz frequencies.

4. Results and discussions

- The end of the result of tests on AlSi10 foam materials and CFRP coated AlSi10 foam materials, natural frequency of AlSi10 foam was found as 2830 rad/s and also natural frequency of CFRP coated AlSi10 foam was found as 3223 rad/s. According to natural frequencies of materials, natural frequency of CFRP coated AlSi10 foam is higher than natural frequency of AlSi10 foam. According to results, CFRP coated AlSi10 foam materials can operate up to higher frequencies without going into resonance. So using areas of the material is widen.
- As a result of the experiment, damping factor of AlSi10 foam materials found as $1{,}1 \times 10^{-2}$, damping factor of carbon-fiber coated AlSi10 foam material found as $4{,}1 \times 10^{-2}$. According to results, damping factor of carbon-fiber coated AlSi10 foam is almost 3,7 times higher than AlSi10 foam material's damping factor. So, it is understood that if AlSi10 foam is coated by CFRP, damping factor of material increases about 3,7 times. After the evaluation, it can be indicated that CFRP coated AlSİ10 foams could be used under vibrated environments.
- Because of mechanical properties and vibration damping capacity of AlSİ10 foam materials, they can't be used without coated by other materials [11]. They can only be used where properties of lightness and good vibration damping, when they are supported by some kind of materials. Especially, excess vehicle weight which one of the biggest problem of aircraft could be fixed by using these coated AlSi10 foams.
- Natural frequency values are found also vibration damping factors. Only one natural frequency value is found range between 0-1000 Hz. According to this result, it could be easy to avoid catching resonance.
- Utilizing equations given at section 3.3, mathematical model and vibration response of the material is identified. Examining vibration response, how the material vibrates can be predicted under impulse forces.

References

[1] Ucun I and Colakoglu M 2004 Değişik kesitli profil malzemelerdeki doğal titreşim frekanslarının ölçülmesi ve hesaplanması Teknoloji **3** 7 pp 435-44
[2] Türker M 2009 Toz Metalurjisi Yöntemi İle Alüminyum Köpük Üretimi 5. Uluslararası İleri Teknolojiler Sempozyumu (IATS'09) pp 13-5 Mayıs Karabük, Türkiye.
[3] Zhang C, Feng Y and Zhang X 2010 Mechanical properties and energy absorption properties of aluminum foam-filled square tubes *Transactions of Nonferrous Metals Society of China* **20** 8 pp 1380-6
[4] Crupi V, Kara E, Epasto G and Guglielmino E 2015 Prediction model for the impact response of glass fibre reinforced aluminium foam sandwiches *International Journal of Impact Engineering* **77** pp 97–107
[5] Crupi Vand Montanini R 2007 Aluminium foam sandwiches collapse modes under static and dynamic three-point bending *International Journal of Impact Engineering* **34** pp 509–21
[6] Baumgärtner F, Duarte Iand Banhart J 2000 Industrialisation of P/M foaming process *Advanced Engineering Materials* **2** 4 pp 168-74 ISSN 1438-656
[7] Degischer H P and Kriszt B 2002 *Handbook of Cellular Metals* Wiley-VCH, 2002, ISBN 3-527-30339-1 Weinheim
[8] Xie Z K, Tane M, Hyun S K, Okuda Y and Nakajima H Vibration-damping capacity of lotus-type porous magnesium
[9] Uyar E 2011 Sistem Dinamiği ve Otomatik Kontrol, Dokuz Eylül Üniversitesi Mühendislik Fakültesi Yayınları **313** 2 pp 144-51
[10] Altintas Y 2011 Manufacturing Automation (Cambridge University Press issue 2) pp 104-10
[11] Baumeister J, Monno M, Goletti M, Mussi V and Weise J Dynamic behavior of hybrid APM (advanced pore morphology foam) and Aluminum foam filled structures

Inventory transparency for agricultural produce through IOT

S.P. Srinivasan [1], D. Sorna Shanthi[2] and Aashish V Anand[3] *

[1]Professor and Head, Department of Mechanical Engineering, Rajalakshmi Engineering College, India
[2]Associate Professor, Department of Computer Science &Engineering, Rajalakshmi Engineering College, India
[3]Under Graduate, Department of Computer Science &Engineering, Rajalakshmi Engineering College, India

*Email: aashishvanand@gmail.com

Abstract. Re-structuring the practices of traditional inventory management is becoming more essential to optimize the supply chain transparency and accuracy of agricultural produce. A flexible and transparent inventory management system is becoming the need of any agricultural commodity. It was noticed that the major setback for the farmers who are the suppliers of the farm produce is due to poor supply chain integration. The recent advent technologies and IT explosion can bring up a greater impact in the process of storing, tracking, distributing and monitoring perishable agriculture produce of day to day life. The primary focus of this paper is to integrate IoT into inventory management and other inbound logistics management of agriculture produce. The unique features of agricultural produce like a prediction of supply, demand, the location of warehouses, distribution and tracking of inventory can be integrated through IoT. This paper proposes a conceptual framework for inventory management transparency involved in the supply chain of agriculture produce.

1. Introduction

India with 127, 0272105 people is the second largest populous country in the world. India's total gross area of 3287260 KM^2 and total cropped area is about 19202 million hectares, and the net sown area is 140 million acre. The supply chain of agricultural produce has grabbed a full consideration as most of the agricultural produce is ordained for human nurturing. Cultivation of vegetables in peri-urban or urban areas is recognized for its potential role in increasing food security, poverty alleviation, employment and income generation, community resource development, waste management and environmental sustainability [1].The eclectic changes in the current human era, where much importance is given to health, consumers are mandated to be concerned with all processes like cultivation, storing, distribution and marketing of the agricultural produce [2]. To note another, the farmers who are the suppliers of the agricultural products falls out with a financial set-back, due to poor supply chain integration [3]. All this put together creates a must essential transparency during the inventory management of the agricultural produces especially perishable produces with reasonable lead time and very less shelf life in the warehouse or at retailer's occupancy.

The recent advent technologies and IT explosion can bring up a greater impact in the process of storing, tracking, distributing and monitoring perishable agriculture produce of day to day life. The use of Information and Communication Technology (ICT) in the supply chain provides significant benefits in the process of involving information exchange to centralized data storage cloud storage [4]. The availability of information will be helpful in the decision-making process of planning and forecasting supply and demand for agricultural produce [5].The current Internet technologies can also be used for the operational management of food supply chain [6][7].

The primary focus of this paper is to integrate IoT (Internet of Things) into inventory and other inbound logistics management of agriculture produce. The integration of IoT layers, like the data transmission layer, the storage layer in the cloud and visibility of data collected through mobile applications can provide the inventory transparency between different actors in the supply chain. As a consequence, the use of IoT in the supply chain of agricultural produce will bring out the required transparency in inventory and logistics level.

2. Methodology

2.1. Traditional inventory practices issues

Availability of rail, road and air transport, cold storage, processing units, export houses and well-established market network, will be helpful for taking up vegetable cultivation in peri-urban areas of metro cities in an organized way. A Huge quantity of solid waste generated during handling and marking of fresh vegetable produce in metro cities, creating health and environmental hazards, can be used or recycled to vermicompost for use in organics vegetable cultivation and distribution. Vegetable quality needs to be maintained since harvest. Growers must take care of harvesting, storing, packing and transporting fresh produce to prevent damage and maintain quality. Compared to Manufacturing products Agriculture products supply is unpredictable, inelastic demand and price and inventory are uncertain. The decision on how much to sell is a major issue in the agriculture produce.

Leafy vegetables kept in the shade will maintain their quality longer than those exposed to full sun. Many growers put their newly harvested leafy greens in large polyethylene bags, during the transport the leaves may be bruised or crushed. Bamboo baskets are not suitable for packing soft fruit vegetables such as tomato due to the rough surface may damage the produce. Rigid containers with smooth surfaces reduce crushing, bruising and damage. Many cities cannot cope with the massive growth of its population which leads to a decrease in urban shelter and security of basic services, increasing inequality and segregation, degradation of the urban environment and growth in poverty, malnutrition and food insecurity. The transparency in Cultivation, storage, and distribution is becoming a need of the hour to avoid these issues [8]. The problems related to this perishable agricultural produce is mentioned in Figure 1.

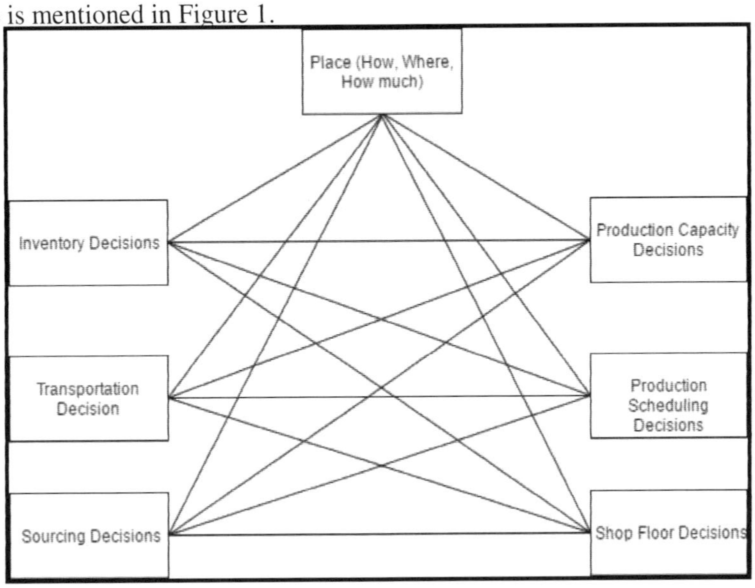

Figure 1. Inventory issues.

2.2. Inventories

There are two different strategies related to finished goods inventory holding at manufacturing sites: make-to-stock (MTS) versus make-to-order (MTO) [9]. MTS decisions refer to decisions to produce finished goods to stock based on planning and forecasts. Under make-to-order production starts when a customer order arrives [10]. In regards to MTS vs. MTO decisions, there are some variables related to them such as obsolescence, perishability, lead time ration, as well as the coefficient of variation of sales state that insufficient information sharing and transparency is a key issue for companies that strive to increase coordination and integration in their supply chains [11] IT facilitates transparency and coordination for efficient implementation of various supply chain processes. There was even no Central logistics department that would be in charge of coordination of inventory management and optimizing finished goods inventory stocks at the company's subsidiaries. And there was in general lack of knowledge of inventory management and lack of educated people within logistics [12].

2.3. Inventory methodology

Researchers have developed a variety of robust optimization techniques for inventory problems when the distribution is not known [13]. In this model a container based approach is proposed which is suitable for IoT-based inventory management.

Number of Containers to maintain smooth operations

Number of container units = demand in the cycle size of each container

$$K= D*(TP+TD)/C \tag{1}$$

C = number of units held in each container

TP = time container spends in production part of a cycle (waiting, being filled and moving to the Store of work in progress)

TD = time container spends in demand part of a cycle (waiting, being emptied and moving to the Store of work in progress).

Total cycle length = TP + TD

$$K<D*(TP+TD) * (1+SF)/C \tag{2}$$

Number of container units with safety factor SF = safety factor (generally less than 0.1)

Maximum stock of work in progress

Maximum stock level = K × C = D × (TP + TD) × (1 + SF)

2.4. Inventory transparency issues in agriculture field

The inventories should meet the need for transparency, accuracy, completeness, consistency and comparability. It can be labelled and transformed by microscopic electronic devices, genetic markers for agricultural products. The barcodes which can read with smartphones combine with the internet technology can contemplate more sophisticated ways to monitor and reveal the agricultural produce. RFID (Radio Frequency Identification) tags were smaller, cheaper and flexible. Near Field Communication (NFC) is a short-range wireless technology which is based on the existing standards of RFID A tag can hold a unique identifier which acts as a basis for a vast amount of web-based supporting data. The mobile apps in which the consumers can readily access this internet of things (IoT) gathering provenance information not just at the generic level of the item category but for the particular item.

Providing online verification codes to end users is an easy way to make the supply chain network transparent. The NFC readers provided by the mobile phone manufacturers are allowing customers to link the data to know about produce origin, inventory, certification, price and distributor details. Visual reader technologies in smartphones are more common nowadays.

3. Internet of things and cloud platform

Internet of Things and cloud computing together brands a new revolution of the Internet, whose goal is to enable things/objects to get connected using any wired or wireless connections without concerning the place, time and the object/anyone of communication [14] . IoT makes communication possible through the Internet where data exchange happens using MQTT (Machine to Machine Authentication Protocol) protocol, which makes the IoT devices to act smart. The flow of data between the objects through various platforms supports the tracking, control, and management of them. Deployment of Cloud computing IOT framework can resolve many issues that get raised during the supply chain management. Efficient knowledge management of agricultural information can support the decision-making process of farmers and consumers.

An automated and highly interactive system for an agriculture produce system planning on the operational level is being developed for use in the Mobile phone. This provides both distributor and farmers enables to design and quickly construct demand and supply framework.

As the capacity planning of the warehouse was developed as a part of a whole system to access the scenarios for cost reductions in harvesting and transport. This proposed model measures and impacts the Harvesting, Storage and time to harvest. This shown in figure1 would help avoid many obstacles in the system which are experienced by harvesters requiring deliveries at inefficient times, the start of the delivery and the best overall coverage of demand and supply.

3.1. Near field communication

Near Field Communication is a short-range wireless technology which is based on the existing standards of RFID. Compared to RFID, in NFC there is no strict difference between the reader and transponder. NFC capable device can read data from an NFC tag and can also write data into NFC tag. The reusability and the cost effective factor of NFC have influenced its usage in a wide range of applications.

3.2. Android device

An Internet-enabled Android device equipped with NFC reader can use any GSM/cellular network to communicate information through the Internet, which can be made accessible to the user through any platform. The Android device with GPS antenna can be used to communicate the vehicle/truck tracking information to the cloud storage database.

4. System description: architecture and components

The system architecture explains the four different stages of this inventory management system which includes farmer-supplier unit, truck–tracking and monitoring unit, Cloud storage and processing unit and Storage facility/Retailers unit as illustrated in figure 2.

4.1. Android application

The agricultural produce harvested by the farmers are loaded with the delivery boxes each of which is tagged with an NFC tag. The NFC tags are then stored with the following information: Agricultural produce a name, date of harvest, box number, through the NFC reader embedded in the Android device with the help of the android application as shown in figure 3. The farmer details, agricultural produce details, location information, are also sent to the cloud backend through the

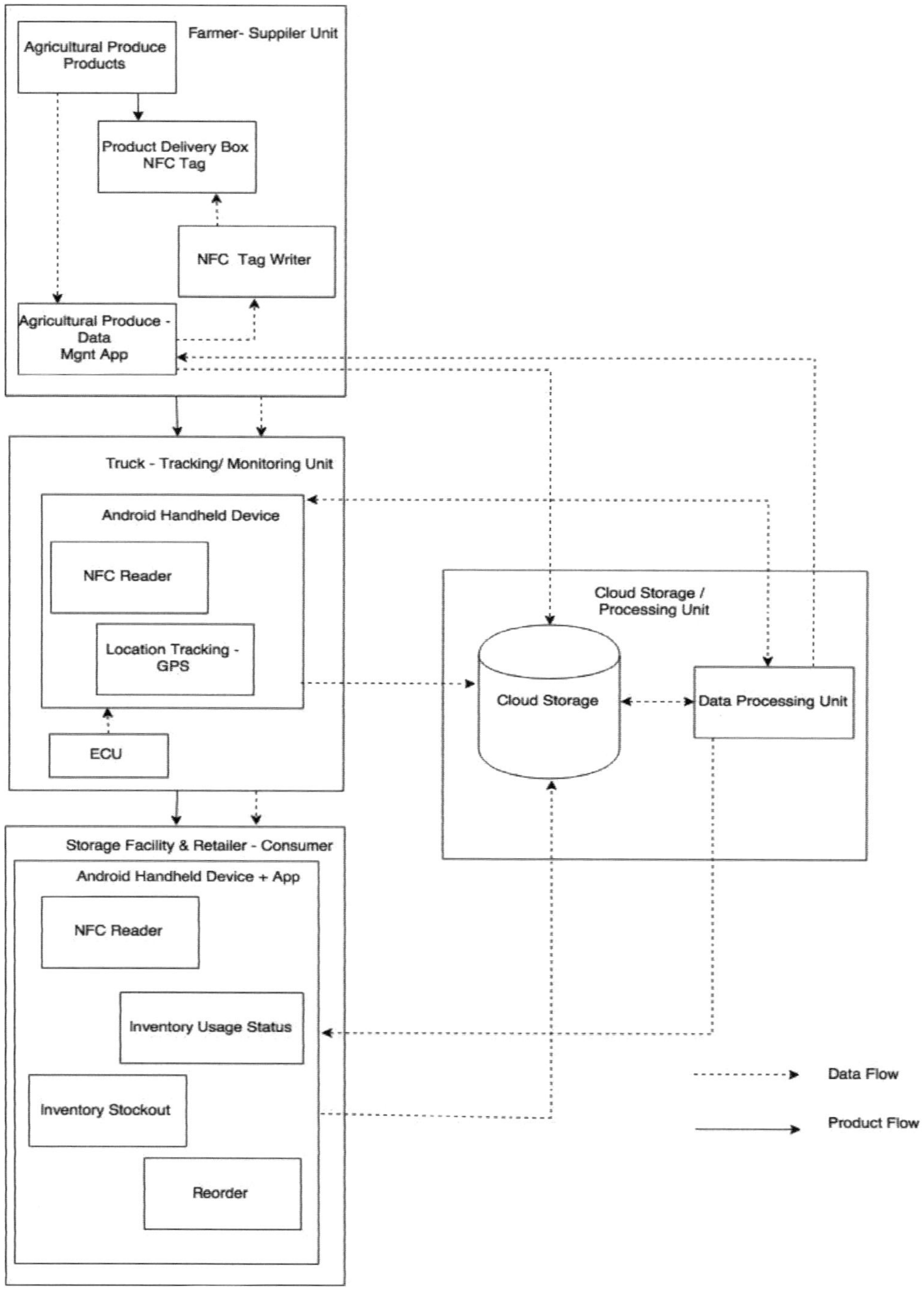

Figure 2. Architecture of proposed system.

Android application as shown in figure 3, through a mobile internet connectivity. The demand of the consumers managed in the cloud storage will be invoked as orders to the farmers, who then initiate the delivery of the agricultural produce through the transportation provided by the warehouses.

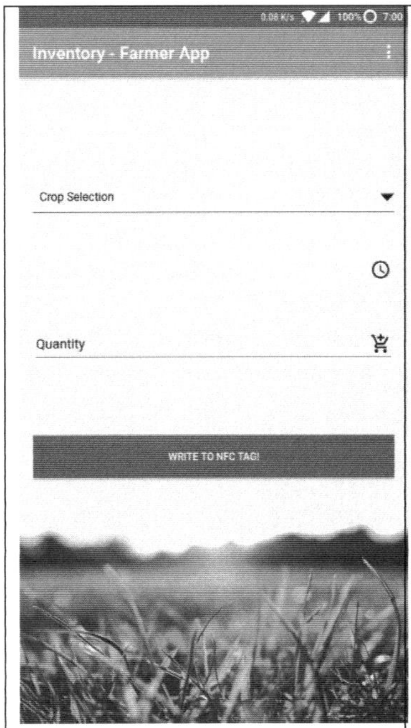

Figure 3(a). Android app screen1.

Figure 3(b). Android app screen2 (regional language hindi).

Figure 3(c). Android app screen3.

4.2. Identification and tracking through NFC

The Box identification is mapped with the truck through the NFC reader enabled Android device present in the vehicle. The ECU (Engine Control Unit) is paired with an Android device, to fetch the ECU data. The GPS data that provides the location of the truck along with the ECU data are sent to the cloud backend to ensure the security, visibility of the agricultural produces on track. Figure 4 shows how the goods in the truck are tracked with the help of GPS.

4.3. Data storage and processing- cloud platform

The instance of the cloud platform provides with a database service (server) which stores the IoT data (supply chain data) which can be accessed from anywhere, anytime and by any actor of the supply chain, by which the visibility and transparency of the inventory process are maintained. The web service (server) handles the request coming from the outside world to redirect it to the application and database server. The truck tracking data stored in the cloud will be used be used to track the product in transport. The delivery time of the orders can be predicted, and Optimized supply based on cost and distance parameters for the required demand is anticipated through Machine learning algorithms, and further notification of the rules is responded to the Android application.

5. Conclusion

This proposed model of Inventory modeling using IoT is a part of the whole system of supply chain network planning. This reduces the unnecessary costs & time in the agricultural supply chain. The agriculture produce spends time waiting for inventory storage space, locomotives, Transport system, etc., can be addressed transparently. Through linking with grower's data and end users data the agriculture produce inventory model will show the consequences across the full industry value chain of proposed inventory information arrangements. The model is data sensitive, the use of geographical information systems and vehicle tracking system has made it possible to generate the model inputs accurately and show solutions in the simpler way for farmers, distributors, and resellers in their smartphones.

References

[1] Stelvia Matos and Jeremy Hall 2007 Integrating sustainable development in the supply chain: The case of life cycle assessment in oil and gas and agricultural biotechnology *Journal of Operations Management* **25** pp 1083–102

[2] O Ahumada and Villalobos J R 2009 Application of planning models in the agri-food supply chain: A review *European Journal of Operational Research* ***196*(1)** pp 1-20

[3] S P Srinivasan and P Malliga 2003 An Optimal Jatropha Seed Warehouse Location Decision Using Myopic and Exchange Heuristics of P Median *Proceedings of 2012 3rd International Asia Conference on Industrial Engineering and Management Innovation (IEMI2012)* pp 439-49

[4] Auramo J, Kauremaa J and Tanskanen K (2005) Benefits of IT in supply chain management: an explorative study of progressive companies *International Journal of Physical Distribution and Logistics Management* Vol 35 No 2 pp 82-100.

[5] Yandra Rahadian Perdana 2012 Logistics Information System for Supply Chain of Agricultural Commodity Procedia - Social and Behavioral Sciences **65** pp 608 – 613

[6] C N Verdouw, A J M Beulens, H A Reijers and J G A J v d Vorst 2015 A control model for object virtualization in Supply chain management *Computer Industry* **68** pp 116–31

[7] C N Verdouw, J Wolfert, A J M Beulens and Rialland 2016 Virtualization of food supply chains with the internet of things *Journal of Food Engineering* **176** pp 128-36

[8] Widodo K H, H Nagasawa, K Morizawa and M Ota 2006 A periodical flowering–harvesting model for delivering agricultural fresh products *European Journal of Operational Research* **170** pp 24–43

[9] Wanke P F and Zinn W 2004 Strategic logistics decision making *International Journal of Physical Distribution and Logistics Management* **34(6)** pp 466-478

[10] Zaerpour N, Rubanni M, Gharehgozli A H and Tavakkoli-Moghaddam R 2008 Make-to-stock or make to-order decision by a novel hybrid approach *Advanced Engineering Informatics* **22(2)** pp 186-201

[11] Dreyer H C, Alfnes E, Strandhen J O and Thomassen M K 2009 Global supply chain control systems: a conceptual framework for the global control centre *Production Planning and Control* **20(2)** pp 147-157

[12] AJ Higgins and LA Laredo 2006 Improving harvesting and transport planning within a sugar value chain ,Journal of the Operational Research Society 57 pp 367–76

[13] C J M Ondersteijn, J H M Wijnands, R B M Huirne and O van Kooten 2006 *Quantifying the agri-food supply chain*, Springer pp 117-32

[14] Ovidiu Vermesan, SINTEF Norway and Peter Friess 2013 EU, Belgium, Internet of Things: Converging Technologies for Smart Environments and Integrated Ecosystems ISBN: 9788792982735

IOP Conf. Series: Materials Science and Engineering **211** (2017) 012010 doi:10.1088/1757-899X/211/1/012010

Optimization of dynamic soaring maneuvers to enhance endurance of a versatile UAV

Imran Mir[1,*], Adnan Maqsood[1] and Suhail Akhtar[2]

[1]Research Center for Modeling and Simulation, National University of Sciences and Technology, H-12 Campus, Islamabad, Pakistan
[2]College of Aeronautical Engineering, National University of Sciences and Technology, Risalpur, Pakistan

*Email: imranmir56@yahoo.com

Abstract. Dynamic soaring is a process of acquiring energy available in atmospheric wind shears and is commonly exhibited by soaring birds to perform long distance flights. This paper aims to demonstrate a viable algorithm which can be implemented in near real time environment to formulate optimal trajectories for dynamic soaring maneuvers for a small scale Unmanned Aerial Vehicle (UAV). The objective is to harness maximum energy from atmosphere wind shear to improve loiter time for Intelligence, Surveillance and Reconnaissance (ISR) missions. Three-dimensional point-mass UAV equations of motion and linear wind gradient profile are used to model flight dynamics. Utilizing UAV states, controls, operational constraints, initial and terminal conditions that enforce a periodic flight, dynamic soaring problem is formulated as an optimal control problem. Optimized trajectories of the maneuver are subsequently generated employing pseudo spectral techniques against distant UAV performance parameters. The discussion also encompasses the requirement for generation of optimal trajectories for dynamic soaring in real time environment and the ability of the proposed algorithm for speedy solution generation. Coupled with the fact that dynamic soaring is all about immediately utilizing the available energy from the wind shear encountered, the proposed algorithm promises its viability for practical on board implementations requiring computation of trajectories in near real time.

1. Introduction

In-spite of being efficient platform, Small Unmanned Aerial Vehicles (SUAVs) are severely limited in their endurance due to on-board energy storage limitations. Without human on board, endurance in SUAVs is mainly dictated by the amount of fuel UAV can carry, which in turn is governed by UAV size and weight constraints. Besides, the smaller size comes with reduced aerodynamic efficiency by flying at lower Reynolds number and the scaling effect does not play in favor of long endurance. It is therefore highly desirable to improve the range, endurance and speed of such small platforms to give them performance parameters matching to that of a platform several times larger. Of recently, great emphasis is being made in using dynamic soaring [1, 2] as means to acquire energy from atmosphere vertical wind gradients for sustained powerless flight. To optimize the energy gain from the atmospheric wind shear, numerical methods are subsequently employed to formulate optimum trajectories for dynamic soaring. Majority of the research work done in formulating optimal trajectories for dynamic soaring involves techniques which are iterative in nature and therefore require substantial time for trajectory optimization. In this research, dynamic soaring optimal trajectories are computed in near real time environment utilizing pseudo spectral technique [3], which is a particular class of direct collocation methods. Pseudo spectral methods which are global in nature have distant advantages over other methods, the most important of which is its fast spectral rate of convergence. This makes such methods the predominant choices in applications that require implementation in near real time environment. In these methods, the original optimal control problem is directly discretized to formulate a nonlinear programming problem (NLP), which is then solved numerically using a sparse nonlinear programming solver to find approximate local optimal solutions. The optimal trajectories formulated in this research are subsequently analyzed to validate different aspects of dynamic soaring and its suitability for real time applications.

2. Problem setup & description

2.1. UAV dynamics

UAV dynamics is represented by a three dimensional point mass model in this work. The dynamic modeling is aligned with other studies [4] for describing system dynamics while formulating dynamic soaring trajectories. The highly coupled non-linear equations of motion utilized in this study are elaborated in equation (1).

$$
\begin{aligned}
&\dot{V}_t = \frac{1}{m}[T - D - mg\sin\gamma - m\dot{V}_{W_x}\cos\gamma\sin\psi] \\
&\dot{\psi} = \frac{1}{mV_t\cos\gamma}[L\sin\varphi - m\dot{V}_{W_x}\cos\psi] \\
&\dot{\gamma} = \frac{1}{mV_t}[L\cos\varphi - mg\cos\gamma + m\dot{V}_{W_x}\sin\psi\sin\gamma] \\
&\dot{h} = V\sin\gamma, \dot{N} = V\cos\gamma\cos\psi, \dot{E} = V\cos\gamma\sin\psi + V_W
\end{aligned}
\tag{1}
$$

where V_∞ is true velocity, γ, ψ and ϕ are flight path, heading and roll angles, V_W is the wind velocity as a function of altitude, T and D are thrust and drag, N and E are the distances along north and east directions and h is the altitude. The geometrical parameters for the baseline UAV utilized in this study are taken from Gao et al [5]. Utilizing the parametric values, prototype UAV was subsequently developed in Solid Works environment. In line with the work done by Mir et al [6], optimal dynamic soaring trajectories for the developed model were than subsequently implemented using rigorously optimized flight parameters.

2.2. Wind shear modeling

Dynamic soaring acquires energy from atmospheric wind shear and is therefore necessary to accurately determine the wind shear conditions. In order to ascertain the estimates for wind shear, two approaches are conventionally followed (a) approximating the wind shear with some known model once the wind conditions are known (b) online estimation of wind shear in unknown wind conditions. Reasonably stable atmospheric conditions exist close to the sea surface with the wind velocity nearly zero at the surface and increasing gradually with altitude. Therefore, the mean velocity profile of actual wind gradients can be approximated using exponential,logarithmic or linear wind models. In this research, linear wind shear model (refer equation (2)), is used to optimize the loiter trajectory.

$$V_W = \frac{\partial V}{\partial h} h = \beta h \tag{2}$$

where β is the linear wind shear parameter

2.3. Optimal control problem formulation

In order to formulate optimal trajectories for dynamic soaring, different type of numerical solvers developed for solving optimal control problems are utilized. Some of the software / algorithm developed are NPSOL [4], ICLOCS [7], GPOPS [5], IDVD [8], PSOPT[9] , PROPT[10], SOCS[11],

AMPL[12] and so on. In this research work, optimal control software GPOPS is utilized to solve the dynamic soaring optimal control problem. GPOPS is a general purpose software developed in Matlab® environment for solving optimal control problems utilizing pseudo spectral techniques. Pseudo spectral methods belong to the class of direct methods and more specifically direct collocation methods. In pseudo spectral methods, the optimal control problem is transcribed to a nonlinear programming problem (NLP) by parameterizing the state and control using global polynomials (basis function are Chebyshev or Lagrange polynomials) and collocating the differential-algebraic equations using nodes obtained from a Gaussian quadrature.

Utilizing three degree of freedom point mass model for UAV dynamics (refer equation (1) and linear wind profile (refer equation (2)), dynamic soaring trajectory optimization problem is configured as an optimal control problem with state and control vectors defined in equation (3) & equation (4) and performance measure in equation (5).

$$\text{State Vector} = [V_t, \psi, \gamma, x, y, h] \tag{3}$$

$$\text{Control Vector} = [C_L, \phi, T] \tag{4}$$

$$J = \min \text{ (wind shear parameter, p)} \tag{5}$$

Different modes of dynamic soaring (basic, loiter and forward flight) are implemented through terminal constraints. Boundary conditions for implementing loiter mode of dynamic soaring is depicted at equation (6).

$$V(t_f) = V(t_0);\ \psi(t_f) = \psi(t_o);\ \gamma(t_f) = \gamma(t_o);\ h(t_f) = h(t_0) + \Delta h; x(t_f) = x(t_0); y(t_f) = y(t_0) \tag{6}$$

Total air relative normalized energy is given in equation (7)

$$E_{norm} = gh + 0.5 * v^2 \tag{7}$$

Initial state condition and path constraints for state/ control are depicted in Table 1. In this problem, initial airspeed, flight path angle and heading angle are left unconstrained, so as to permit the optimization process to determine their optimal values for sustainable soaring flights.

Table 1. Initial and path constraints.

S No	Parameter	*Value*	S No	Parameter	*Value*
1	East distance (x) ranges	[-500, 500] m	7	Lift coefficient range [C_{Lmin}, C_{Lmax}]	[0, 1.5]
2	North distance (y) range [y_{min} , y_{max}]	[-500, 500] m	8	Bank angle range [ϕ_{min} ,ϕ_{max}]	[-70°,70°]
3	Altitude (z) range [z_{min} , z_{max}]	[0, 1000] m	9	Initial State [x_0,y_0,z_0, γ_0,ψ_0,v_0]	[0,0,0, $\gamma_{min} \leftrightarrow \gamma_{max}$, $\psi_{min} \leftrightarrow \psi_{max}$, $v_{min} \leftrightarrow v_{max}$]
4	Velocity range [V_{min} , V_{max}]	[3, 100] m/s	10	State Path Constraints	[$x_{min} \leftrightarrow x_{max}$, $Y_{min} \leftrightarrow Y_{max}$, $\psi_{min} \leftrightarrow \psi_{max}$, $v_{min} \leftrightarrow v_{max}$]
5	Flight path angle range [ψ_{min} ,ψ_{max}]	[-60°, 60°]	11	Control Path Constraint	[$C_{Lmin} \leftrightarrow C_{Lmax}$, $\phi_{min} \leftrightarrow \phi_{max}$]
6	Heading angle range [γ_{min}, γ_{max}]	[-360°, 90°]	12	Terminal Constraints [x_f ,,y_f, z_f γ_f, ψ_f, v_f]	[$x_f = x_o$,,$y_f = y_o$, $\gamma_f = \gamma_o$,$z_f = z_o$, $\psi_f = \psi_o$,$v_f = v_o$]

3. Results & discussion

3.1. Simulation results

Dynamic soaring optimal control problem formulated in GPOPS environment was subjected to extended simulations under different operating conditions to formulate optimal trajectory for loiter mode. The results achieved greatly qualified the optimization criteria and showed promising results. Figure 1 depicts the graphical 3 dimensional view of the optimal trajectory for the loiter mode of dynamic soaring with a cycle time of 24 sec. The normalized energies (total, potential and kinetic energies) are shown in Figure 2, in which the kinetic energy is exchanged for the potential energy as the UAV gains height. After reaching maximum altitude, the UAV takes the high altitude turn and start its downwind flight where the potential energy is traded for the kinetic energy.

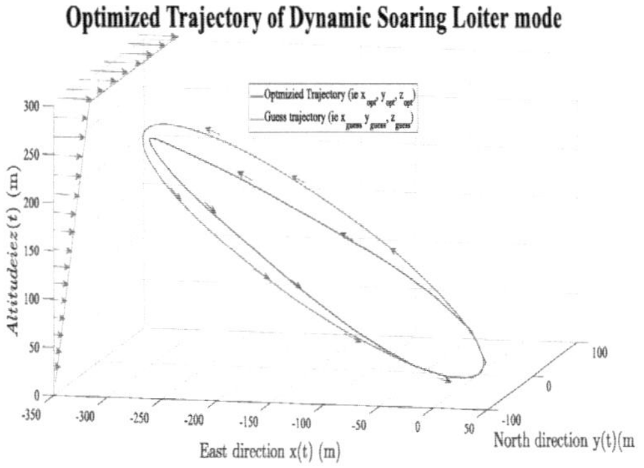

Figure 1. 3D Optimized trajectory.

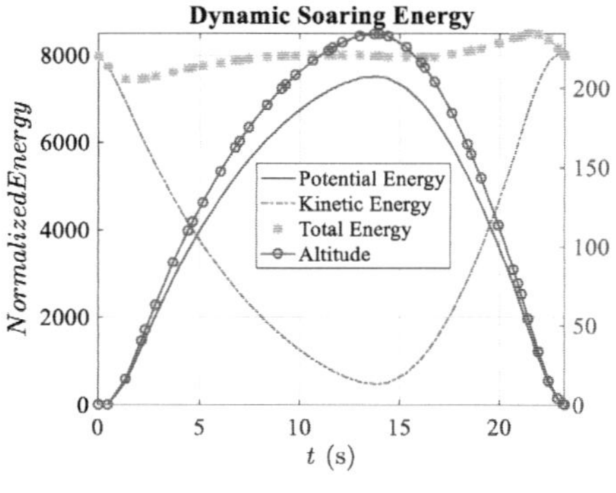

Figure 2. Variation in UAV energy.

Figure 3 reflects UAV attitude showing the flight path, roll and heading angles during various phases of the maneuver. The angles are optimized to implement the desired trajectory.

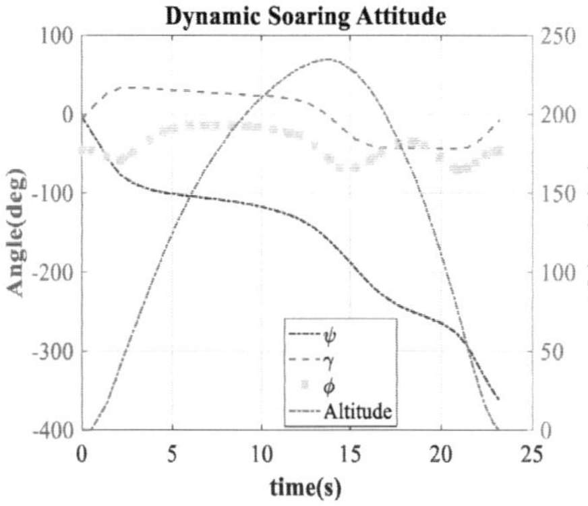

Figure 3. Dynamic soaring attitude.

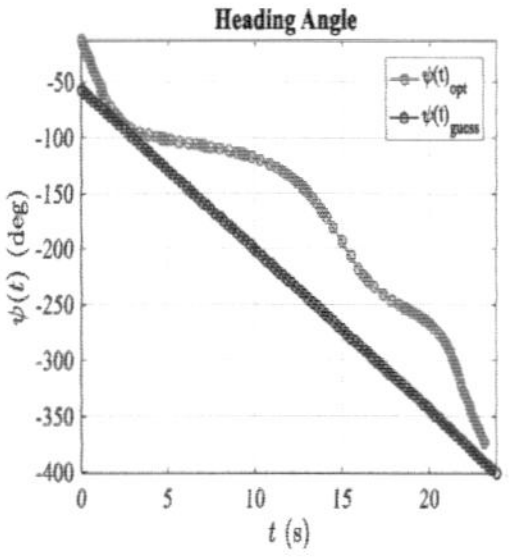

Figure 4. Heading angle.

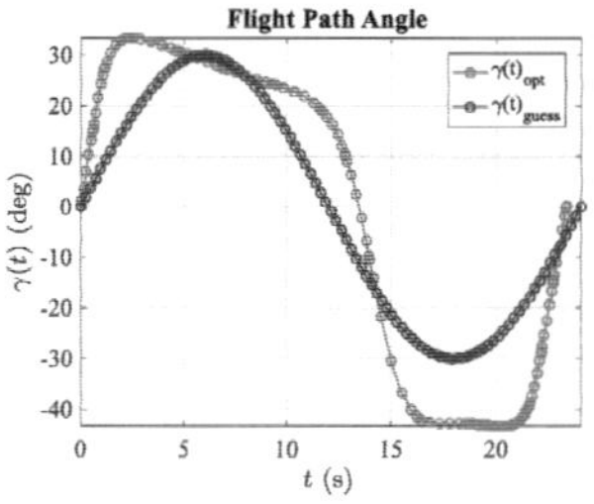

Figure 5. Flight path angle.

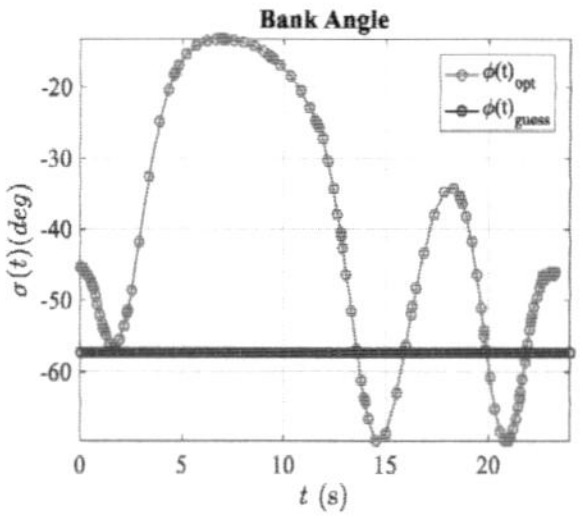

Figure 6. Bank angle.

Optimized trajectory along with guess trajectories for individual UAV states and control variables are depicted from Figure 4 to Figure 6.

It can be seen from Figure 7 that the lift coefficient begin to increase during the windward climb phase and gets to maximum at the highest point (high altitude turn phase) where the flight direction changes from windward climb to downwind decent. Lift coefficient then decreases along the downwind descent phase, goes to the minimum at the lowest point of the maneuver (low altitude turn phase) and then the cycle repeats. Similarly UAV velocity decreases during the wind ward climb phase, reaches to the minimum at the highest point of the maneuver and then increases during the downwind descent phase. Similar trend is also encountered in the bank angle and other states during dynamic soaring cycle. UAV in the start moves away from the shear wind and loses altitude to achieve higher wind shears available near the surface. Subsequently, it then takes a low altitude turn, faces the head wind and gains altitude and comprises kinetic energy from the gained kinetic energy. This smooth shifting between the speed gained while descending and altitude gained during the climb phase is shown in Figure 8.

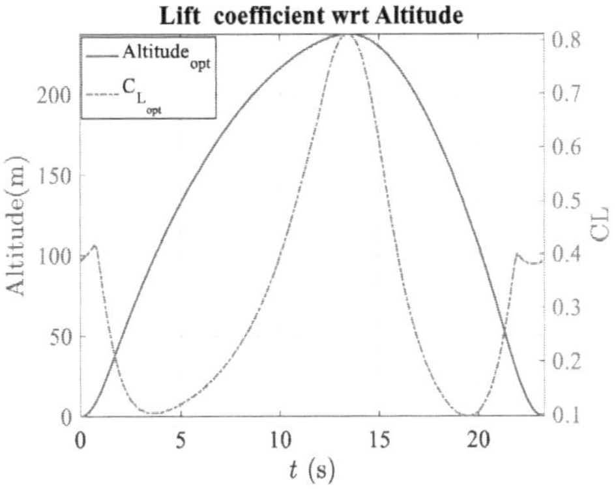

Figure 7. Lift coefficient variation.

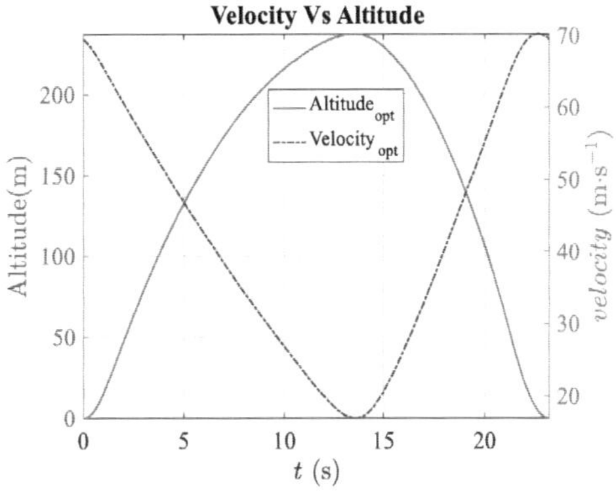

Figure 8. Velocity variation.

3.2. Qualitative comparison with other techniques

The problem of generating optimal dynamic soaring trajectories for UAV in near real time was an area not worked upon by many researchers. In past, majority of the technical studies used off-line numerical optimization techniques such as ALTOS and GESOP [13-15], NPSOL [4, 16] and so on for wind shear calculation and trajectory optimization. Such trajectory optimization software's have a typical computation time of about 100 to 1000sec depending on performance measure. Reinforcement learning technique to acquire speedy results was utilized for such problems due to the inherent goal of balancing exploration and exploitation, but it suffered from the problem of state space complexity and slow learning rates [17, 18]. Akhtar et al. [19] developed trajectory tracking algorithm for powered sailplane that could be implemented in near real time. The technique was based on the "Direct method of Taranenko" [20] could produce optimized trajectories in near real time environment with optimization time of about eight sec. In another study, Akhtar [8, 21] formulated energy-saving trajectories for powered sailplane in near real time by utilizing another method called "IDVD" (Inverse dynamics in the virtual domain). IDVD is a non-linear constrained optimization method, where reference polynomials are determined by the boundary conditions. This method again resulted in considerably rapid solution generation in about eight sec. Similarly , Gao [5] utilizing GPM of GPOPS ([3, 22]) developed dynamic soaring trajectories in real world environment with a computational time of about 15 sec. In this research work, which is a combination of optimization software implemented in GPOPS environment with NLP solver IPOPT, rigorously optimized parameters and guess trajectories, a computational time of below 5 sec is achieved. Generation of optimized trajectories for dynamic soaring within a time frame of less than 5 sec from a nominal core

I-7 platform depicts the effectiveness of the proposed strategy and its viability for practical on board implementations.

4. Conclusion

This paper aims to demonstrate a viable algorithm which can be implemented in near real time environment to formulate optimal trajectories for dynamic soaring maneuvers. The algorithm based upon state of the art optimal control solver and diverse knowledge of dynamic soaring heuristics produces optimal trajectories in a short time of under 5 sec. This optimization time is much lesser than the times achieved earlier. The advantage lies in its characteristics of simplicity as well as rapid trajectory length and parameter computation. Coupled with the fact that dynamic soaring is all about immediately utilizing the available energy from the wind shear encountered, the proposed algorithm promises its viability for practical on board implementations requiring computation of trajectories in near real time.

References

[1] Liu D-N, et al 2016 Bio-inspired energy-harvesting mechanisms and patterns of dynamic soaring *Bioinspiration & Biomimetics*

[2] Liu D-N, Z-X Hou and X-Z Gao 2016 Flight modeling and simulation for dynamic soaring with small unmanned air vehicles *Proceedings of the Institution of Mechanical Engineers Part G: Journal of Aerospace Engineering* p 0954410016641440

[3] Rao A V, et al 2010 Algorithm 902: Gpops, a matlab software for solving multiple-phase optimal control problems using the gauss pseudospectral method *ACM Transactions on Mathematical Software (TOMS)* **37(2)** p 22

[4] Zhao Y Y J and Y C Qi 2004 Minimum fuel powered dynamic soaring of unmanned aerial vehicles utilizing wind gradients *Optimal Control Applications & Methods* **25(5)** p 211-33

[5] Gao X Z et al 2014 Analysis and design of guidance-strategy for dynamic soaring with UAVs *Control Engineering Practice* **32** p 218-26

[6] Mir I A Maqsood and S Akhtar 2017 Optimization of Dynamic Soaring Maneuvers for a Morphing Capable UAV *AIAA Information Systems-AIAA Infotech@ Aerospace* p 0678

[7] Liu Y S Longo and E C Kerrigan 2013 Nonlinear predictive control of autonomous soaring UAVs using 3DOF models *Control Conference (ECC) European IEEE*

[8] Akhtar N, J F Whidborne, and A K Cooke 2012 Real-time optimal techniques for unmanned air vehicles fuel saving *Proceedings of the Institution of Mechanical Engineers Part G-Journal of Aerospace Engineering* **226(G10)** p 1315-28

[9] Becerra and V M 2010 Solving complex optimal control problems at no cost with PSOPT *2010 IEEE International Symposium on Computer-Aided Control System Design*

[10] Rutquist, P E and M M Edvall 2010 Propt-matlab optimal control software *Tomlab Optimization Inc* p 260

[11] Betts J and I Kolmanovsky 2002 Practical methods for optimal control using nonlinear programming *Applied Mechanics Reviews* **55** p 68

[12] Vanderbei and R J 2015 *Linear programming* Springer

[13] Sachs G 2005 *Minimum shear wind strength required for dynamic soaring of albatrosses* Ibis **147(1)** p 1-10

[14] Sachs G and O da Costa 2003 Optimization of dynamic soaring at ridges *AIAA Atmospheric Flight Mechanics Conference and Exhibit Austin Texas*

[15] Sachs G and M Mayrhofer 2001 Shear wind strength required for dynamic soaring at ridges *Technical Soaring* **25(4)** p 209-15

[16] Zhao YJ 2004 Optimal patterns of glider dynamic soaring *Optimal control applications and methods* **25(2)** p 67-89

[17] Wharington J and I Herszberg 1998 Control of a high endurance unmanned air vehicle *Proceedings of the 21st ICAS Congress*

[18] Kyle JA 2006 *Optimal soaring by a small autonomous glider*
[19] Akhtar N, J F Whidborne and A K Cooke 2009 Wind shear energy extraction using dynamic soaring techniques *American Institute of Aeronautics and Astronautics AIAA* p 734
[20] Yakimenko O A 2000 Direct method for rapid prototyping of near-optimal aircraft trajectories *Journal of Guidance, Control, and Dynamics* **23(5)** p 865-75
[21] Akhtar N 2010 *Control system development for autonomous soaring*
[22] Rao A, D Benson, and G Huntington 2010 Constructing an optimal control problem in GPOPS *A MATLAB software for solving multiple-phase optimal control problem using pseudospectral methods* (University of Florida Gainesville) p 18-39

IOP Conf. Series: Materials Science and Engineering **211** (2017) 012011 doi:10.1088/1757-899X/211/1/012011

Incipient loose detection of hoops for pipeline based on ensemble empirical mode decomposition and multi-scale entropy and extreme learning machine

Xiaowei Li[1]*, Qin Wei[1], Yongzhi Qu[2] and Lin Cai[1]

[1]School of Information Engineering, Wuhan University of Technology, Wuhan, Hubei, China
[2]School of Mechanical and Electronic Engineering, Wuhan University of Technology, Wuhan, Hubei, China

*Email: 742745426@qq.com

Abstract. Hoops are very important fittings in hydraulic pipeline, incipient loose detection of hoops will help to prevent hydraulic piping system from breaking down. Since the vibration signals of fluid pipe are non-stationary and of great complexity, multi-scale entropy(MSE), a method characterized by evaluating complexity and irregularity of time series on multiple scales, is used for extracting feature vectors from the vibration signals. In order to obtain components related to system characteristics, ensemble empirical mode decomposition(EEMD) is applied to reconstruct the original signals before the procedure of MSE. Extreme learning machine(ELM) is a new machine learning algorithm characterized by high accuracy and efficiency. In this paper, ELM is introduced as a classifier to identify the different conditions of hoops according to feature vectors extracted by EEMD and MSE algorithms. Thus a novel loose detection method combining with EEMD-MSE and ELM is put forward. The analysis and experimental results demonstrate that the proposed loose detection and feature extraction method for hydraulic pipeline is effective with high performance.

1. Introduction

Hydraulic system pipeline usually works in complex environment such as high temperature, high pressure, strong vibration and fluid-structure interaction and so on. Once frequencies of exciting signals are approximately equal to intrinsic frequency of pipeline, resonance occurs[1]. Resonance phenomena will lead to vibration damage of pipeline. Therefore, it is of great importance to restrain the vibration of hydraulic piping system effectively. One of the most common and economical way is applying hoops to hydraulic pipeline[2]. However, if hoops get loose due to improper assembly or long-time running of hydraulic piping system, it will seriously affect the stability and service life of the hydraulic pipeline. Thus it is necessary for real-time incipient loose detection of hoops in hydraulic piping system, so as to discover hoop failure in time and prevent accidents from happening. In recent years, however, study of hoop mainly focus on modal analysis[3] and optimization of the hoop layout on pipeline[2], relatively few studies on hoop condition monitoring.

Different piping vibration states caused by conditions of hoop can be detected by a vibration sensor, vibration signals under which condition are usually complex non-stationary signal submerged by strong noise[1-3]. As typical signal in field of fault diagnosis, non-stationary vibration signals have been analyzed in time-domain, frequency-domain, and time-frequency domain respectively for many years. Among which empirical mode decomposition(EMD) is a typical one. Although EMD can decompose nonlinear signals into terms of different mode, it still shows mode mixing and end effects problems. To alleviate those problems, ensemble empirical mode decomposition(EEMD) is proposed by Wu and Huang in 2009[4]. In 2015, EEMD method is applied to diagnostics of gear deterioration by Yang et al[5]. It enhanced the accuracy of gear fault diagnosis.

In recent years, feature extraction method based on the statistical analysis has been increasingly introduced into fault diagnosis and condition monitoring of machinery, of which entropy analysis is an important one. In 2015, a fault diagnosis method combined with LMD, sample entropy(SampEn) and energy ratio for roller bearings was presented by Minghong Han[6]. SampEn was an improvement of approximate entropy(ApEn), since ApEn shows apparent disadvantages like the heavy dependence on data length and lack of relative consistency. SampEn shows good traits such as data length independence and trouble-free implementation compared to ApEn. Although SampEn is better and stable than ApEn, it still reflects only single scale information of the time sequence. On this condition, multi-scale entropy(MSE) was proposed by M.Costa et al[7]. MSE actually calculates SampEn on different time scales, it was introduced as a method to evaluate the complexity of time series under different scales. MSE is earliest used in physiological time series analysis, but in recent years, many researchers applied it to mechanical signal analysis and achieved some pretty good results [8]. Unfortunately, there has been only a few study about application of entropy methods in field of hydraulic piping system fault diagnosis thus far.

Generally, after feature extraction of vibration signals for hoop conditions, an intelligent classifier is applied to complete automatic condition monitoring. Automated condition monitoring can make reliable decision quickly and reduce the risk of operation on the running condition of hydraulic piping system. Traditionally, the most widely used intelligent techniques are backward propagation neural network(BPNN) and support vector machine(SVM) and so on. However, gradient-based learning algorithms such as BPNN may face several issues like local minima, improper learning rate and overfitting, etc[9]. SVM has difficulty in large-scale training set because of its huge amount of calculation, and it is also not easy for SVM to make real-time prediction since it normally generate much number of support vectors(computation units). In order to avoid these issues, extreme learning machine(ELM) is proposed by Huang[10], ELM is a new learning algorithm for single-hidden layer feed forward neural networks (SLFNs) which randomly chooses hidden nodes, it tends to reach solutions straightforward without many trivial issues, it has better generalization performance than BPNN and much faster than SVM. The ELM method has been used in many applications[9], in this paper, ELM algorithm is introduced in loose detection of hoops for hydraulic pipeline.

According to the hydraulic piping dynamics theory, different hoop conditions show different impacts on dynamic characteristics of pipeline[1-3]. Therefore the complexity of signals under different hoop conditions is different, as well as its entropy value. Thus MSE method is applied to loose detection of hoops for hydraulic piping system and the entropy values are used as feature vector to evaluate different loose conditions.

In this paper, a method based on EEMD, MSE and ELM is proposed for loose detection of hoops installed on hydraulic pipeline. EEMD-MSE algorithm is used for feature extraction and ELM is introduced as an intelligent classifier to identify the different conditions of hoops. In Section 2, the methods of EEMD, MSE and the feature extraction procedures based on EEMD-MSE are presented simply, as well as ELM algorithm. Experimental analysis is expressed in Section 3. In this Section, method based on EEMS-MSE and ELM is conducted for loose detection of hoops. Moreover, ELM model is compared with BPNN and SVM in performance of loose detection of hoop based on experimental data. Finally, the discussion and conclusion are provided in Section 4.

2. Methods

2.1. Ensemble Empirical Mode Decomposition

The Empirical Mode Decomposition (EMD), proposed by Huang[11], is an adaptive time–frequency data analysis method designed for nonlinear and non-stationary signal analysis. In the EMD approach, the data $x(t)$ is decomposed into terms of Intrinsic Mode Functions($IMFs$), c_j, i.e.,

$$x(t)=\sum_{j=1}^{n} c_j + r_n \tag{1}$$

where r_n is the residue of original data $x(t)$ after n number of $IMFs$ are extracted. Among which $IMFs$ should satisfies two conditions[11]:

(1) Throughout the whole length of a single IMF , the number of extremes and the number of zero-crossings must either be equal or differ at most by one.

(2) At any data location, the mean value of the envelope defined by the local maxima and the envelope defined by the local minima is zero.

However, one shortcoming of the EMD method is the mode mixing problem, which is defined as a single IMF including oscillations of dramatically disparate scales, or a component of a similar scale residing in different $IMFs$ .When mode mixing occurs, physical meaning of individual IMF becomes unclear, what's more, it may even suggest wrong time-frequency distribution.

To solve the problem of mode mixing in EMD, the EEMD method is proposed by Wu and Huang[9]. It based on the fact that the white noise could provide a uniformly distributed scale in the time–frequency space. The intrinsic oscillations in the signal with different scales would automatically associate with the similar scales of reference ridings provided by white noise. The detailed procedures of EEMD are described as follows[4]:

(1) Add a white noise series to the targeted data;

(2) Decompose the data with added white noise into IMFs using EMD method;

(3) Repeat step (1) and step (2) again and again, but with different white noise series each time;

(4) Obtain the ensemble means of corresponding $IMFs$ of the decompositions as the final result.

Through the above steps, problem of mode mixing is solved by the elegant use of noise in EEMD.

2.2. Sample Entropy and Multi-scale Entropy

SampEn[12] is proposed as an unbiased estimator of the conditional probability that two similar sequences of m consecutive data points (m is the embedded dimension) would remain similar when one more consecutive point is included. It represents the complexity and irregularity of time series on single scale. However, acceleration signals of hydraulic pipeline usually own correlation in different time scales. In order to indicate the correlation and complexity of time series under different scales, MSE method is introduced in this paper.

MSE is improvement of SampEn, it actually calculates SampEn on different time scales. Its algorithm process contains two parts, the first part is a "coarse-graining" transform of the time series, while the other part is calculation of SampEn . The specific algorithm procedures is as follows:

Consider the discrete time series $\boldsymbol{x}=\{x_1,x_2,...,x_L\}$ of length L , conduct "coarse-graining" transform of the time series $\boldsymbol{x}$ and get new series $\boldsymbol{y}=\{y_1^{(\tau)},y_2^{(\tau)},...,y_N^{(\tau)}\}$ under scale τ , each element of y is calculated according to Eq.(2):

$$y_j^{(\tau)}=1/\tau\sum_{i=(j-1)\tau+1}^{j\tau} x_i \qquad (1\le j\le L/\tau) \tag{2}$$

where $\tau=1,2,...$ represents the scale factor, $N=L/\tau$, represents length of each "coarse-graining" series y . If scale $\tau=1$, the new series is original sequence.

For each time series corresponding scale τ , given the embedding dimension m , and threshold r ,calculate the sample entropy $SampEn(\tau,m,r)$ [12] of them.

Multi-scale entropy is defined as a collection of sample entropy in multiple scales. So MSE of the time series is as follows:

$$MSE = \{\tau \mid SampEn(\tau, m, r)\} \tag{3}$$

Multi-scale entropy indicates the complexity of time series under different scales. Generally higher entropy value means higher complexity. Therefore, if the entropy values of a time series are higher than another in most of the scales, the former is considered more complicate than the latter. If the entropy values of a time series progressively decrease as the time scales increase, the sequence is considered to be with simple structure and there is more information on lower scales. Otherwise, if the entropy values of a time series progressively increase as the time scales increase, it indicates that there is more information on higher scales.

2.3. The Procedures of Method Based On EEMD-MSE

In this paper, a method based on EEMD and MSE to extract features of hydraulic pipeline is proposed. The whole processing procedure includes reconstructed procedure use EEMD and feature-obtaining procedure with MSE. The detailed procedure is described as follows:

(1) Set the parameter k, calculate k $IMFs$ of original signal $x(n)$ using EEMD method;

$$x(n) = \sum_{j=1}^{k} c_j(n) + r \tag{4}$$

(2) Calculate the Pearson product-moment correlation coefficient r_j of $x(n)$ with each IMF, $c_j(n)$:

$$r_j = \frac{\sum_{i=1}^{n}(x(i)-\bar{x})(c_j(i)-\bar{c_j})}{\sqrt{\sum_{i=1}^{n}(x(i)-\bar{x})^2}\sqrt{\sum_{i=1}^{n}(c_j(i)-\bar{c_j})^2}} \quad (j = 1, 2, \dots, k) \tag{5}$$

where $\bar{x} = \frac{1}{n}\sum_{i=1}^{n} x(i)$ and $\bar{c_j} = \frac{1}{n}\sum_{i=1}^{n} c_j(i)$.

(3) Sort $r_1, r_2, \dots, r_k$ in descending order, choose $IMFs$ whose correlation coefficients with $x(t)$ are greater than threshold λ, add all of them together as the reconstructed signal $y(n)$. Generally $\lambda = 0.4$, when $r_j > 0.4$, it means $c_j(n)$ is significant correlation with $x(n)$.

(4) Calculate MSE of the reconstructed signal $y(n)$, the entropy values of which are treated as feature vector.

2.4. Extreme Learning Machine

Extreme learning machine(ELM) is a new learning algorithm for single-hidden layer feed forward neural networks (SLFNs) which randomly chooses hidden nodes and analytically determines the output weights of SLFNs[10]. In theory, this algorithm tends to provide good generalization performance at extremely fast learning speed.

In ELM model, the input layer is connected to the input feature vector $\mathbf{x}$.The number of input nodes is denoted as N.

At the hidden layer, the number of hidden nodes is denoted as L.The output of a hidden node indexed by i is denoted as $g(\mathbf{x}; \mathbf{w}_i, b_i) = (\mathbf{x} \bullet \mathbf{w}_i + b_i)$, where g is the activation function, $\mathbf{w}_i$ is the input weight vector between the hidden node and all input nodes, b_i is the bias of the node and $i = 1, \dots, L$.

At the output layer, the number of output nodes is denoted as M, each output node represents a traffic sign class. The output weight between the ith hidden node and jth output node is denoted as $\beta_{i,j}$, where $j = 1, \dots, M$. The value of a output node indexed by i is denoted as:

$$f_j(\mathbf{x}) = \sum_{i=1}^{L} \beta_{i,j} \times g(\mathbf{x}; \boldsymbol{\omega}_i, b_i) \tag{6}$$

Thus, the output vector for the input vector $\mathbf{x}$ can be written as:

$$f(\mathbf{x}) = [f_1(\mathbf{x}), ..., f_M(\mathbf{x})] = \mathbf{h}(\mathbf{x})\boldsymbol{\beta} \tag{7}$$

where

$$\boldsymbol{\beta} = \begin{bmatrix} \boldsymbol{\beta}_1 \\ \vdots \\ \boldsymbol{\beta}_L \end{bmatrix} = \begin{bmatrix} \beta_{1,1} & \cdots & \beta_{1,M} \\ \vdots & \vdots & \vdots \\ \beta_{L,1} & \cdots & \beta_{L,M} \end{bmatrix} \tag{8}$$

Class label of input vector $\mathbf{x}$ is determined as $\max f_j(\mathbf{x})$, where $j = 1, ..., M$.

For a supervised training with N training sample pairs, each input feature vector corresponds to a ground truth class label vector $\mathbf{t}_k = [t_{k,1}, ..., t_{k,M}]$ where $k = 1, ..., N$. All labels form a matrix denoted as $\mathbf{T} = [\mathbf{t}_1, ..., \mathbf{t}_N]^T$.

Suppose vector $\mathbf{y}_k$ represents actual output for the input feature vector $\mathbf{x}_k$,taking all training samples into equation (12), can form a linear representation:

$$\mathbf{H}\boldsymbol{\beta} = \mathbf{Y} \tag{9}$$

where

$$\mathbf{H} = \begin{bmatrix} \mathbf{h}(\mathbf{x}_1) \\ \vdots \\ \mathbf{h}(\mathbf{x}_N) \end{bmatrix} = \begin{bmatrix} g(\mathbf{x}_1; \mathbf{w}_1, b_1) & \cdots & g(\mathbf{x}_1; \mathbf{w}_L, b_L) \\ \vdots & \vdots & \vdots \\ g(\mathbf{x}_N; \mathbf{w}_1, b_1) & \cdots & g(\mathbf{x}_N; \mathbf{w}_L, b_L) \end{bmatrix} \tag{10}$$

And

$$\mathbf{Y} = \begin{bmatrix} \mathbf{y}_1 \\ \vdots \\ \mathbf{y}_n \end{bmatrix} = \begin{bmatrix} y_{1,1} & \cdots & y_{1,M} \\ \vdots & \vdots & \vdots \\ y_{N,1} & \cdots & y_{N,M} \end{bmatrix} \tag{11}$$

Aim of the training process is to minimize the training error which is defined as $\| \mathbf{T} - \mathbf{H}\boldsymbol{\beta} \|^2$ and obtain the norm of output weight $\boldsymbol{\beta}$.

Giving a set of training feature vectors $\left\{ (\mathbf{x}_k, \mathbf{y}_k)_{k=1,...,N} \right\}$, activation function g and number of hidden node L the training routine for the ELM-based classifier is described as follows[10]:

(1) Randomly generate hidden neuron parameter such as input weight vector $\boldsymbol{\omega}_i$ and bias b_i, where $i = 1, ..., L$.

(2) Calculate the hidden layer output matrix $\mathbf{H}$ through formula(10);

(3) Calculate the minimum norm least squares solution $\boldsymbol{\beta}$ of $\mathbf{H}\boldsymbol{\beta} = \mathbf{Y}$, therefore, $\boldsymbol{\beta}$ is the estimation of the output weight $\boldsymbol{\beta}$.

Thought procedures described above, a classifier based on ELM with great performance is established.

3. Experiment Analyses

3.1. Experiment Introduction

To confirm the effectiveness of the algorithm, hoop experiments are carried out on fluid flow pipeline. The experimental setup is depicted in Figure 1. It consists of hydraulic station, a straight metal pipe, two hoops, one accelerometer, valves and fittings and so on. Parameters of the pipe in experiment are shown in table 1. Filled with hydraulic oil, the tested pipe is clamped on a fixed support platform and excited by fluid and various external vibrations. The two hoops are installed on the surface of the pipe. Installation positions of the hoops are near the joints on each end of the pipe, as shown in Figure 1.

The accelerometer is installed at the entrances of the fluid flow pipeline to monitor vibration states of the pipe, acquisition of the accelerometer data is conducted by LMS Test.Lab 14A, a high speed multi-channel data acquisition equipment produced by LMS company. The hydraulic station is equipped with various sensors such as pressure sensor, temperature sensor and flow sensor so as to monitor the system status in real time. Control and acquisition platform is depicted in Figure 2. The hydraulic control console is applied to control pressures and flow rate in pipeline, which can be collected through the data acquisition card and can be shown on laptop-A. The LMS equipment is used for acquisition of the accelerometer data, which can be shown on laptop-B in real time.

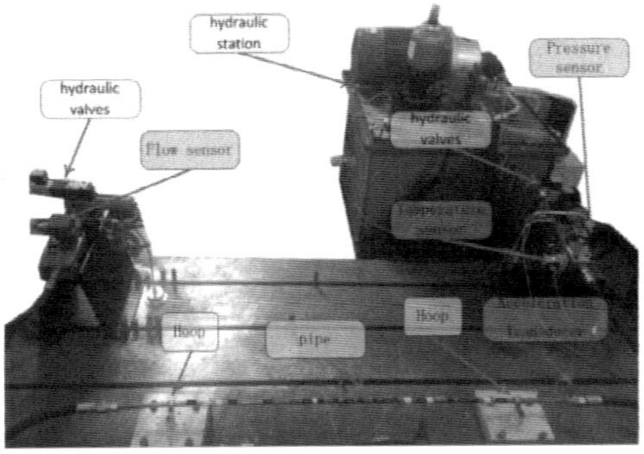

Figure 1. Experimental setup.

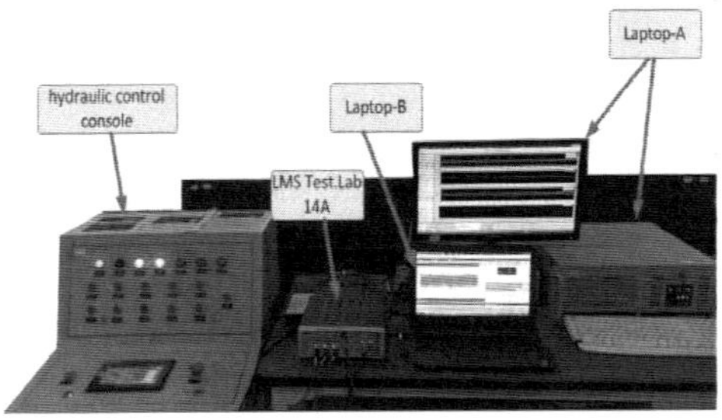

Figure 2. Control and data acquisition platform.

Table 1. Parameters of pipe in experiments.

Length (m)	Outside diameter(mm)	Wall thickness(mm)	Material
1.2	15	1	Steel

Experiments are conducted in fluid flow condition with pump and valve opening. In order to verify the robustness of the algorithm, three experiments are conducted on different days. In each experiment, experimental conditions keep similar. For example, sampling frequency of the accelerometer is 4096Hz, fluid flow rate is controlled at 98mL/s and keep basically stable, pressures in the pipeline varies from 3MPa to 15MPa with interval 3MPa. Both of fluid flow rate and pressure in pipeline can be set through the hydraulic control console and showed in laptop-A. There are two conditions of the hoop, one is normal assembly and the other is loose condition, for loose condition, it is divided into left side loose and right side loose.

Accelerometer data is collected under different pressures and hoop conditions. Sampling frequency is 4096Hz. Experiments are repeated three times, from which 210 samples are collected in total. Duration of each sample is 3 seconds. Number of samples of each hoop state is shown in Table 2.

Table 2. Number of samples of each hoop conditions.

Hoop Condition	Number of Samples
Normal Condition	70
Left Side Loose	70
Right Side Loose	70

3.2. Experimental Results

Three working conditions of hoops for hydraulic pipeline are analyzed through our method. Time domain waveforms and frequency spectrums of the original acceleration signals under normal and loose conditions of hoops are shown in Figure 3. Among which loose conditions consist of left side loose and right side loose respectively. Obviously, it is very difficult to distinguish three hoop conditions just through waveforms of time domain or frequency domain.

To reconstruct the original signals, EEMD is applied to decompose the acceleration signals and then correlation coefficients are used to select fault related. The Pearson product-moment correlation coefficients of each with its original signal under three hoop conditions are shown in Figure 4. In this paper, the threshold of correlation coefficient is 0.4, that is to say, whose correlation coefficients with the original signals exceed 0.4 are chosen to reconstruct the signal while the others are filtered since they show low correlation with system characteristics. The reconstructed signals are shown in Figure 5. Compared to original signals, high frequency components have been filtered and low frequency parts are more pure in reconstructed signals. All of which can be found in frequency spectrums.

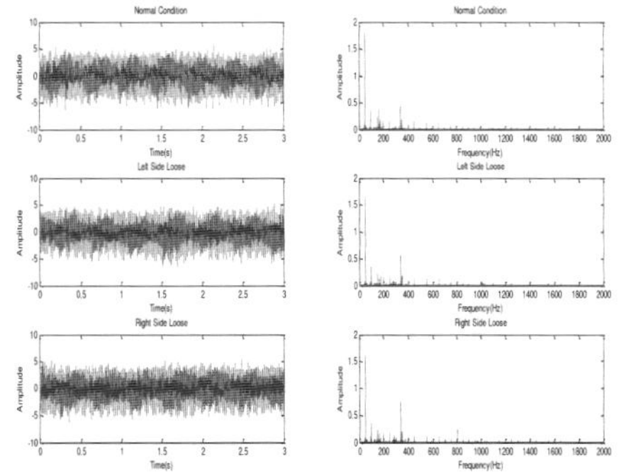

Figure 3. Acceleration signals of three hoop conditions.

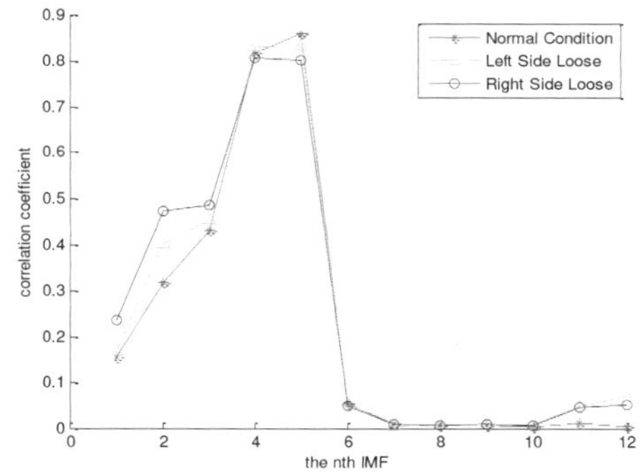

Figure 4. Correlation coefficients of each *IMF* with its original signal.

After reconstruct of the detected signals, the three different conditions of hoops for hydraulic pipeline, such as normal condition, left side loose and right side loose are described by MSE method quantitatively. Figure 6 gives the MSE over 20 scales of the reconstructed signals under different pressures and hoop conditions. It is apparent that different hoop conditions show different entropy values under each pressure. Therefore, hoop conditions can be distinguished by entropy values of MSE over 20 scales regardless of different pressures.

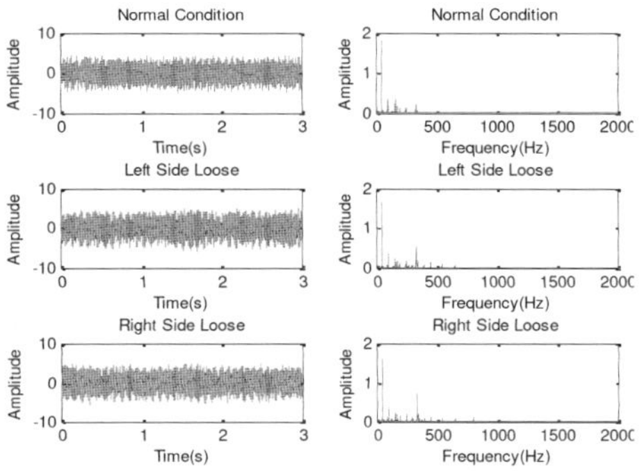

Figure 5. Reconstructed signals of three hoop conditions.

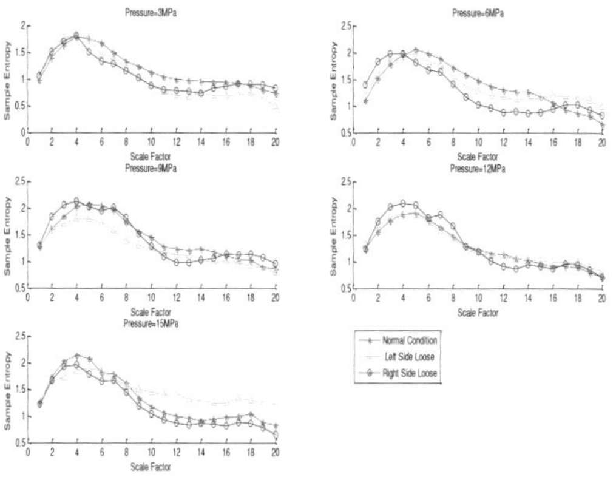

Figure 6. MSE over 20 scales of the reconstructed signals under different pressures and hoop conditions.

However, it may not enough just through the curves of MSE over different scales to identify the different three hoop conditions. In this paper, 20 entropy values of MSE are considered as a feature vector and ELM is used for intelligent condition monitoring. According to experiment introduction, each hoop conditions obtain 70 sets of feature vectors, as is shown in Table 2. In order to simplify the problem and improve the recognition accuracy, firstly detect whether the hoop is loose or not. On this condition, left side loose and right side loose belong to the same class. And then on the premise of loose condition, to detect loosening position. namely left side loose or right side loose.

Firstly detect whether the hoop is loose or not, at this point there are only two output states: normal condition and loose condition. Feature vectors with a total of 210 sets, of which normal condition 70 sets and loose condition 140sets (As shown in Table 2). Randomly select 60% of feature vectors namely 126 sets as the training dataset, the remaining 40% namely 84 sets feature vectors as testing dataset. ELM algorithm is used as classifier. The process is repeated 100 times, each time the classification accuracy and run time are shown in Figure 7. Among which average accuracy is $92.02\% \pm 3.11\%$ and average run time is $0.0127s \pm 0.0047s$.

As contrast, results of classification method based on BPNN and SVM in detecting whether hoop is loose or not are shown in Figure 8 and Figure 9, respectively.

Three parameters are selected to evaluate performance of the algorithms, accuracy, efficiency and stability. Accuracy is reflected by average classification accuracy over 100 times training and testing, efficiency is measured by average run time over 100 times training and testing, and stability is measured by standard deviation of classification accuracy and run time. Performance contrast of three classifier method in detecting whether hoop is loose or not is shown in Table 3.

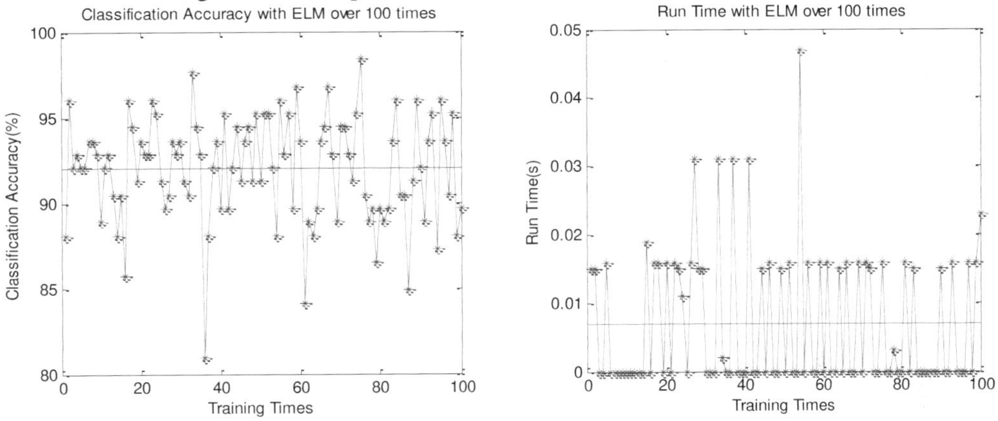

(a) classification accuracy. **(b)** run time.

Figure 7. Classification accuracy and run time in detecting whether the hoop is loose or not based on ELM.

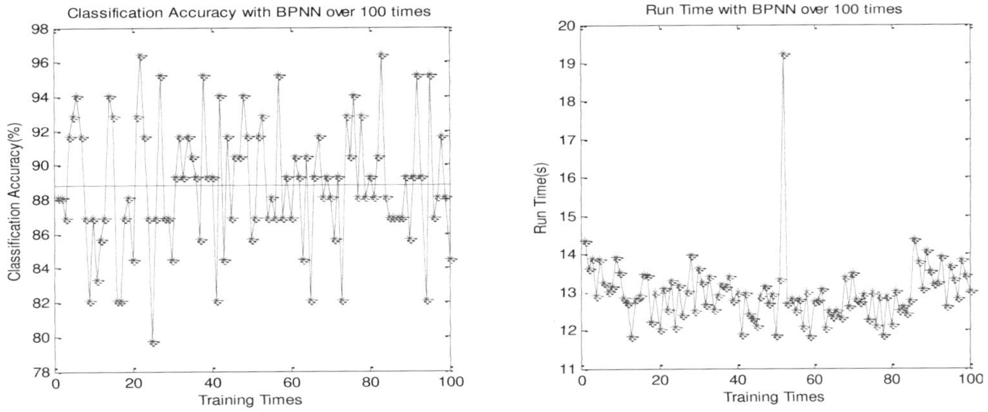

(a) classification accuracy. **(b)** run time.

Figure 8. Classification accuracy and run time in detecting whether the hoop is loose or not based on BPNN.

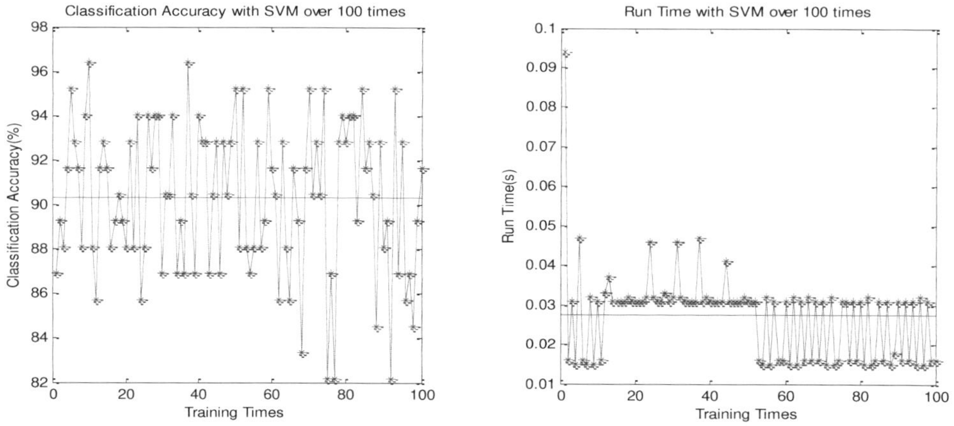

(a) classification accuracy **(b)** run time

Figure 9. Classification accuracy and run time in detecting whether the hoop is loose or not BASED ON SVM.

Table 3. Performance contrast of three classifiers in detecting whether hoop is loose or not.

Classifier	Features	Average Accuracy(%)	Average Run Time(s)
		Mean ± Std	Mean ± Std
ELM	EEMD-MSE	92.02±3.11	0.0127±0.0047
BPNN	EEMD-MSE	88.81±3.63	12.9931±0.8541
SVM	EEMD-MSE	90.34±3.41	0.0275±0.0109

Note: mean represents average value, and std represents standard deviation.

In Table 3, it is obviously that the maximum of average accuracy is 92.02% in ELM and the minimum of average run time is 0.0127s in ELM. Moreover, the minimum standard deviation of classification accuracy is 3.11% in ELM and the minimum standard deviation of run time is 0.0047s in ELM. It indicates that ELM classifier behaves better than BPNN and SVM both in accuracy, efficiency and stability in detecting whether the hoop is loose or not.

When it comes to detecting loosening position, in this case the output two states are left side loose and right side loose, feature vectors with a total of 140 sets, of which left side loose 70 sets and right side loose 70 sets(As shown in Table 2). Randomly select 60% of feature vectors namely 84 sets as the training dataset, the remaining 40% namely 56 sets as testing dataset. ELM algorithm is used as classifier. It is also repeated 100 times and among which average accuracy is $98.49\% \pm 1.26\%$ and average run time is $0.0051s \pm 0.0008s$.

As mentioned before, accuracy, efficiency and stability are selected to evaluate performance of three algorithms, accuracy is reflected by average classification accuracy, efficiency is measured by average run time and stability is measured by standard deviation of classification accuracy and run time. Performance contrast of three classifier method in detecting which side loose is shown in Table 4.

Table 4. Performance contrast of three classifier methods in detecting which side loose.

Classi fier	Features	Average Accuracy(%)	Average Run Time(s)
		Mean ± Std	Mean ± Std
ELM	EEMD-MSE	98.49+1.26	0.0051+0.0008
BPNN	EEMD-MSE	96.12+2.71	12.5961+0.6232
SVM	EEMD-MSE	95.98+3.64	0.0143+0.0084

Note: mean represents average value, and std represents standard deviation.

In Table 4, it is obviously that the maximum of average accuracy is 98.49% in ELM and the minimum of average run time is 0.0051s in ELM. Moreover, the minimum standard deviation of classification accuracy is 1.26% in ELM and the minimum standard deviation of run time is 0.0008s in ELM. It also indicates that ELM classifier behaves better than BPNN and SVM both in accuracy, efficiency and stability in detecting which side loose.

4. Conclusion

In this paper, a method based on EEMD-MSE and ELM is proposed for loose detection of hoops installed on hydraulic pipeline. EEMD is applied to reconstruct the original vibration signals and MSE is used for extracting feature vectors from the reconstructed signals. ELM algorithm is introduced as classifier. In order to simplify the problem and improve the detection accuracy, firstly detect whether the hoop is loose and then detect loosening position. The experimental results indicate that, feature extraction method based on EEMD and MSE is effective in loose detection of hoop, and classification method based on ELM behaves better than BPNN and SVM both in accuracy, efficiency and stability. Therefore, method based on EEMD-MSE and ELM for loose detection of hoops is available with high performance, especially in real-time condition monitoring.

References

[1] X Wei, H Zhang, C Lu and Z Lu 2012 Analysis on vibration signal of pipeline for hydraulic exciting system *Proceedings of SPIE - The International Society for Optical Engineering* **8334** pp 833424-5
[2] Z Tang, Z Lu, D Li and F Zhang 2011 Optimal design of the positions of the hoops for a hydraulic pipelines system *Nuclear Engineering & Design* **241** pp 4840–55
[3] P X Gao, J Y Zhai, Y Y Yan, Q K Han, F Z Qu and X H Chen 2015 A model reduction approach for the vibration analysis of hydraulic pipeline system in aircraft *Aerospace Science & Technology* **49** pp 144-53
[4] Z WU and N E HUANG 2009 ENSEMBLE EMPIRICAL MODE DECOMPOSITION: A NOISE-ASSISTED DATA ANALYSIS METHOD *Advances in Adaptive Data Analysis* **1** pp 1-41
[5] C Y Yang and T Y Wu 2015 Diagnostics of gear deterioration using EEMD approach and PCA process *Measurement* **61** pp 75-87
[6] M Han and J Pan 2015 A fault diagnosis method combined with LMD, sample entropy and energy ratio for roller bearings *Measurement* **76** pp 7-19
[7] M Costa 2002 Multiscale entropy analysis of complex physiologic time series *Physical Review Letters* **89** pp 705 - 708
[8] J L Lin, Y C Liu, C W Li, L F Tsai, and H Y Chung 2010 Motor shaft misalignment detection using multiscale entropy with wavelet denoising *Expert Systems with Applications* **37** pp 7200-4
[9] Z Huang, Y Yu, J Gu and H Liu 2016 An Efficient Method for Traffic Sign Recognition Based on Extreme Learning Machine *IEEE Transactions on Cybernetics*
[10] G B Huang, Q Y Zhu and C K Siew 2006 Extreme learning machine: Theory and applications *Neurocomputing* **70** pp 489-501
[11] N E Huang et al 1998 The empirical mode decomposition and the Hilbert spectrum for nonlinear and non-stationary time series analysis *Proceedings of The Royal Society A: Mathematical, Physical and Engineering Sciences* **454** pp 903-95
[12] J S Richman and J R Moorman 2000 Physiological time-series analysis using approximate entropy and sample entropy *American Journal of Physiology Heart & Circulatory Physiology* **278** pp H2039-49

Acknowledgment
The authors wish to thank the support of the National Nature Science Fund (No.51475347).

IOP Conf. Series: Materials Science and Engineering **211** (2017) 012012 doi:10.1088/1757-899X/211/1/012012

Progressive failure of composite wind blades with a shear-web spar subjected to static testing

T Y Kam* and Y H Chiu

Mechanical Engineering Department, National Chiao Tung University
Hsin Chu 300, Taiwan

*Email: tykam@mail.nctu.edu.tw

Abstract. Composite wind blades of 1m long comprising glass-fabric/epoxy skins and a sandwich plate-type spar were designed and fabricated for static testing. In the composite wind blades, the spar supports the top and bottom skins to form the airfoil shape of NACA4418. The blades were tested to failure and the failure modes were identified at different loading stages. A structural failure analysis method which consists of a geometrically nonlinear finite element (FE) model and appropriate phenomenological failure criteria is used to study the progressive failure behaviours of the blades subjected to different types of quasi-static loads. The experimental load-displacement curves as well as failure loads and locations for different failure modes are used to validate the suitability of the proposed failure analysis method.

1. Introduction

In a wind turbine, the wind blades are used to convert wind energy to electrical power. To attain high efficiency, the structure of a wind blade is required to be light and have small mass moment of inertia so that they can response swiftly to wind direction change and have low cut-in wind speed. During the operating life of a wind turbine, the wind blades have to survive the attack of severe environment. In particular, a wind blade has to be able to sustain the extreme wind loads that may occur during its lifetime. Such load carrying capacity requirement has made the failure analysis of a wind turbine blade become an important topic of research, especially the determination of the failure behaviour of the blade under an extreme wind load. Recently, many papers [1-5] have been devoted to study the failure behaviour of wind turbine blades subjected to quasi-static loads. In the previous studies, the failure modes and their criticality have been studied in the failure analysis of the wind blades. Due to their different structural configurations and loading conditions, large and small composite wind blades may have distinct mechanical behaviours and failure modes. Therefore, to assure the reliability of small wind blades, it is desired to have a deeper understanding about the actual blade failure behaviour. In this paper, the progressive failure behaviour of 1m long composite wind blades comprising a sandwich spar is studied via both experimental and theoretical approaches. The small composite wind blades were subjected to either load- or stroke-control testing. A structural failure analysis procedure is used to analyse the nonlinear failure behaviour of the blades. The experimental results are used to validate the suitability of the proposed failure analysis method.

2. Design and fabrication of composite wind blade

The small composite wind blade shown schematically in Figure 1 consists of 3 main parts, namely, top skin, bottom skin, and longitudinal sandwich spar. The skins are fabricated with glass fabric/epoxy laminate. The top and bottom skins which are adhesively connected to the spar to form the aerodynamic profile of the blade are adhesively bound to two glass fiber/epoxy stiffeners at, respectively, the leading and trailing edge to prevent delamination or opening failure mode from occurring at these edges. The root end of the blade is also adhesively bound to a rigid metallic connector. The blade orients in the x-direction with its root end located at x = 0. Each skin layer has one fiber direction coincident with the span-wise (x) direction and the other with the chord (y) direction. To save weight, the numbers of glass fabric/epoxy layers in the skins vary along the span of the blade. The blade cross section starts to change its shape gradually from airfoil at x = 0.14m to circle at x = 0 m. The spar is made of two flanges and a shear web. The shear web of the spar is a glass fabric composite sandwich panel and the flanges of the spar are Balsa wood panels. The width and thickness of each spar flange are 30 and 2mm, respectively. For the composite sandwich spar web, each glass fabric face sheet comprises 3 glass fabric/epoxy layers and the core is a 2 mm thick Balsa wood panel. The blade is equally divided into ten stations. The geometric properties such as chord length *c*, skin thickness *t*, and twist angle β_T are listed in Table 1. The wind blade has an efficiency of 0.42 for the designed tip speed ratio of 5. The hand layup technique was used to make the parts of the wind blade. Regarding the fabricating process, the moulds used to manufacture the top and bottom blade skins are shown in Figure 2. The glass fabric sheets coated with epoxy resin were placed in the mould and the vacuum forming technique was used to cure the glass fabric/epoxy layers. The parts of the blade including skins, spar, and stiffeners are shown in Figure 3. The assembled wind blade with painted surface is shown in Figure 4. Glass fabric/epoxy panels were also manufactured to determine the material constants of the composite lamina experimentally. The material constants of the glass fabric/epoxy lamina are listed in Table 2.

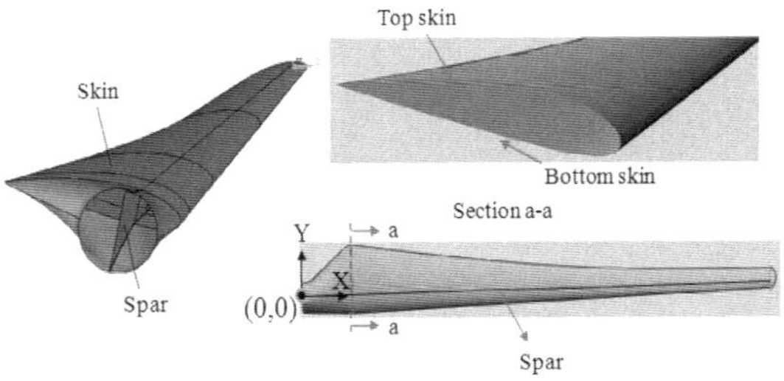

Figure 1. Schematic description of wind blade.

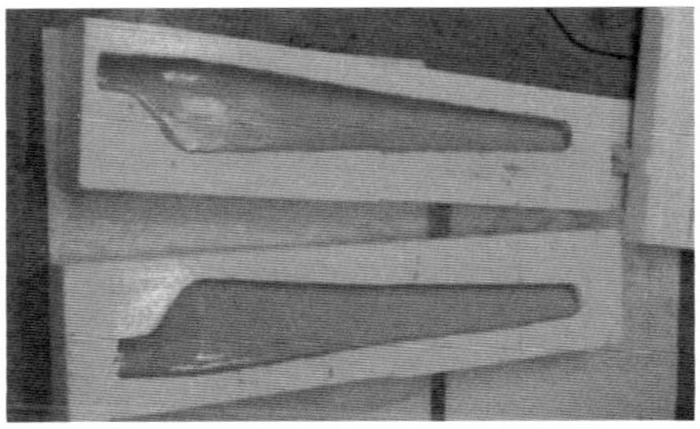

Figure 2. Mould for blade fabrication.

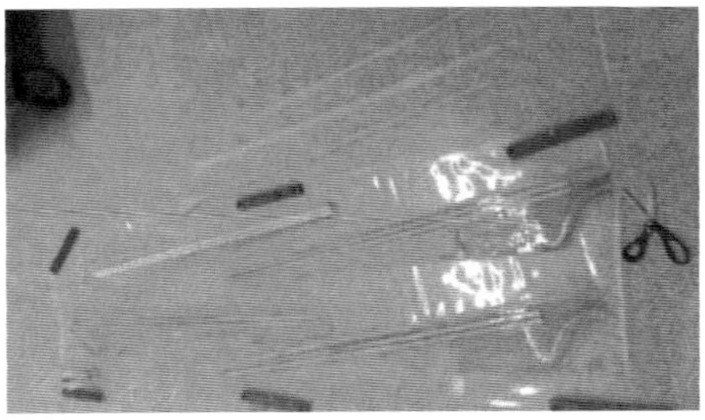

Figure 3. Parts for fabricating wind blade.

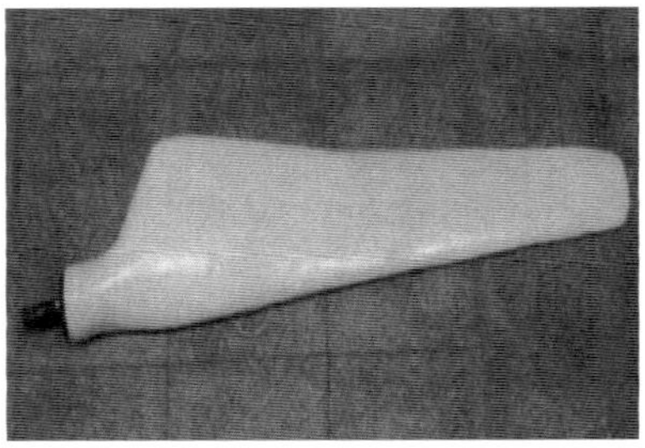

Figure 4. Assembled wind blade.

Table 1. Cross-sectional properties of blade at different stations.

Blade length (m)	Property / Station location (mm)	Twist angle β_T (Degree)	Chord length C (mm)	Skin thickness t (mm) [Layer range (mm)] Number of layers
1	0	-	-	Top skin 0.75

120	20.00	180.14	[0≦X≦350] 3 Layers Bottom skin 1.00 [0≦X≦350] 4 Layer
210	17.01	167.62	
300	13.75	155.76	
390	10.41	144.70	0.50 [350≦X≦1080] 2 Layers
480	7.26	134.44	
570	4.61	124.84	
660	2.64	115.65	
750	1.30	106.67	
840	0.51	97.78	
930	0.14	88.92	
1020	0.00	80.06	

Table 2. Properties of blade materials.

Material constant	Glass fabric/epoxy	Balsa wood	Strength parameter	Glass fabric/epoxy (MPa)
E_x	20GPa	3.7GPa	X_T	252
E_y	20GPa	55MPa	X_C	305
$G_{xy} = G_{xz}$	2.637GPa	55MPa	Y_T	252
G_{yz}	2.637GPa	8.33MPa	Y_C	305
υ_{xy}	0.15	0.02	S_{xy}	64
υ_{yz}	0.212	0.2	$S_{xz} = S_{yz}$	64

3. Experimental investigation

Several wind blades were subjected to load- and stroke testing for experimental study. In the load-control test as shown in Figure 5, the composite blade with the bottom skin facing upward was fixed at the root end. Several strain gages were mounted on the blade surfaces to measure the blade surface strains during testing. In the double load test, with same increment the total weights of the steel balls in the buckets were increased incrementally. For each blade under load-control testing, when the load was increased to a certain value, the incipient buckling of the front (top) skin of the blade would occur by inducing several dimples on the skin as shown in Figure 6. The location of each dimple was determined by measuring the distance between the dimple center and the blade root. The load that caused the incipient buckling of the blade was identified via the observation of the sign changes of the measured strains. As the load was kept increasing, the failure of blade materials such as skin glass fabric/epoxy lamina (first-ply failure) or adhesive at the interface between skin and spar (skin-spar interfacial debonding) might occur before the total fracture occurs at the blade root. On the other hand, for the blade subjected to stroke-control test as shown in Figure 7, with the front (top) skin facing upward, the blade was fixed at the root end while the blade tip was clamped using a rigid gripper in the form of the blade airfoil shape. The gripper was connected to the MTS material testing machine via a rigid link which consisted of three steel rings in chain. The link with its two ends working as two pin joints, respectively, transmitted an axial force from the testing machine to the blade tip. The

IOP Conf. Series: Materials Science and Engineering **211** (2017) 012012 doi:10.1088/1757-899X/211/1/012012

vertical movement of the top end of the link was controlled using the testing machine which was increased incrementally at a speed of 3mm/min. A load cell was used to measure the load associated with the assigned vertical displacement of the link top. The vertical displacements of the link top at different stroke stages together with the associated loads were recorded by a data acquisition system to construct the load-displacement curve. When the blade tip vertical displacement was increased to a certain value, the incipient buckling of the front (top) skin of the blade would occur by inducing a dimple on the skin. The location of the first dimple on the front blade skin and the strains were recorded to determine the incipient buckling load. As the stroke was kept increasing, additional dimples would be developed and a crack would be formed due to the failure of the blade material at the edge of the first dimple as shown in Figure 8. The extension of the crack across the blade skin would finally lead to the total failure of the blade.

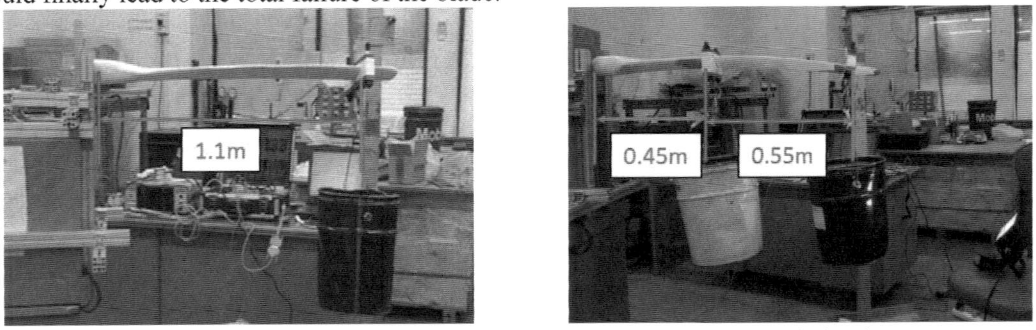

(a). Single load test. **(b).** Double load test.

Figure 5. Load-control test of wind blade.

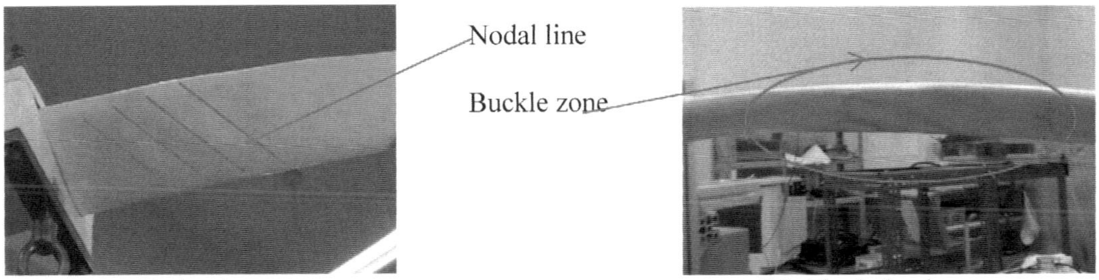

(a). Double load. **(b).** Single load.

Figure 6. Incipient buckling of blades under load-control testing.

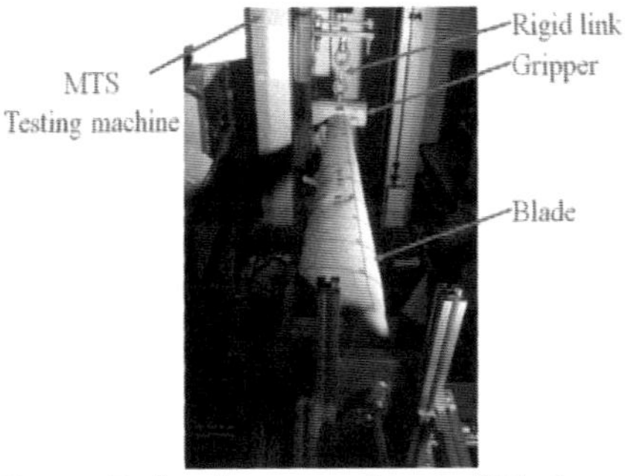

Figure 7. Stroke-control test of blade.

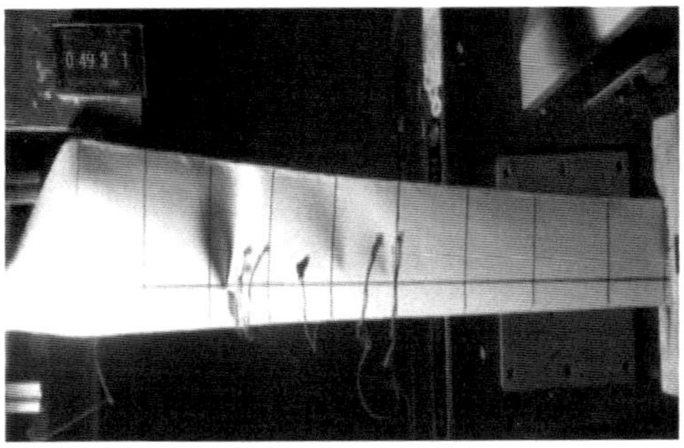

Figure 8. Crack developed from material failure.

4. Experimental and theoretical results

IOP Conf. Series: Materials Science and Engineering **211** (2017) 012012 doi:10.1088/1757-899X/211/1/012012

The failure of a small composite wind blade is susceptible to the effects of geometrical nonlinearity. Therefore, in the proposed failure analysis procedure, the geometrically nonlinear FE method of ANSYS [6] together with the Tsai-Wu and quadratic failure criteria for identifying material and interfacial failure, respectively, is used to analyse the failure behaviour of the small composite wind blades which have been tested. In the FE formulation, the composite wind blade is modelled using the laminated composite Shell91 elements of which each has eight nodes and four integration points. To bind the skin and spar together, the nodes of the skin elements are located at the top surfaces of the elements while those of the spar elements at the middle surfaces of the elements. At the critical interfacial regions, a row of thin epoxy elements is used to bind the skin and spar. In the blade failure analysis, the stresses at each layer of an element and the interface between skin and spar are used to determine if material or interfacial failure has occurred. Regarding material degradation due to material failure, the material constants of a lamina in an element are treated as zero when material failure in that lamina is detected to occur. Similarly, when failure occurs in an interfacial element, the material constants of that element become zero. In the nonlinear finite element analysis with the consideration of material degradation, the load-displacement curve is constructed using an incremental displacement procedure in which the tip displacement is increased incrementally via a stroke-control approach. Whenever material failure in the blade is detected, the material constants of the failed element(s) are modified in accordance with the proposed material degradation model and the blade with such new material constants is analysed again starting from the initially undeformed configuration via the incremental displacement approach. The experimental and theoretical load-displacement curves are shown in Figure 9. The experimental results have shown that the bottom skin of the blade buckles at the tip stroke of 66mm (load is 143.9N) with the first dimple at x = 45mm; material failure occurs at x = 45mm, the edge of the first dimple, when the tip stroke is 104mm (load is 212.2N); the applied load drops to 199.4N at the tip stroke of 105mm due to the enlargement of the dimple size; the applied load further drops to 168.9N at the tip stroke of 120mm due to the extension of the crack initiated at the material failure point; the bottom skin is torn apart to make the blade fail totally at the tip stroke of 165mm (load is 59.4N). It is noted that the theoretical load-displacement curve predicted using the material degradation method closely matches the experimental one. The difference between the theoretical and experimental ultimate loads is less than 4%. Therefore, it is appropriate to adopt the proposed material degradation method for progressive failure analysis of small composite wind blade.

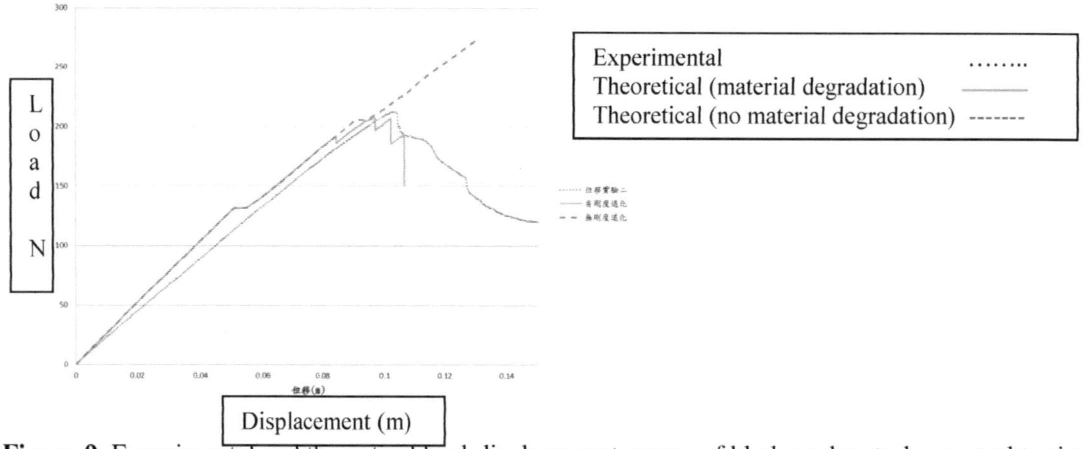

Figure 9. Experimental and theoretical load-displacement curves of blade under stroke-control testing.

5. Conclusion

The progressive failure of small composite wind blade has been studied via both experimental and theoretical approaches. A number of wind blades have been subjected to load- and stroke-control testing to observe the failure behaviour of the blades. A nonlinear failure analysis method with or without considering material degradation has been proposed. It has been shown that the failure analysis method with the consideration of material degradation is appropriate for progressive failure analysis of small composite wind blades. The difference between the theoretical and experimental ultimate loads is less than 4% for the blade subjected to stroke-control testing.

References

[1] Overgaard L C T, Lund E and Thomsen O T 2010 Structural collapse of a wind turbine blade Part A: Static test and equivalent single layered models *Composites Part A-Applied Science and Manufacturing* **41 (2)** pp 257-70
[2] Overgaard L C T and Lund E 2010 Structural collapse of a wind turbine blade. Part B: Progressive interlaminar failure models *Composites Part A-Applied Science and Manufacturing* **1 (2)** pp 271-83
[3] Sierra-Perez J, Torres-Arredondo M A and Gueemes A 2016 Damage and nonlinearities detection in wind turbine blades based on strain field pattern recognition. FBGs, OBR and strain gauges comparison *Composite Structures* **135** pp 156-66
[4] Haselbach P U, Bitsche R D and Branner K 2016 The effect of delaminations on local buckling in wind turbine blades *Renewable Energy* **85** pp 295-305
[5] Kim S H, Bang H J, Shin H K and Jang M S 2014 Composite Structural Analysis of Flat-Back Shaped Blade for Multi-MW Class Wind Turbine *Applied Composite Materials* **21** pp 525-39
[6] ANSYS 2010 **12** 1 (ANSYS Inc. USA)

Acknowledgment

This research work was supported by the Ministry of Science and Technology under MOST 105-2221-E-009-149.

IOP Conf. Series: Materials Science and Engineering **211** (2017) 012013 doi:10.1088/1757-899X/211/1/012013

A fast learning method for large scale and multi-class samples of SVM

Yu Fan[1 2 *] and Huiming Guo[2 3]

[1]China Changfeng Science Technology Industry Group Corp, Beijing, China
[2]The Second Institute of China Aerospace Science & Industry Corp, Beijing, China
[3] Beijing Aerospace Changfeng Co. Ltd, Beijing, China

[*]Email: alex3936@163.com

Abstract. A multi-class classification SVM(Support Vector Machine) fast learning method based on binary tree is presented to solve its low learning efficiency when SVM processing large scale multi-class samples. This paper adopts bottom-up method to set up binary tree hierarchy structure, according to achieved hierarchy structure, sub-classifier learns from corresponding samples of each node. During the learning, several class clusters are generated after the first clustering of the training samples. Firstly, central points are extracted from those class clusters which just have one type of samples. For those which have two types of samples, cluster numbers of their positive and negative samples are set respectively according to their mixture degree, secondary clustering undertaken afterwards, after which, central points are extracted from achieved sub-class clusters. By learning from the reduced samples formed by the integration of extracted central points above, sub-classifiers are obtained. Simulation experiment shows that, this fast learning method, which is based on multi-level clustering, can guarantee higher classification accuracy, greatly reduce sample numbers and effectively improve learning efficiency.

1. Introduction

Support vector machine (SVM) is a machine learning method proposed by Vapnik[1]. It is mainly used to solve the problem of small samples, nonlinear and high dimensional pattern recognition. A single SVM can only solve two classification problems. For multi-classification, two kinds of classifiers are combined according to certain combination rules. In order to realize multi-class separable, scholars around the world have done a lot of research on multi-classification. The basic algorithms are "one to one" and "one to many" [1], and an improved algorithm for the two types. Among them, the SVM multi-classification method based on binary tree is a kind of effective method to solve the multi-class recognition problem in [2-4]. The binary tree construction method is used to solve the problem of multi-class SVM to classification. However, there is a problem still unavoidable, which is: in the realization of SVM multi-classification, if the training set is large, the SVM learning speed will be very slow, and even the memory is too small to cause the program to execute.

In order to improve the learning speed of SVM, researchers have done a lot of works: (1) To reduce the learning time by reducing the size of the solution of the two programming neutron problem without changing the size of the training set. Such as SMO[5]. (2) Selecting the training samples of support vector regression[6-8]. (3) Constructing a multi strategy decision tree based on multi strategy fusion. In addition, in order to make the two classification SVM able to learn quickly, a new method based on particle distribution and support vector machine is proposed in literature [9].

Literature [10] considered the local cluster, and that paper proposed a method to construct effective relative distance matrix from binary tree. And the learning classifier, which is based on that method, had fast learning speed and fast classification speed. Inspired by literature [9], combined with the bottom-up construction method in literature [10], this paper proposed a fast learning method for multi-classification of SVM. The effective relative distance matrix based on the bottom-up construction of binary tree, while training each sub classifier utilizing samples by selective reduction method of multilayer clustering. The simulation results showed that this algorithm took into account the distribution information of the sample points, and improved the learning speed of multi-classification SVM based on the reduction of the training samples.

2. Design of binary tree SVM hierarchy

As for the N classification problem, compared with other multi-classification methods, this binary tree SVM multi-classification method constructs the least number of classifiers at the learning stage. When the sample is tested, it is not necessary to traverse all the classifiers. There will be no case where a test sample belongs to multiple classes at the same time, and there will not be a case where the test sample has no category. The construction of the binary tree hierarchy consists of the following two parts: the construction of the effective relative distance matrix and the design of the structure of the binary tree based on the effective relative distance matrix.

2.1. Construction of effective relative distance matrix

Binary tree SVM hierarchical structure using the relative distance between the two types of samples to measure the degree of aliasing. The relative distance is determined by two factors: the absolute distance between two class centres, and distribution range between two class centres. Set the samples contain N category, c_i is the centre of category i, x_i is the sample of i category. The relative distance is defined between each of the two categories, which is as follow: $D_{ij} = \frac{d_{ij}}{R_i + R_j}$ $(D_{ij} \in [0, \infty))$, and $d_{ij} = \|c_i - c_j\|$ means the Euclidean distance of the sample centre between i and j. R is the smallest hypersphere radius of each sample: $R_i = \max\{\|c_i - x_i\|\}$.

The steps to build an effective relative distance matrix are as follows:

(1) Calculate the relative distance of each of the two samples, and arrange according to the distance. Forming N * (N-1) /2 column and 3 rows of the relative distance matrix D1, the matrix of the first 2 columns for the category, the third column is the relative distance between the two categories.

(2) Select effective relative distance. Determine the rank of D1 is greater than 0, if more than 0, said the matrix contains a valid category of choice, the first line of the two categories saved to a valid relative distance in D_{valid} matrix, and the larger category value assigned to max.

2.2. Binary tree hierarchy design steps

After the effective relative distance matrix formed, the following steps are used to design the hierarchical structure.

Step 1: Select the first row of the two categories from the valid relative distance D_{valid} matrix. A new non-leaf-node is generated by combining the two, and the left and right branches are set. Save new non-leaf-nodes, left and right node information.

Step 2: Determine the second line of the two categories. If the two categories in the previous node have not appeared, then the combination of these two categories into a new non-leaf-node.

Step 3: Repeat Step2 until the last line. Binary tree structure is completed.

The following is the four classification of the structure of the tree as an example, the paper describes the design process of the binary tree hierarchy. In a certain four classification problem, the category is 1,2,3,4, and the effective relative distance matrix is $D_{valid} = \begin{bmatrix} 1 & 2 & 0.1 \\ 3 & 4 & 0.2 \\ 1 & 3 & 0.5 \end{bmatrix}$, so the design procedure of the binary tree hierarchy is as follows:

Step 1: Choose two categories from D_{valid}, the first line has 1 and 2. Set 1 and 2 are combined to form a 1V2 node, left is 1, right is 2.

Step 2: Judge the second line of the two categories 3 and 4. All nodes traversal formed before, found 3 and 4 node does not appear in any non-leaf-node, then setting 3 and 4 nodes are combined to form a 3V4 node, the 3 left, 4 right.

Step 3: Judge the third lines of the two categories of 1 and 3. All nodes traversal before the formation .Finding 1 has appeared in the 1V2 node, 3 has appeared in the 3V4 node. 1V2 node and 3V4 node will then combine to form 1,2V3,4 node ,and set left is 1,2, right is 3,4.

So far, the four classification of the tree structure has been designed over, and its structure is shown as figure 1.

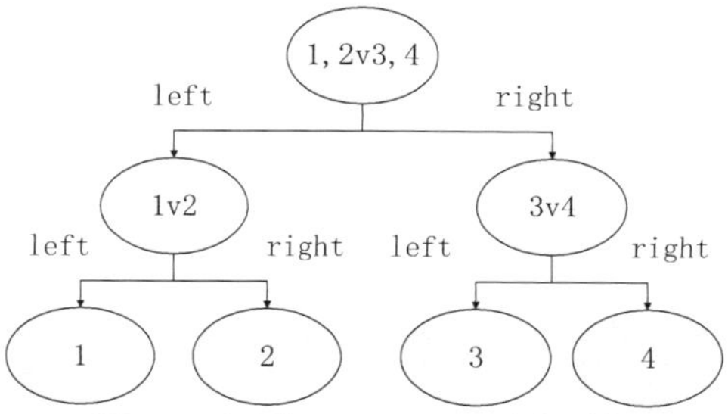

Figure 1. The structure of the binary tree of the four classification.

3. Multi-layer clustering fast learning method

Each non-leaf-node in the binary tree corresponds to a sub classifier. This paper mainly introduces the steps of the fast learning method.

Step 1: In the training samples, we extract the same samples of all the left and right branches in a non-leaf-node. The left the category of the sample for the +1, the right branch of the category of the sample for -1. The two types of samples are combined, and the K-Medoids method is used to cluster the samples for the first time.

Step 2: If all the samples in the first I class cluster are the same class, the cluster is a pure class cluster, and only the center points of the clusters are extracted and reserved. If it is not for the same category, the cluster is a mixed class cluster, go to step 3.

Step 3: Use formula $mix = 1 - \frac{|n_1 - n_2|}{n_1 + n_2}$,(n1 is the number of positive samples of mixed class clusters, n2 is the number of negative samples of mixed class clusters), and to determine the degree of mixture of positive and negative samples of each mixed class cluster .If $mix \geq 1/3$, it means that the sample of the cluster is more balanced, go to step 4. If $mix < 1/3$, it means the cluster sample is not balanced, go to step 5.

Step 4: The reduction factor of positive and negative samples is s, $s \in (0,1]$. The positive and negative of samples in the mixed cluster are clustered by K-Medoids method, and the number of clusters is s * n1, s * n2, respectively, and the center points of all positive and negative clusters are extracted and reserved, and then transferred to step 6.

Step 5: the reduction factor of the positive and negative samples in the class of the unbalanced mixed cluster is t, $t \in (0,1]$. Assuming that the minority class samples are positive, most of the samples are negative, and the positive and negative of samples are clustered by K-Medoids method.

Step 6: repeat step 2 to step 5 until K class cluster analysis is complete.

Step 7: the central point of the first clustering in step 2 is merged with the central points retained in step 4, step, and re clustering in. The training sample set is constructed, and the sample set is studied.

4. Test results

The simulation software used in the experiment is Matlab R2009a, the support vector machine toolbox is LSSVM, the experimental data were selected from UCI data set, as shown in table 1. Waveform was 21 dimensional data, sleep used the training set of the first 10000 data to learn, and tested with 10000 data in the test set. Shuttling used the first 5 types to learning and testing, removed the smallest proportion of the sixth and the seventh class.

Operating environment: 3.3GHz, 1.97GB memory. In the simulation process, the RBF kernel function was used, the parameters needed to be set such as kernel parameters, penalty factor, reduction factor s and t. In this paper, the method of grid verification was used to get the optimal parameter values for each data set, as shown in table 2. At the same time to simplify the calculation, so that s=t=0.6.

Table 1. Experimental Data Sets.

Data	Class number	Feature dimension	Train set	Test set
waveform	3	21	4700	300
sleep	5	13	10000	10000
shuttle	5	9	43483	14494

Table 2. Optimal Parameters.

Data set	Kernel parameter	Penalty factor
waveform	0.2	21
sleep	100	2
shuttle	100	12

4.1. Analysis of multi cluster training sample reduction

Taking the waveform data set as an example, the paper discusses the reduction of training samples by multi-level clustering. Figure 2 is a hierarchical structure of the waveform, which is obtained by the steps described before. As can be seen from Figure 2, the 1V2 and 1,2V3 corresponding to the training samples for the first time after the cluster reduction, the number of unbalanced mixed clusters were 15, 21. Because of the same training sample reduction method of node 1V2 and 1,2V3, we only take the unbalanced hybrid clusters corresponding to the node 1V2 as an example, and then use the literature [9] method and this paper method to reduce the cluster again. The number of samples obtained after the reduction of each class of unbalanced clusters is accumulated. Because the number of clustering number obtained by this method will be less than literature [9] method, center points then extracted, namely the number of samples of each cluster after the reduction, it must be less than the reduction method of [9] after the number of samples. Therefore, with the increase in the number of samples, the gap will gradually increase. It can be seen that this method can greatly reduce the sample.

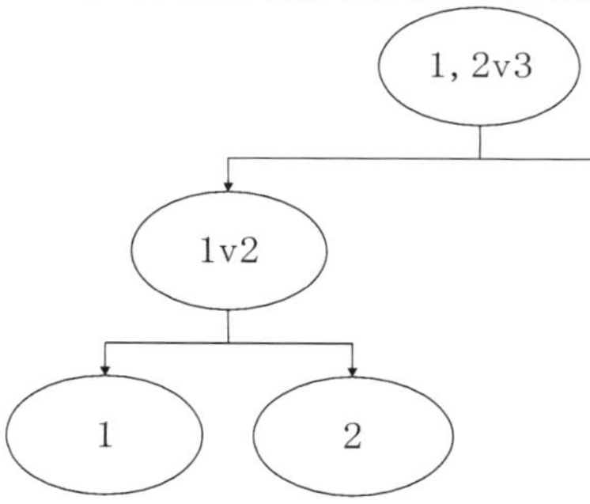

Figure 2. Waveform three classification of the structure of the binary tree.

4.2. Analysis of learning results in data sets

Table 3, 4, 5 respectively using waveform, sleep, shuttle data sets, based on the different methods of the SVM multi-classification learning results were compared. By literature [10], [9] method and the method of comparative analysis, each classifier of [10] method in learning before the reduction of the training samples, the method in reference [9] were used in section third of all mixed cluster 4 reduction training samples, the method of balanced mixed clusters by step in the third quarter of 4. The equilibrium mixed cluster by step 5 reduction of training samples.

It can be seen from the experimental results that the training samples, the number of support vectors and the learning time of the single classifier are significantly reduced by using the method of this paper and the [9] method. However, the reduction of learning time is the least. The classification accuracy of each classifier was observed and compared with literature [10], the method of classification method[9] which accuracy drops less than 2%, it shows that using this method, the training samples are greatly reduced at the same time.

Table 3. Comparison of learning results in three waveform data sets.

Waveform	Sub classifier	Trainin g sample number	Reduced training sample number	Cluster ing time /s	Learning time/s	Support vector number	Classificati on accuracy /%
Literature [10] method	$SVM_{1\text{-}2}$	3114	--	--	2.42	3114	93.68
	$SVM_{1,2\text{-}3}$	4700	--	--	6.50	4700	90.67
Literature [9] method	$SVM_{1\text{-}2}$	3114	1013	1.92	0.17	1013	92.11
	$SVM_{1,2\text{-}3}$	4700	1704	3.03	0.50	1704	93.33
This paper method	$SVM_{1\text{-}2}$	3114	738	1.70	0.14	738	92.11
	$SVM_{1,2\text{-}3}$	4700	1221	2.93	0.34	1221	92.67

Table 4. Comparison of learning results in sleep data set five.

Shuttle	Sub classifier	Training sample number	Reduced training sample number	Clustering time /s	Learning time/s	Support vector number	Classification accuracy /%
Literature [10] method	SVM4-1	2842	2842	--	3.60	2842	72.00
	SVM2-4,1	7848	7848	--	122.32	7848	83.51
	SVM5-4,1,2	8995	8995	--	206.57	8995	90.49
	SVM3-4,1,2,5	10000	10000	--	188.75	10000	93.43
Literature [9] method	SVM4-1	2842	1637	2.19	0.45	1637	73.53
	SVM2-4,1	7848	4235	7.47	3.58	4235	85.42
	SVM3-4,1,2,5	10000	2492	7.59	1.23	2492	94.13
This paper method	SVM4-1	2842	1344	2.64	0.17	1344	73.81
	SVM2-4,1	7848	2480	5.90	0.61	2480	83.62
	SVM5-4,1,2	8995	1430	5.35	0.20	1430	92.01
	SVM3-4,1,2,5	10000	1131	5.73	0.15	1131	94.04

Table 5. Comparison of learning results in shuttle data set five.

Shuttle	Sub classifier	Training sample number	Reduced training sample number	Clustering time /s	Learning time/s	Support vector number	Classification accuracy /%
Literature [10] method	SVM2-5	2495	2495	--	1.75	2495	99.76
	SVM3-4	6880	6880	--	25.62	6880	99.68
	SVM2,5-3,4	9375	9375	--	65.98	9375	99.67
	SVM1-2,5,3,4	43483	43483	--	2276.95	43483	99.73
Literature [9] method	SVM2-5	2495	181	2.31	0.00	181	100.00
	SVM3-4	6880	445	7.66	0.03	445	99.23
	SVM2,5-3,4	9375	683	16.39	0.16	683	99.83
This paper method	SVM2-5	2495	172	4.59	0.00	174	100.00
	SVM3-4	6880	197	10.11	0.01	168	99.82
	SVM2,5-3,4	9375	209	23.83	0.01	203	99.40
	SVM1-2,5,3,4	43483	1075	96.06	0.62	1620	98.59

4.3. Overall performance analysis

Contrast the "one to many", the literature [10], [9] methods, as shown in table 6 for comparison, overall performance of the four methods on three data sets: tx means the study time, tc means the test time. According to the classification accuracy, method [9] and this paper method results can be affected by the influence that each learning result is different, so take the average of the five learning as its final learning results, "one to many" and the method of [10] due to the use of all the samples, and did not use clustering, so only take one result of the previous study.

For waveform, sleep, shuttle three data sets, compared to the method [9], the classification accuracy of this method is slightly lower. This is because this method is equivalent to the class of mixed uneven cluster based on the method [9] for further reduction, so that the sample is reduced, and the classification accuracy rate will be slightly lower. This method makes the classification accuracy a little higher than the "one to many", method [10] classification accuracy is slightly higher. For the shuttle data set, we can find that the classification accuracy of the data set is up to 99.72%, which shows that there is almost no mixed sample at the junction of class and class in the data set of [10]. the sub classifier learning, using multi-layer clustering reduction training samples, the number of participants in training less support vector class at the junction at the same time will also be reduction, classification learning are not accurate enough. So this method of classification accuracy compared to the "one to many" literature, [10] method classification accuracy is slightly lower.

Table 6. Comparison of the overall performance of the four methods.

	Literature [10] method			Literature [9] method			This paper method		
	tx/s	tc/s	acc/%	tx/s	tc/s	acc/%	tx/s	tc/s	acc/%
wavefo rm	6.06	1.24	86.67	5.69	4.47	87.67	4.63	0.47	87.04
sleep	521.26	336	67.84	34.08	318.56	72.06	20.78	280	70.98
shuttle	2370.31	672	99.72	164.02	118.8	98.50	135.25	88.91	98.06

5. Conclusion

In this paper, a multi-class SVM fast learning method based on multi-level clustering is proposed. Using the relative distance matrix to construct the binary tree structure. Learning while using each non-leaf-node, reduce the positive and negative training samples by multi-layer clustering. And get the reduction mode by the type of cluster. Then use the training samples to learn and to improve the learning speed. The simulation results of three large data sets showed that the proposed method can greatly shorten the learning time under the premise of ensuring high classification accuracy. In the future research work, we should combine this method with other methods of pre extracting support vectors, which can be used to reduce the training samples and further improve the learning speed.

References

[1] Vapnik V N 1996 The nature of statistical learning theory *Technometrics* **8(4)** p 1564

[2] Sahbi H and Geman D 2006 A hierarchy of support vector machines for pattern detection *Journal of Machine Learning Research* **7(2)** pp 2087-123

[3] Fan B C, Wang J Y and Yu-Ming B O 2010 Binary tree svm multi-class classification algorithm with feature selection *Computer Engineering & Design* **31(12)** pp 2823-25

[4] Scholten E L, Semaan R, Illei P, Mallow C, Arias S and Feller-Kopman D, et al 2017 Stylet use does not improve diagnostic outcomes in endobronchial ultrasonographic transbronchial needle aspiration: a randomized clinical trial *Chest* **151(3)** p 636

[5] Lkopf B, Burges C J C and Smola A J 1999 Advances in kernel methods: support vector learning (MIT Press)

[6] Li X R, Zhao G Z and Zhao L Y 2008 Design of decision-tree-based support vector machines multi-class classifier based on vector projection *Control & Decision* **23(7)** pp 745-50

[7] Yang X S, Deb S and Fong S 2012 Accelerated particle swarm optimization and support vector machine for business optimization and applications *Communications in Computer & Information Science* **136(13)** pp 53-66

[8] Zhang Zhao, Fei Yinan, Song Lin and Wang Suozhu 2008 Research of multi-class algorithm based on fuzzy support vector machine[J] *Communications in Computer & Information Science* **28(7)** pp 1681-3

[9] Yu Z, Wenjian W, et al 2013 Support vector machines accelerated training method based on grain distribution *Journal of Nanjing University (NATURAL SCIENCE)* **49(5)** pp 644-9

[10] Lang Yuning 2010 Research and application of multi-classification methods based on support vector machines (*Doctoral, dissertation, Southwest Jiao Tong University*)

Modeling of ultrasonic processes utilizing a generic software framework

P Bruns*, **J Twiefel and J Wallaschek**

Institute of Dynamics and Vibration Research, Gottfried Wilhelm Leibniz Universität Hannover, Appelstraße 11, D-30167, Hannover, Germany

*Email: bruns@ids.uni-hannover.de

Abstract. Modeling of ultrasonic processes is typically characterized by a high degree of complexity. Different domains and size scales must be regarded, so that it is rather difficult to build up a single detailed overall model. Developing partial models is a common approach to overcome this difficulty. In this paper a generic but simple software framework is presented which allows to coupe arbitrary partial models by slave modules with well-defined interfaces and a master module for coordination. Two examples are given to present the developed framework. The first one is the parameterization of a load model for ultrasonically-induced cavitation. The piezoelectric oscillator, its mounting, and the process load are described individually by partial models. These partial models then are coupled using the framework. The load model is composed of spring-damper-elements which are parameterized by experimental results. In the second example, the ideal mounting position for an oscillator utilized in ultrasonic assisted machining of stone is determined. Partial models for the ultrasonic oscillator, its mounting, the simplified contact process, and the workpiece's material characteristics are presented. For both applications input and output variables are defined to meet the requirements of the framework's interface.

1. Introduction

Ultrasonically assisted processes are in the focus of researchers since decades and numerous applications are already state of the art in industry. For investigation and research activities as well as for designing, accurate models are needed. The utilized oscillators and tools as well as the respective processes and loads span a wide range and can be very different from each other. Thus customized models for the various applications usually are developed. A wide variety of occurring loads, utilized oscillators, and modeling approaches have been studied in the past. Doumanidis et al. [1] e.g. present both an analytical and numerical model for ultrasonic welding of thin metal foils to analyze the process mechanics. Liu et al. [2] present a finite element model to allow investigations on ultrasonic surface rolling. Ultrasonically assisted drilling processes are modeled by Potthast et al. [3] as well as Heisel et al. [4]. The investigations usually concentrate on the influence of a superimposed ultrasonic vibration on process forces. For all given modeling approaches the main distinguishable feature, however, is their complexity. Both different physical domains and different size scales need to be considered to sufficiently represent the respective processes and systems. Usually either simplified models or a division into partial models are used to overcome this. The former can results in insufficient accuracy whereas the latter leads to customized and mostly complex coupling of the partial models. Usually approaches like co-simulations [5] are utilized for modeling such complex systems. However, these approaches are mostly application-oriented and tailored for specific problems. Thus a generic but simple software framework is presented permitting an easy to implement, expandable, and well comprehensible modeling. It is capable of including different physical domains and size scales and thus permits modeling of complex processes without any significant restrictions.

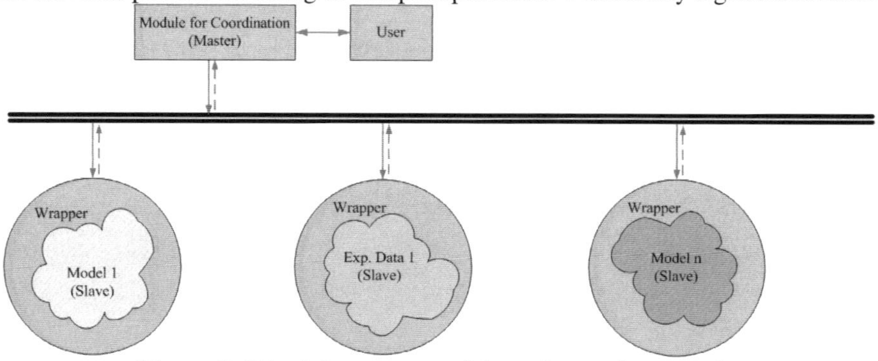

Figure 1. Principle structure of the software framework.

2. Generic software framework

The developed software framework's basic concept is depicted in Figure 1. It consists of a master module and any number of generic slave modules (wrappers) permitting implementation of the respective partial models. The parts are structured as simply as possible and implemented in the commercial software MathWorks MATLAB. The master module is essentially responsible for coordination of the calculations and simulations and thus for the execution of the slave modules in correct order. The explicit order needs to be set by the user. Slave modules can be performed once per calculation step or within an inner iteration loop (cf. Figure 2). An adaptable convergence examination is implemented permitting the selection of any parameter from the implemented partial models to be checked for convergence with a defined accuracy. Furthermore, the amount of calculation steps to be performed is determined in the master module. Hence the default abort criterion for the calculations is a manual defined number of runs through the overall model and not a model internal value. This can be adapted in the developed framework though. The definition of global variables that are not altered during the calculations and simulations in terms of a structure array is a further function of the master module. With the number of considered partial models not being limited a large amount of data possibly can be produced. Hence an appropriate data and storage management is implemented.

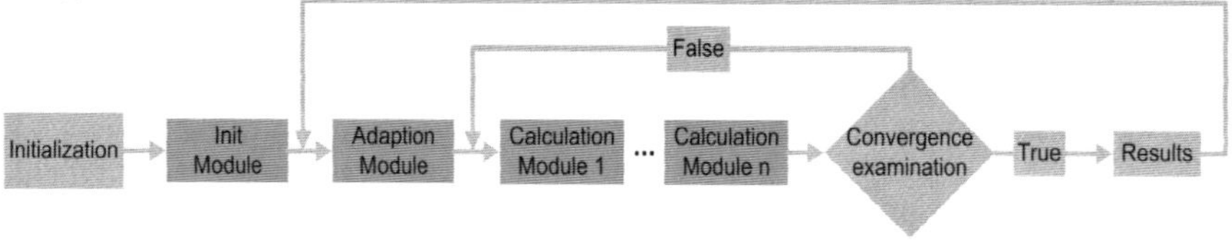

Figure 2. Exemplary flow chart of the developed master module.

The wrappers' generic character allows not only an implementation of any desired model but also of experimental, interpolated, or precalculated data. This generic character is achieved by the wrappers' well defined interfaces in terms of formatting rules for input and output parameters whereat a very simple structure is achieved (see Figure 3). Only defined parameters and variables (as well as the global structure array) are handed over from the master module to the individual wrappers and vice versa. A distinction between the respective wrappers' first and following executions is made permitting the one-off declaration of input and output variables in the respective wrapper. Specific parameters not being altered can be defined either in the respective wrapper itself or in the master module and the defined global structure array respectively. The latter permits centrally collected data of model settings and calculation results.

```
function task = wrapper_i(task,GlobalVariable)

%% Declaration and transfer of variables
switch strcmp(task.status, 'init')
   case 1     %Declaration of input and output variables in first call
      task.variable = [];
      variable = value;
   otherwise     %Transfer of variables in any other call
      variable = task.variable(end);
end

%Declaration of model-specific variables in every call
parameter1 = value;
parameter2 = GlobalVariable.parameter2;

%% Integration of model

%% Postprocessing/Output
%Generation of output
task.variable(end+1) = variable;
```

Figure 3. Basic structure of generic slave modules.

3. Examples of application

The described framework is utilized for modeling two different ultrasonic processes. First parameterization of a simplified load model for ultrasonically-induced cavitation is presented. The model is intended to be used for investigations on the impact of cavitation and respective operating parameters on the ultrasonic oscillator. Second the hybrid process of ultrasonic assisted machining of stone is modeled with the objective to determine a process-optimized mounting position. Keeping the focus on the software framework the respective process of material removal is simplified.

3.1. Ultrasonic-Induced Cavitation

Beside ultrasonic-induced cavitation there are only few processes with so many different physical domains intermixing with each other. It includes inter alia a piezoelectric part with electrical and mechanical domain as well as the cavitation part with acoustics, fluid dynamics, and thermodynamics. Due to this complexity just a brief description of cavitation and few application examples are given in this context. Further information can be found in well-known literature, for example, [6], [7], [8] and [9].

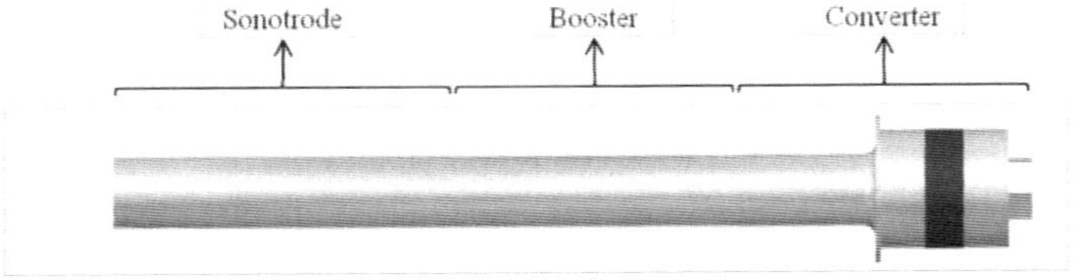

Figure 4. Ultrasonic oscillator for cavitation induction.

Figure 4 shows a model of a piezoelectrically driven ultrasonic oscillator for inducing cavitation. It consists of a converter, a booster, and a sonotrode. Applying a voltage and a current with a specific frequency in the ultrasonic range to the piezoelectric stack, results in the excitation of a longitudinal vibration mode. The converter here is an interface between the electrical and mechanical domain. Utilizing the inverse piezoelectric effect, it converts the electrical to mechanical oscillation and transfers it to the booster which modifies the amplitude [6], [10]. The sonotrode is the only part of the oscillator in contact with a liquid medium and, therefore, causes the cavitation due to its ultrasonic vibration.

The cavitation itself can be basically defined as the breakdown of a liquid medium under very low pressures [11]. It can occur in both hydrostatic and hydrodynamic liquids. In the contribution at hand a process with ultrasound being applied to a liquid stored in container is investigated. Thus only hydrostatic cavitation and acoustic cavitation respectively is considered. In accordance with [11] the effect of cavitation is comparable to boiling with the main difference lying in the driving mechanism.

For cavitation it is a drop of pressure while for the boiling it is an increase in temperature. However, due to the applied ultrasonic vibration along with the changing pressure field, pulsing bubbles are generated in the liquid and implode inter alia in dependence of their size and the pressure field. Cavitation is utilized in a variety of applications as e.g. surface cleaning [12], sonochemical processes [6] and [13], food processing [14], and ultrasonic peening [15].

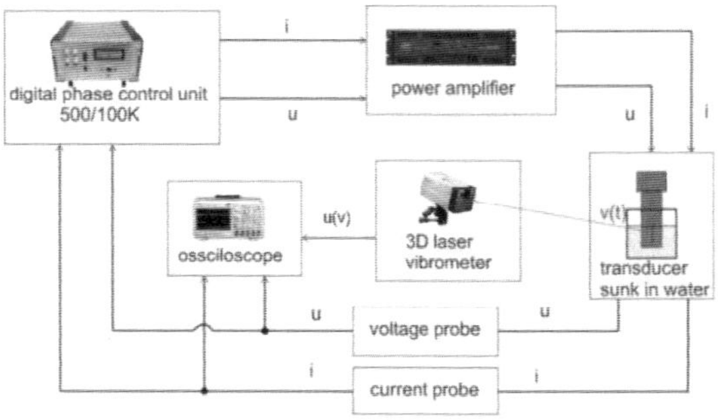

Figure 5. Experimental setup for observation of ultrasonic-induced cavitation.

The explicit models for the cavitation process can be arbitrarily complex. To allow examination of the process' impact on the oscillating system, however, there is no need to model the cavitation process itself in all details. Thus a simplified load model with sufficient accuracy is developed. Therefore, the straightforward process of cavitation is considered and the ultrasonic oscillator depicted in Figure 3 introduced into a container filled with water. The system first is examined experimentally with the setup depicted in Figure 5. Voltage and current probes are utilized to measure the electrical values. The oscillator's vibration amplitude at the front face is measured with a 3D single point laservibrometer. A database of the measured parameters is received by varying the operational parameters as driving voltage, immersion depth, and water level.

In a second step the overall model of the process and ultrasonic oscillator with a simplified load model is developed and parameterized. Therefore, the system is divided in partial models of the ultrasonic oscillator, the respective mounting, the process load, and the vibration behavior. Each partial model is implemented into a wrapper as depicted in Figure 6. In addition, the partial model types are given. The different partial model's input and output parameters are summarized in Table 1. Parameters not defined in the respective wrapper but in the master module's global structure array are not listed here.

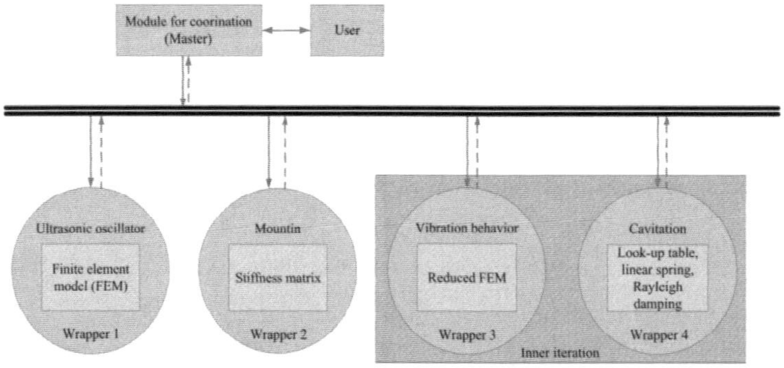

Figure 6. Partial models and respective model type for ultrasonic-induced cavitation.

Operational parameters are implemented into the global structure array in the master module. Furthermore, the execution sequence for the partial models and wrappers respectively is determined here and corresponds to the sequence in Table 1 and Figure 6 (from left to right). Both the wrapper for calculating the vibration behavior and the wrapper with the simplified cavitation model are implemented into the framework's inner iteration loop. First free oscillation characteristics and resulting load characteristics are determined. The calculated load then is used to update the calculation of the vibration behavior and so forth. Constant vibration amplitude at the ultrasonic oscillator's front surface is defined as convergence criterion. The developed partial models are briefly summarized in the following.

Table 1. Input, and output parameters of partial models for ultrasonic-induced cavitation.

	Ultrasonic Oscillator	Mounting	Vibration behavior	Load model
Input	Index of DOF , mass, damping, stiffness matrix	Index of DOF	System of differential equations, mounting stiffness, contact forces	Index of DOF, Vibration amplitudes
Output	System of differential equations	Mounting stiffness	Vibration amplitudes, eigenfrequencies	Process forces, Rayleigh damping matrix

The ultrasonic oscillator is modeled by means of the finite element method (FEM [16]) considering both the mechanical and electrical degrees of freedom (DOF) with the commercial software ANSYS. The mass $\boldsymbol{M}$, damping $\boldsymbol{D}$ and stiffness matrices $\boldsymbol{K}$ are exported and within a first wrapper imported into a system of differential equations. Furthermore the system of differential equations is sorted by mechanical (master) and electrical (slave) DOF in the first wrapper (see equation 1).

$$\begin{bmatrix} \mathbf{M}_{mm} & \mathbf{M}_{ms} \\ \mathbf{M}_{sm} & \mathbf{M}_{ss} \end{bmatrix} \begin{bmatrix} \ddot{\underline{x}}_m \\ \ddot{\underline{x}}_s \end{bmatrix} + \begin{bmatrix} \mathbf{D}_{mm} & \mathbf{D}_{ms} \\ \mathbf{D}_{sm} & \mathbf{D}_{ss} \end{bmatrix} \begin{bmatrix} \dot{\underline{x}}_m \\ \dot{\underline{x}}_s \end{bmatrix} + \begin{bmatrix} \mathbf{K}_{mm} & \mathbf{K}_{ms} \\ \mathbf{K}_{sm} & \mathbf{K}_{ss} \end{bmatrix} \begin{bmatrix} \underline{x}_m \\ \underline{x}_s \end{bmatrix} = \begin{bmatrix} \underline{f}_m \\ \underline{f}_s \end{bmatrix} \tag{1}$$

An idealized mounting setup is defined in a second wrapper. The explicit nodes of the ultrasonic oscillator's FE mesh are determined and a matrix with entries in terms of a defined stiffness at corresponding positions is created as output data. It can be added to the oscillator's stiffness matrix.

Within a third wrapper the vibration behavior (vibration amplitudes and eigenfrequencies) of the ultrasonic oscillator in dependence of the load is determined. Because of the considered electrical DOF leading to zero entries in the mass matrix a static reduction in accordance with [17] and [18] has to be implemented. Furthermore, a modal reduction in accordance with [19], [18] and [20] is utilized to overcome the issue of a big amount of DOF and a high computing time respectively. Output data of this wrapper are the system's eigenfrequencies and deflection amplitudes.

The simplified cavitation model itself consists of spring-damper-elements and is implemented into a fourth wrapper. Damping and stiffness coefficients (k_{cav}, α, and β) according to the selected operational parameters are picked from an implemented look-up table. For calculation of the resulting forces a linear relation for the spring force

$$\underline{f}_m = k_{cav} \underline{x}_m \tag{2}$$

and for the damping impacts occurrence of Rayleigh damping [21] only

$$\underline{C}_R = \alpha\, \mathbf{M}_{mm} + \beta\, \mathbf{K}_{mm} \tag{3}$$

are assumed. The spring-damper-elements are connected to the nodes of the oscillator's FE mesh (cf. Figure 7). Applied forces are assumed to be the same at each node regardless of their position in x- and y-directions. Thus vectors and matrices with entries at the corresponding positions are created as the wrapper's output data.

IOP Conf. Series: Materials Science and Engineering **211** (2017) 012014 doi:10.1088/1757-899X/211/1/012014

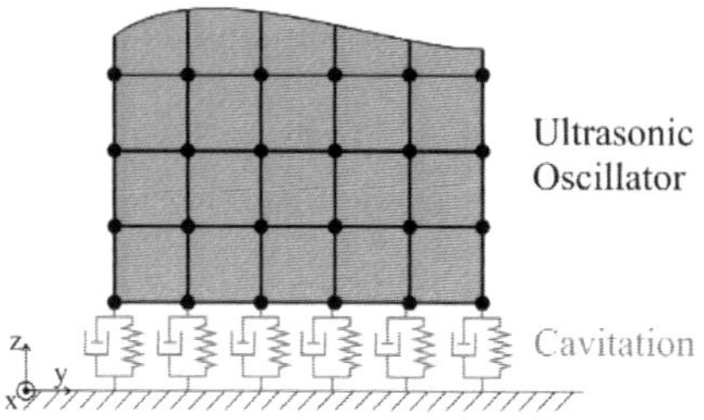

Figure 7. Schematic representation of load model for ultrasonic-induced cavitation.

The developed overall model permits a parameterization of the load model based on the conducted measurements. Therefore, the complete framework is implemented into an iteration loop. Damping and stiffness coefficients for respective operational parameters are determined iteratively and used for adapting the look-up table for the load model. The corresponding measured eigenfrequency and deflection amplitude are defined as convergence criteria.

Table 2. Comparison of experimental and numerical determined cavitation impacts on the oscillator.

Operating parameters			Experimental results		Numerical Results	
Current [A]	Immersion depth [mm]	Water level [mm]	Frequency [Hz]	Amplitude [μm]	Frequency [Hz]	Amplitude [μm]
0.7	10	27	20416	4.36	20414	4.35
1.1	10	27	20513	7.23	20515	7.21
0.7	30	47	20423	7.42	20425	7.43
0.7	30	76	20475	1.94	20747	1.92

Numerical and experimental results for a representative selection of operational parameters are given in Table 2. It becomes apparent that the developed overall model allows very good prediction of the cavitation influences on the ultrasonic oscillator. Average error regarding the eigenfrequency is 0.3 % and regarding the vibration amplitude 0.2 %. The developed overall model thus allows investigations on load influences on the oscillator's vibration shape.

3.2. Ultrasonic Assisted Machining of Stone

The second process modeled with the developed framework is ultrasonic assisted machining of stone. An ultrasonic vibration superimposed to an actual machining process reduces the occurring process forces [22], [23], improves the surface quality [24], and has a positive effect on the chip breaking [25]. However, implementing the ultrasonic system into the process increases the modeling complexity significantly. In addition to the actual feeding movement within the centimeter range and velocities around 100 m/min the ultrasonic vibration of several micrometers and at a frequency above 20 kHz has to be considered. Furthermore, the ultrasonic oscillator typically is driven by a piezoelectric converter thus the electrical domain has to be implemented into the overall model. Due to these challenges no model of the process in combination with the ultrasonic oscillator and the interactions between these two parts has yet been developed. Such an overall model though would offer distinct benefits like an improved process understanding or process optimized design of the utilized tools and mountings inter alia. The latter is selected as application example in the contribution at hand. Therefore, a simplified contact model is utilized to remain the focus on the software framework whereat the material removal itself will be neglected.

IOP Conf. Series: Materials Science and Engineering **211** (2017) 012014 doi:10.1088/1757-899X/211/1/012014

Table 3. Partial model type, input, and output parameters for ultrasonic assisted machining of stone.

	Ultrasonic Oscillator	Mounting	Workpiece characteristics	Vibration behavior	Contact process
Input	Index of DOF , mass, damping, stiffness matrix	Index of DOF	-	System of differential equations, mounting stiffness, contact forces	Index of DOF, Vibration amplitudes, young's modulus
Output	System of differential equations	Mounting stiffness	Young's modulus	Vibration amplitudes, eigenfrequencies	Contact forces

The system is divided into partial models for a given ultrasonic oscillator, the mounting, the workpiece characteristics, the simplified contact process, and for calculating the vibrational behavior. The input and output parameters are summarized in Table 3. Again parameters not defined in the respective wrapper are not listed here. These operational parameters are defined within the master module where the execution sequence for the partial models and wrappers respectively again is determined (cf. Table 3 from left to right). The partial model for calculating the vibration behavior and the partial model of the contact forces are implemented into the inner iteration loop and repeated iteratively whereat constant vibration amplitude at the ultrasonic oscillator's front surface is defined as convergence criterion.

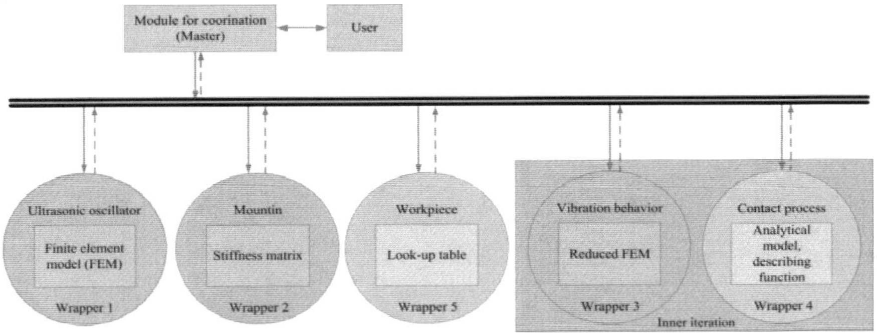

Figure 8. Connection between partial models for ultrasonic assisted machining of stone.

One main benefit of the software framework can be exploited for developing partial models for the oscillator, the mounting implementation, and for calculating the vibration behavior. Since respective partial models already have been developed for ultrasonic-induced cavitation those wrappers can be adapted, if necessary, and easily implemented into the new overall model. Thus the previous model's first wrapper for defining a system of differential equation similar to equation 1, the second wrapper for implementing idealized mounting conditions into a respective stiffness matrix, and the third wrapper for calculating the vibration behavior are slightly modified. Again static reduction in accordance with [17] and modal reduction in accordance with [19] are implemented. However, partial models of the contact process and workpiece characteristics need to be developed (cf. Figure 8).

The complex contact process is simplified and therefore divided into axial forces resulting from the micro hammering effect (cf. [22] and [26]) and forces in the machining plane resulting from frictional processes. Both non-linear contact forces can be described in terms of piecewise-linear stiffness as depicted in Figure 9. A linearization is inter alia possible with the describing function and harmonic balance method [27] permitting an implementation in appropriate analytical models. Considering the friction contact the resulting forces (see also Figure 9 c)) can be written as:

$$F_a = \begin{cases} -N_0 & \text{for} \quad \hat{v}_e < -c, \\ \dfrac{N_0 - N}{c}\hat{v}_e - N & \text{for} \quad -c < \hat{v}_e < 0, \\ 0 & \text{for} \quad \hat{v}_e = 0, \\ \dfrac{N_0 - N}{c}\hat{v}_e + N & \text{for} \quad 0 < \hat{v}_e < c, \\ N_0 & \text{for} \quad c < \hat{v}_e \end{cases} \tag{4}$$

and linearized to

$$B(\hat{v}_e, N_0) = \frac{2(N_0 - N)}{\pi\, c}\arcsin\left(\frac{c}{\hat{v}_e}\right) + \frac{1}{\pi\, \hat{v}_e}\left[2c(N_0 - N)\sqrt{1-\left(\frac{c^2}{\hat{v}_e^2}\right)} + 4N\right] \tag{5}$$

The determined describing function (equation 5) is equivalent to a replacement stiffness that depends on the velocity amplitude and can be easily implemented into the constitutive contact model and spring elements respectively. The described calculation of the friction forces as well as the analogous calculation of the impact forces (cf. [20]) is implemented into an appropriate wrapper. Analogous to the partial model of the mounting the explicit nodes of the ultrasonic oscillator's FE mesh at the front face are determined and a matrix with entries at corresponding positions is created. Outputs of the wrapper thus are a stiffness matrix depending on the calculated vibration amplitudes and a damping force vector. The workpiece characteristics in terms of Young's modulus are gained from literature whereat a database for different materials and probe dimensions is implemented into a further wrapper.

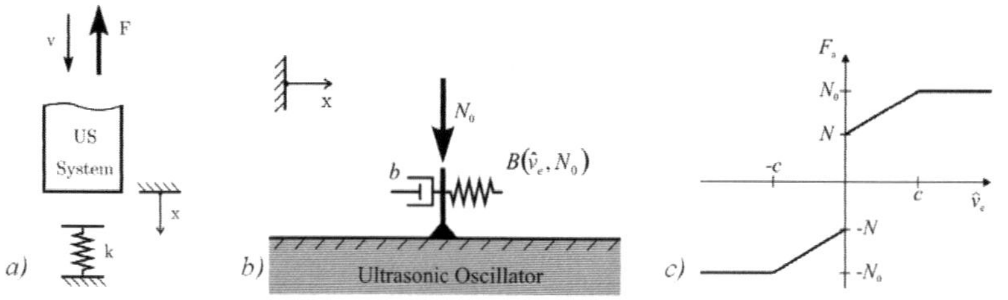

Figure 9(a). Analogous models for the impact process; **Figure 9(b).** analogous model for the frictional process, and **Figure 9(c).** related force-deflection graph.

Permitting a comparison of the resulting vibration shapes and achievable deflection amplitudes the framework with the described overall model is executed for different mounting position. Starting directly next to the piezoelectric stack the mounting here is coupled to distinct positions on the converter and moved towards the sonotrode. The possible mounting positions and the minimum step size depend directly on the FE mesh and node distribution respectively. Figure 10 a) shows the calculated mean displacement amplitudes at the oscillator's front face for different mounting distances to the piezoelectric stack. A clear maximum is distinguishable thus a process-optimized mounting position located as close as possible to the piezoelectric stack can be identified. Vibration shapes of the oscillator with original and optimized mounting positions can be easily calculated with the developed overall model and are depicted in Figure 10 b). Again the increase in displacement amplitudes is apparent. In addition, a load impact in terms of a shift of the maximum displacement away from the front face can be noticed. In the region of the piezoelectric stack load and mounting impacts are negligibly small.

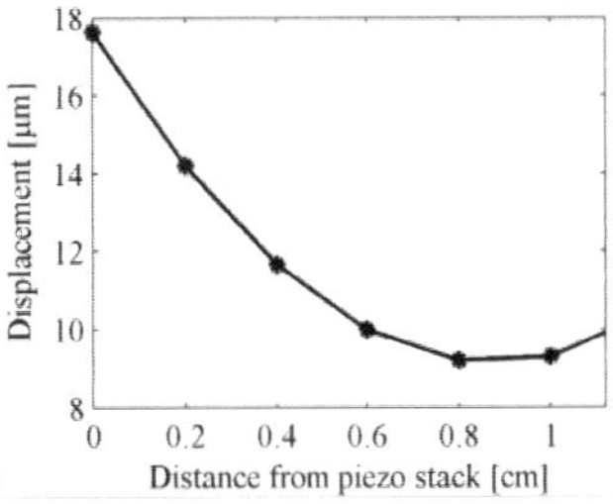

a)

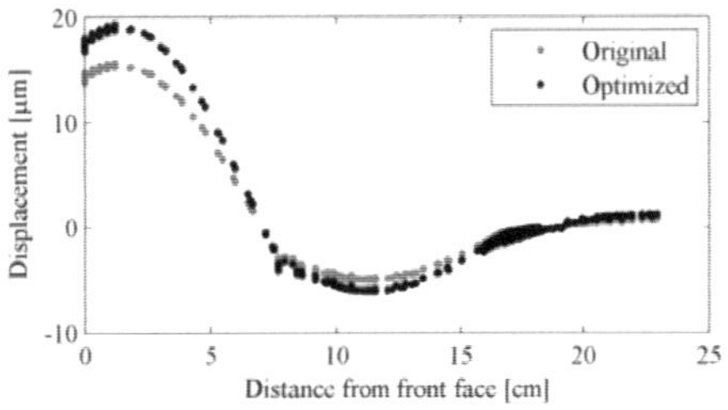

b)

Figure 10(a). Mean displacement amplitudes at oscillator's front face for different mounting positions.

Figure 10(b). vibration shapes of original and optimized oscillator.

4. Conclusion

In the contribution at hand a generic but simple software framework implemented in MathWorks MATLAB is presented. By dividing the respective overall model into partial models it is applicable for processes and systems of almost any complexity. The framework consists of a master module for coordination and any number of slave modules with well-defined interfaces permitting the integration of the partial models. The defined format of the respective input and output parameters leads to the frameworks generic character and an easy coupling of the wrappers. Furthermore, there are no significant restrictions in terms of partial model type to be included (e.g. experimental data, interpolated data, analytical models, or numerical models). The framework is demonstrated by means of two explicit application examples and tasks respectively that could be satisfactorily solved. Considering the cavitation process and the presented simplified load model an experimental based parameterization could be performed. The developed overall model is capable of predicting the cavitation influences on an ultrasonic oscillator without consideration of complex descriptions of the cavitation process itself. In addition, a sufficiently high accuracy can be reached. A process-optimized mounting position could be identified by modeling the process of ultrasonic assisted machining of stone in terms of process loads, ultrasonic oscillator, and respective interactions. Integrating the overall model into the developed framework, furthermore, permitted the determination of load impacts on the utilized oscillator's vibration shape.

In conclusion, the easy integration of partial models into the developed framework and the possibility of adapting existing partial models for new modeling tasks became apparent in particular. Further benefits of the developed framework are the arbitrary expandability and the good comprehensibility of the overall models due to the clearly defined but still simple coupling of clearly separated partial models.

References

[1] Doumanidis C and Gao Y 2004 Mechanical modeling of ultrasonic welding *Welding Journal-New York* **83** 4 pp 140-S
[2] Liu Y, Lijun W and Dongpo W 2011 Finite element modeling of ultrasonic surface rolling process *Journal of Materials Processing Technology* **211** 12 pp 2106-13
[3] Potthast C, Twiefel J and Wallaschek J 2007 Modeling approaches for an ultrasonic percussion drill *Journal of Sound and Vibration* **308** 3 pp 405-17
[4] Heisel U et al 2011 Kinematic Model for Ultrasonic-Assisted Manufacturing of Bore Holes with Undefined Cutting Edges *Advanced Materials Research* (Trans Tech Publ) **223**
[5] Tudoret S, et al 2000 Co-simulation of hybrid systems: Signal-Simulink *Int. Symp. on Formal Techniques in Real-Time and Fault-Tolerant Systems* (Springer Berlin Heidelberg)
[6] Crawford A E 1955 Ultrasonic Engineering (Butterworths, London)
[7] Flynn H G 1964 Physics of acoustic cavitation in liquids *In Physical acoustics* ed W P Mason (New York: Academic) IB p 157

[8] Franc J P and Michel J M 2004 Fundamentals of Cavitation (London: Kluwer Academic Publishers, Dordrecht, Boston)
[9] Brennen C E 2014 Cavitation and Bubble Dynamics (Cambridge: Cambridge University Press)
[10] Ikeda T 1990 Piezoelectricity (New York: Oxford university press)
[11] Franc J P and Michel J M 2004 Fundamentals of Cavitation (London: Kluwer Academic Publishers, Dordrecht, Boston)
[12] Shchukin D G, et al 2011Ultrasonic cavitation at solid surfaces *Advanced Materials* **23** 17 pp 1922-1934.
[13] Suslick K S 1989 The chemical effects of ultrasound *Scientific American* **260** 2 pp 80-6
[14] Knorr D, Zenker M, Heinz V and Lee DU 2004 Applications and potential of ultrasonics in food processing *Trends in Food Science & Technology* **15(5)** pp 261-6
[15] Mordyuk B N and Prokopenko G I 2007 Ultrasonic impact peening for the surface properties' management *Journal of Sound and Vibration* **308** Jg Nr 3 S pp 855-66
[16] Knothe K and Wessels H 2008 Finite Elemente (Springer Verlag Berlin Heidelberg)
[17] Guyan R J 1965 Reduction of stiffness and mass matrices *The American Institute of Aeronautics and Astronautics Journal* **3** 2 p 380
[18] Bruns P and Twiefel J 2015 Influences of Non-axial Process Loads on the Transducer and the Associated Mounting *Ultrasonic Machining. Energy Harvesting and Systems* ISSN (Online) pp 2329-8766 ISSN (Print) pp 2329-8774
[19] Weaver V, Timoshenko S P and Young D H 1990 *Vibration problems in engineering* 5th ed (Fong and Sens Printers Ltd)
[20] Bruns P, Twiefel J and Wallaschek J 2016 Experimental and Theoretical Investigation of a Vibro-Impact Process Utilizing an in situ Capacitive Distance Measurement *Proceedings of ISMA2016 International Conference on Noise and Vibration Engineering, Löwen*
[21] Craig R R and Kurdila A J 2006 Fundamentals of structural dynamics (John Wiley & Sons)
[22] Heisel U, Eisseler R, Eber R, Wallaschek J, Twiefel J and Huang M 2011 Ultrasonic-assisted machining of stone *Prod. Eng. Res. Devel. 5* pp 587–94
[23] Stehle T, Eisseler R, Drewle K, Weiland S, Bruns P and Twiefel J 2015 Investigation of Process Forces of Ultrasonically Assisted Scratch Tests *Proc. of the 3rd International Conference on Stone and Concrete Machining (ICSCM)* Bochum (Germany), pp 137-45
[24] Silberschmidt V V, Mahdi S M A, Gouda M A , Naseer A, Maurotto A and Roy A 2014 Surface-roughness improvement in ultrasonically assisted turning *Procedia CIRP* **13** pp 49-54
[25] Babitsky, et al 2003 Ultrasonically assisted turning of aviation materials *Journal of Materials Processing Technology* **(132)** pp 157–67
[26] Twiefel J, et al 2008 Fundamental experiments as benchmark problems for modeling ultrasonic micro-impact processes *Journal of Electroceramics* **20** 3-4 pp 209-214
[27] Zhang W S 1992 Nonlinear Vibrations of Piecewise-Linear Systems by Incremental Harmonic Balance Method *Journal of Applied Mechanics* **59** 1-2 p 153

Acknowledgment
The work was supported by the German Research Foundation (DFG).

IOP Conf. Series: Materials Science and Engineering **211** (2017) 012015 doi:10.1088/1757-899X/211/1/012015

Kalman filter implementation for small satellites using constraint GPS data

Elmahy M Wesam *, **Xiang Zhang, Zhengliang Lu and Wenhe Liao**

Nanjing University of Science and Technology, China

*Email: Wesam_elmahy@yahoo.com

Abstract. Due to the increased need for autonomy, an Extended Kalman Filter (EKF) has been designed to autonomously estimate the orbit using GPS data. A propagation step models the satellite dynamics as a two body with J_2 (second zonal effect) perturbations being suitable for orbits in altitudes higher than 600 km. An onboard GPS receiver provides continuous measurement inputs. The continuity of measurements decreases the errors of the orbit determination algorithm. Power restrictions are imposed on small satellites in general and nanosatellites in particular. In cubesats, the GPS is forced to be shut down most of the mission's life time. GPS is turned on when experiments like atmospheric ones are carried out and meter level accuracy for positioning is required. This accuracy can't be obtained by other autonomous sensors like magnetometer and sun sensor as they provide kilometer level accuracy. Through simulation using Matlab and satellite tool kit (STK) the position accuracy is analyzed after imposing constrained conditions suitable for small satellites and a very tight one suitable for nanosatellite missions.

1. Introduction

A Global Navigation Satellite System (GNSS) is a system comprising a constellation of satellites capable of providing accurate orbit determination (OD) and time synchronization. It transmits navigation signal required by GNSS users to estimate their Position, Velocity and Time (PVT) [1]. GPS consists of three segments: the control segment, the space segment and the user segment. User segment refers to the GPS receivers that use the satellite-transmitted signals for navigation [2]. The GPS space segment nominally consists of 24 satellites in six orbital planes they are designed to ensure a visibility of at least four satellites at all times. Only four satellites are needed for a navigation fix. The control segment consisting of worldwide ground stations to maintain the operation of the space segment. The system provides very precise positioning information in meter and cm level. The GPS offers an attractive alternative to ground-based tracking systems. It provides very comparable accuracy in an autonomous way or nearly autonomous as it depends on the GPS satellites. CubeSats have evolved from purely educational tools to a standard platform for technology demonstration and scientific instrumentation and application in less than a decade [3]. Due to their low cost in development, they are spreading widely and becoming the lead in the list of choices for satellite manufacturing. With the availability of light and low power receivers lead a direct way to the usage of GPS navigation in nanosatellites. In [4], it is found that 16 % of 94 studied pico- and nanosatellites carried a GPS receiver, not all have been successful. The CanX-2 nanosatellite was carrying a GPS receiver among other payloads. It provided promising results for GPS navigation on nanosats missions as it was the first mission to successfully deliver GPS navigation fixes and raw measurements [5]. The measurements obtained from the GPS receiver can be applied for precise orbit determination (POD) in two ways either for post processing where a least squares technique is applied or in real- time onboard navigation where a Kalman filter is used [6]. An Extended Kalman Filter (EKF) is a filter suitable with nonlinear systems [8]. It allows for accurate prediction of a given state without requiring expansive past data to be stored and computed, which allows for efficient on-board computation. Accuracy at the decimeter level can only be achieved under the best cases: post processing dual-frequency data with continuous receiver operation providing accuracy in cm level [7]. Unfortunately, such requirements may not be accessible in nanosats as they have to operate with low power. Single-frequency receivers are more suitable for CubeSats. Several experiments especially atmospheric ones need very accurate positioning information in meter level to ensure their success. This high accuracy can not be accomplished using autonomous sensors like magnetometers and sun sensors as they provide kilometer level accuracy. The GPS accurate navigation information opens the door for more experiments to be carried onboard the CubeSat. This makes the nanosats perfect candidates for more experiments and challenges. As a result of these power restrictions, the GPS will be shut off most of the mission and open when performing experiments. In this paper, several constrained conditions are imposed on the availability of GPS data to see how this affects the corrected states accuracy. Thus, the remaining of current article is organized as follows: firstly, Section 2 outlines the OD process consisting of its three main parts (dynamical model, estimation technique and measurement model). Secondly, Section 3 shows the simulation scenario using simulated GPS measurements with different test cases and the results are also analyzed. Finally, Section 4 declares conclusions and future work.

2. The OD process using GPS data

To accomplish an OD process using GPS data a satellite orbit can be determined using a dynamical method of analysis, where a force model is used to represent the most effective forces acting on the satellite (depends on its altitude). With an initial state vector, the numerical integration of these accelerations will give the position of the satellites at discrete-time intervals, resulting in the so called predicted orbit. The dynamics then allow the propagation of the state forward in time. The GPS measurements and the dynamic information are combined through an EKF for the correction process and hence produce a GPS solution.

2.1. Satellite's Dynamic Model

For a continuous-time system, there are different forms for the state vector representation. Each form depends on the solution procedure and the analysis type. In this work, the position and velocity vectors, in the Earth Centered Inertial (ECI) coordinates frame, are used to represent the state. This form is more suitable for numerical processing and usually easier to deal with during the OD process since they have no singularities. The state is presented as Equation (1):

$$\mathbf{x}(t)=\begin{bmatrix}\mathbf{r}\\ \mathbf{v}\end{bmatrix}_{ECI}=\begin{bmatrix}x & y & z & \dot{x} & \dot{y} & \dot{z}\end{bmatrix}^T \tag{1}$$

The state dynamic model that describes its time changes. The state propagation is in Equation (2):

$$\begin{aligned}\dot{\mathbf{X}}(t)&=f\,(\mathbf{x}(t))+w\,(t)=\begin{bmatrix}\mathbf{v}(t)\\ a(t)\end{bmatrix}+w\,(t)\\ &=\begin{bmatrix}\dot{x} & \dot{y} & \dot{z} & a_x & a_y & a_z\end{bmatrix}^T+\begin{bmatrix}0 & 0 & 0 & w_x & w_y & w_z\end{bmatrix}^T\end{aligned} \tag{2}$$

The order of the total acceleration used in the current work is equal to Newton's acceleration and the acceleration perturbation due to J_2. The w terms represent the modeling errors arising from the limited acceleration order. The acceleration components, a_x , a_y and a_z are represented in the ECI frame, as Equation (3):

$$
\begin{aligned}
a_x &= a_{x_{Newton}} + a_{x_{J2}} = -\frac{\mu x}{r^3} + \frac{\mu J_2 R^2}{2}\left(15\frac{xz^2}{r^7} - 3\frac{x}{r^5}\right) \\
a_y &= a_{y_{Newton}} + a_{y_{J2}} = -\frac{\mu y}{r^3} + \frac{\mu J_2 R^2}{2}\left(15\frac{yz^2}{r^7} - 3\frac{y}{r^5}\right) \\
a_z &= a_{z_{Newton}} + a_{z_{J2}} = -\frac{\mu z}{r^3} + \frac{\mu J_2 R^2}{2}\left(15\frac{z^3}{r^7} - 9\frac{z}{r^5}\right)
\end{aligned} \quad (3)
$$

where: r is the norm vector between the center of the Earth and the satellite, the standard gravitational parameter $\mu = 3.986004418*10^{14}$ m^3s^{-2}, the dimensional coefficient $J_2 = 0.0010827$ has been found from satellite observations and the radius of the Earth $R = 6378.137km$ at the equator [9].

2.2. *Continuous-Time EKF*

The Kalman Filter (KF) is a recursive algorithm which requires only data from the last and current iterations as the predicted state estimate ($\hat{\mathbf{X}}_{k/k-1}$) contains all required information on the system up to time (k-1). The KF generates a new state estimate as new observations become available, thus opening the possibility of real-time estimation. In case of nonlinear systems, Extended Kalman Filter EKF is constructed by linearizing either the system equation or measurement equation or both about the current (predicted) state estimate. This is done using the first two terms of a Taylor series expansion and using the same structure of the linear Kalman filter to provide the next estimate [10]. This produces a constant term and a first derivative term, known as the Jacobian (a matrix of partial derivatives). Most physical systems are represented as continuous-time models while discrete-time measurements are frequently taken for state estimation via a digital processor. The filter is comprised of two main steps: the prediction and the update. In continuous–time form, the dynamics of a linear system (process and measurement models) can be represented by the following first-order vector-matrix differential shown in Equations (4) and (5):

$$\dot{\mathbf{x}}(t) = f(\mathbf{x}(t), \mathbf{u}(t)) + \mathbf{w}(t), \text{ and } \mathbf{w}(t) \sim N(0, \mathbf{Q}(t)) \quad (4)$$

$$\mathbf{z}(t) = h(\mathbf{x}(t)) + \mathbf{v}(t), \text{ and } \mathbf{v}(t) \sim N(0, \mathbf{R}(t)) \quad (5)$$

where as shown in Equation (6):

$$\mathbf{x}_k = \mathbf{x}(t_k), \hat{\mathbf{x}}_{0/0} = E[\mathbf{x}(t_0)], \mathbf{P}_{0/0} = var[\mathbf{x}(t_0)] \quad (6)$$

Suppose that as shown in Equation (7):

$$\phi(t) = \left.\frac{\partial f}{\partial x}\right|_{\hat{\mathbf{x}}(t), u(t)} \text{ and } \mathbf{H}_k = \left.\frac{\partial h}{\partial x}\right|_{\hat{\mathbf{x}}_{k/k-1}} \quad (7)$$

By solving Equations (8) and (9):

$$\dot{\hat{\mathbf{x}}}(t) = f(\hat{\mathbf{x}}(t), u(t)) \quad (8)$$

$$\dot{\mathbf{P}}(t) = \phi(t)\mathbf{P}(t) + \mathbf{P}(t)\phi^T(t) + \mathbf{Q}(t) \quad (9)$$

With Equation (10):

$$\hat{\mathbf{x}}(t_{k-1}) = \hat{\mathbf{x}}_{k/k-1} \text{ and } \mathbf{P}(t_{k-1}) = \mathbf{P}_{k/k-1} \quad (10)$$

The predicted state estimate and predicted covariance matrix will be as Equation (11):

$$\hat{\mathbf{x}}_{k/k-1} = \hat{\mathbf{x}}(t_k) \text{ and } \mathbf{P}_{k/k-1} = \mathbf{P}(t_k) \quad (11)$$

The update equations (12-16) of continuous-time EKF are identical to those of discrete-time EKF given by:

$$\mathbf{\Psi}_k = \mathbf{z}_k - \mathbf{h}(\hat{\mathbf{x}}_{k/k-1}) \quad (12)$$

$$\mathbf{S}_k = \mathbf{H}_k \mathbf{P}_{k/k-1} \mathbf{H}_k^T + \mathbf{R}_k \quad (13)$$

$$\mathbf{K}_k = \mathbf{P}_{k/k-1} \mathbf{H}_k^T \mathbf{S}_k^{-1} \quad (14)$$

$$\hat{\mathbf{x}}_{k/k} = \hat{\mathbf{x}}_{k/k-1} + K_k \, \mathbf{\Psi}_k \quad (15)$$

$$\mathbf{P}_{k/k} = (\mathbf{I} - K_k \mathbf{H}_k)\mathbf{P}_{k/k-1} \tag{16}$$

To prevent the Kalman filter divergence, the following algebraically equivalent covariance correction Equation (17), which is called "Joseph Alternative KF Form", is applied [10]:

$$\mathbf{P}_{k/k} = (\mathbf{I} - K_k \mathbf{H}_k)\mathbf{P}_{k/k-1}(\mathbf{I} - K_k \mathbf{H}_k)^T + K_k \mathbf{R}_k K_k^T \tag{17}$$

Where, the symmetry of matrix multiplication forces $P_{k/k}$ to remain symmetric. The first Jacobian state matrix, Φ, is independent of the type of measurement. Thus, Φ is given by Equation (18):

$$\Phi = \Phi\left(\hat{\mathbf{x}}(t)\right) = \begin{bmatrix} 0_{3x3} & I_{3x3} \\ \Phi_{33} & 0_{3x3} \end{bmatrix}_{\mathbf{x}(t)=\hat{\mathbf{x}}(t)} \quad and \ldots \Phi_{33} = \begin{bmatrix} \frac{\partial a_x}{\partial x} & \frac{\partial a_x}{\partial y} & \frac{\partial a_x}{\partial z} \\ \frac{\partial a_y}{\partial x} & \frac{\partial a_y}{\partial y} & \frac{\partial a_y}{\partial z} \\ \frac{\partial a_z}{\partial x} & \frac{\partial a_z}{\partial y} & \frac{\partial a_z}{\partial z} \end{bmatrix} \tag{18}$$

The partial derivatives are too lengthy to be stated here and are available in [11]. The Runge-Kutta method of fourth order is used to propagate the state and covariance in-between measurements. It provides good performance to a relatively low computational cost even for larger integration steps. The Kalman Filter can be "tuned" by adjusting the covariance of the process noise, w_k, and the covariance of the observation noise, v_k.

2.3. GPS measurement model

The atomic clocks aboard all GPS satellites are synchronized to a time referred to as GPS system time. GPS satellites continuously transmit signals that can be picked up by a GPS receiver. In an ideal case, 3 satellites are enough for a 3-dimensional position fix, but in practice a fourth satellite is needed to correct the receiver's clock as the receiver clock is not usually synchronized with the system time [12].
The vector **r** from the receiver to a GPS satellite can be expressed as Equation (19):

$$\mathbf{r} = \mathbf{S} - \mathbf{u} \tag{19}$$

where $\mathbf{S}$ is the position vector of the GPS satellite and **u** is the position vector of the receiver. The pseudo range p can be expressed as Equation (20):

$$p = r + c.t_u \tag{20}$$

c is the light's speed and t_u is the offset between receiver clock and GPS system time as Equation (21):

$$p - c.t_u = \|\mathbf{S} - \mathbf{u}\| \tag{21}$$

This is actually an equation for a sphere centered on the GPS satellite. We have a set of four unknowns (x_u, y_u, z_u, t_u): the receiver position and the time offset. With four satellites available, the pseudo range is as Equation (22):

$$p_i = \sqrt{(x_i - x_u)^2 + (y_i - y_u)^2 + (z_i - z_u)^2} + c.t_u \ , \ i=1:4 \tag{22}$$

where (x_i, y_i, z_i) are the individual GPS satellite positions.

The receiver velocity can be obtained by either forming an approximate derivative of the position or by studying the Doppler shift of the signal from individual satellites. More advanced techniques include using the carrier phase of the incoming signal to smooth the pseudo range obtained from the code phase to completely eliminate ionospheric error. The GPS receiver output is in the earth centered earth fixed (ECEF) frame. This fix information can be used to correct the on-board orbit propagator in the satellite, and it can also be downloaded from the satellite and used in more accurate orbit propagators. Satellite orbits are calculated in an earth centered inertial (ECI) frame as the equations of motion have to be applied in an inertial frame. The measured state vector is then transformed into ECI frame for compatibility. The observation Jacobian matrix, H is hence a 6x6 identity matrix [13].

IOP Conf. Series: Materials Science and Engineering **211** (2017) 012015 doi:10.1088/1757-899X/211/1/012015

3. Simulation scenario

Several scenarios are carried out with different constrained conditions to see the performance of the filter. The cases are tabulated in table.1 showing simulation times when the GPS receiver is on and off as well as the duty cycle. It has to be noted that the switching time differs from one mission to another depending on their own requirements. Case 5 in particular is suitable for nanosatellite operations.

3.1. The initial orbit and conditions for the EKF

A sun synchronous orbit (700 km) is chosen to test the filter so that the limited acceleration model outlined in the last section is convenient to be applied at that altitude. The reference state is generated from the high-precision orbit propagator (HPOP) in the satellite tool kit (STK). In the configuration, Runge-kutta 7/8 method for numerical integration of equations of motion, the EGM 96 model (up to a degree of 70*70) for perturbation due to the nonsymmetrical geopotential, Harris-Priester model for atmospheric drag, lunar/solar gravitational attraction and solar radiation pressure are also included. The atmospheric drag coefficient is set to 2.2, the solar radiation pressure coefficient is set to 1.0 and the constant area-to-mass ratio is set to 0.0043.

The initial state to start the filter is set as a truth model i.e. with no initial error. The simulated GPS measurements were created using the rand function in Matlab which produces normally distributed pseudo-random numbers, using a chosen error and rate. The error added to the position is 10 meters and to the velocity is 0.01. The Q and R are taken to be 6x6 diagonal matrices. The process noises for acceleration uncertainty are set to 1×10^{-5}. The step size is 1 second and the initial covariance is set to:

$$\mathbf{P}_o = \text{diag}\left[(10\text{ m})^2, (10\text{ m})^2, (10\text{ m})^2, (0.01\text{m/s})^2, (0.01\text{m/s})^2, (0.01\text{m/s})^2\right]$$

R matrix is set to equal $\mathbf{P}_o$ as no initial error is present. When GPS measurements are not available, the solution is based only on the dynamics part, as it is better than not having a solution at all. This causes quick convergence when the measurements are available again. More details are in [7].

Table 1. Availability of GPS data.

Case #	GPS_ON	GPS_OFF	Duty cycle
Case 1	15 Min	15 Min	50 %
Case 2	10 Min	10 Min	50 %
Case 3	5 Min	15 Min	25 %

Case #	GPS_ON	GPS_OFF	Duty cycle
Case 4	5 Min	10 Min	25 %
Case 5	9.8 Min	88.2 Min	10 %
Case 6	-	-	100 %

Simulation results and analysis: The simulations were performed for 1 day, 8hours and 1 orbit to see how this affects the position error as well as to test the filter. After running simulations, the results are presented in table. 2 showing the average root sum square (ARSS) for position, the maximum RSS and the average root sum square AVSS for velocity.

Table 2. Simulation Results.

Case	ARSS [m]			Max RSS [m]			AVSS in [m/sec]		
	1 Day	8 h	1 Orb	1 Day	8 h	1 Orb	1 Day	8 h	1 Orb
Case 1	9.1904	8.1858	10.68	120.1	82.317	67.39	0.0313	0.0278	0.0339
Case 2	4.8439	4.209	4.822	55.114	39.8	41.739	0.0234	0.02	0.0244
Case 3	13.799	12.14	14.58	129.77	107.11	107.29	0.045	0.038	0.049
Case 4	8.2841	7.361	9.163	80.065	55.322	50.778	0.0347	0.0298	0.039
Case 5	348.72	378.51	542.9	1700.0	1340.0	1130.0	0.3569	0.3889	0.562
Case 6	0.5082	0.534	0.738	9.1657	8.823	9.683	0.0063	0.0063	0.0064

As expected, the position error increased as the time decreased. The difference in accuracy between Cases 1 and 2 is relatively high. Case 2 is better by nearly 50% in ARSS and the maximum RSS for all simulation times relative to case 1. Also, Cases 3 and 4 have a similar pattern. Different users maybe prefer one case to another depending on their requirements and what type of small satellite they have.

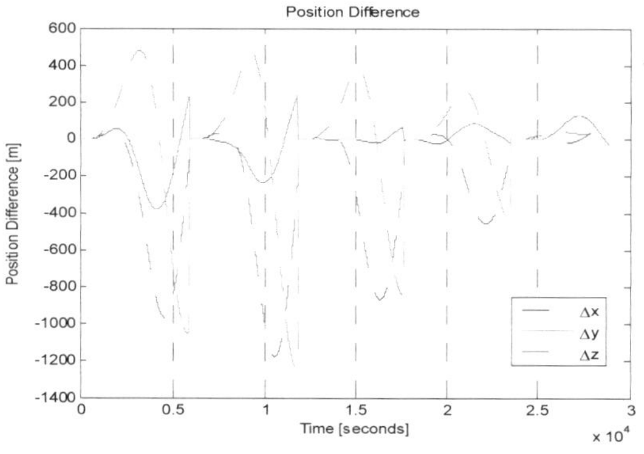

Figure 1. Position diff. 8h simulation Case 5.

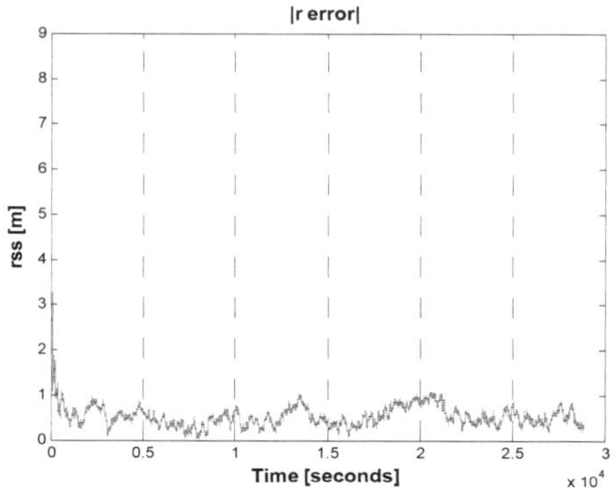

Figure 2. RSS 8h simulation Case 6.

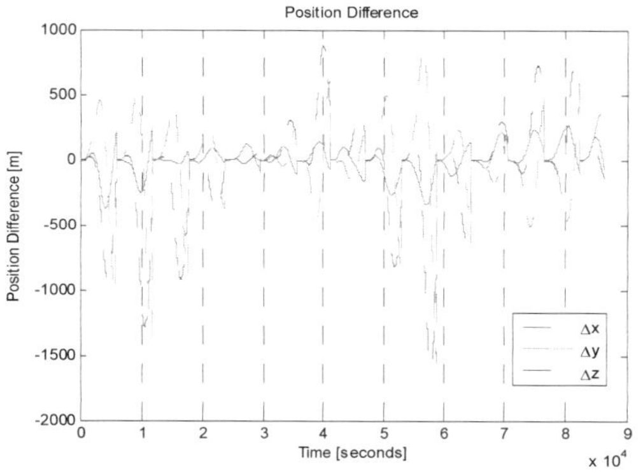

Figure 3. Pos. diff. 1 day simulation Case 5.

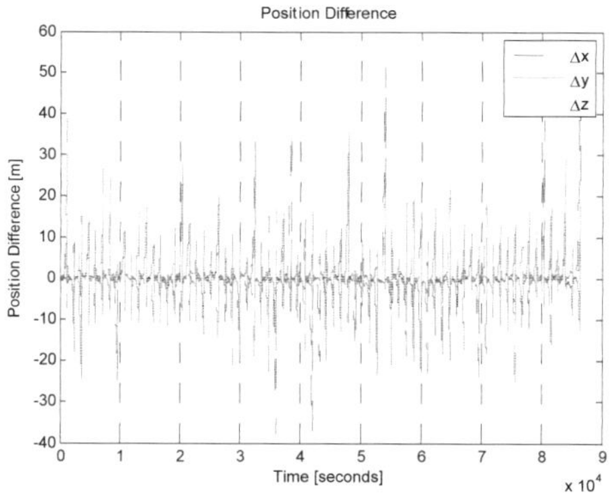

Figure 4. RSS 1 day simulation Case 2.

The microsatellites are more flexible than nanosatellites concerning the power limits making the cases other than Case 5 more suitable. Case 5 with the least duty cycle have the worst results of all cases which were also expected since saving power will result in less accurate results. The maximum error reaches 1.34km for 8h and 1.7km for 1 day simulation. This is still a very good result compared to the coarse accuracy provided by other sensors that can be onboard nanosatellites. Some values listed in table 2 are depicted for more clarity and to show the performance of the filter .Figures 1 and 2 show the position difference for Case 5 and the RSS for Case 6 with 8h simulation, respectively. Figures.3 and 4 show the 1 day simulation for Case 5 and Case 2, respectively.

4. Conclusion and future work

An EKF has been designed to autonomously correct the state using GPS measurements. The GPS measurements are simulated to represent only single-frequency data. Due to power constraints that are imposed on small satellites, especially nanosats, it is impossible for the GPS to be turned on all the time. In fact, in nanosats it will be shut down most of the mission's life time. For that reason, different test cases with different duty cycles and simulation times are applied. As expected, the accuracy degraded as the simulation time and the duty cycle decreased. On average ,a meter level is obtained even on a case with 10% duty cycle which is desirable when accurate on board tests are needed. For future work, more tuning for the filter is hopefully to be tried to reduce the maximum error. With real data available, it will be preferable to see the results in real case situation.

References

[1] Rush John 2000 Current issues in the use of the Global Positioning System aboard satellites *Acta Astronautica* **47** 2-9 pp 377-87

[2] Kaplan Elliott D and Christopher J Hegarty 2006 Understanding GPS: Principles and Applications Second Edition Artech House

[3] Puig-Suari J, Turner C and Twiggs R J 2001 CubeSat: The Development and Launch Support Infrastructure for Eighteen Different Satellite Customers on One Launch *Proceedings of AIAA/USU* Conf. Logan UT pp 1–5

[4] Bouwmeester J and J Guo 2010 Survey of worldwide pico- and nanosatellite missions, distributions and subsystem technology *Acta Astronautica 67* pp 854-62

[5] Kahr E, Montenbruck O, O'Keefe K, Skone S, et al 2011 GPS tracking on a nanosatellite CanX-2 flight experience *International ESA conference on guidance, navigation & control systems* (Karlovy Vary, Czech Republic) pp 5–10

[6] Oliver Montenbruck and Eberhard Gill 2005 Satellites Orbits Models Methods Applications (3rd Edition, Springer Berlin, Heidelberg, NewYork)

[7] Mander, Amandeep 2011 *Constrained GPS-based precise orbit determination of low earth orbiters* MSc. Thesis, York University

[8] Hendy H, Rui X T and Khalil M 2013 An Integrated GPS/INS Navigation System for Land Vehicle *Applied Mechanics and Materials* **336** pp 221-6

[9] D Vallado 2007 *Fundamentals of Astrodynamics and Applications* (3rd edition Springer, Hawthorne, CA)

[10] El-Mahy M K 1994 An Investigation into Kalman Filter Target Tracking Algorithms and their Real Time Parallel Transputer Implementation (Ph.D. Thesis Royal Military College of Science Cranfield University U.K)

[11] Rikard Ottemark 2015 Autonomous Satellite Orbit Determination Based on Magnetometer and Sun Sensor Measurements Design: Test and Verication (MSc. Thesis, Luleå University of Technology, Department of Computer Science, Electrical and Space Engineering)

[12] Hannu Leppinen 2013 Integration of a GPS subsystem into the Aalto-1 nanosatellite (MSc in Technology, School of Electrical Engineering Aalto University)

[13] Elizabeth M Keil 2014 Kalman Filter Implementation to Determine Orbit and Attitude of a Satellite in a Molniya Orbit *MSc in Aerospace and Ocean Engineering*

Acknowledgment

The research was supported by the Research Fund for the Doctoral Program of Higher Education of China (20113219110025). The authors declare that there is no conflict of interest regarding the publication of this paper.

IOP Conf. Series: Materials Science and Engineering **211** (2017) 012016 doi:10.1088/1757-899X/211/1/012016

Low voltage ride-Through strategy based on virtual synchronous generator control

L Guo*, J H Su, L Yu and J D Lai

School of Electrical Engineering and Automation, Hefei University of Technology, Hefei, China

*Email: 710651751@qq.com

Abstract. Inverters controlled in virtual synchronous generator(VSG) method is able to operate in both grid-connected mode and islanding mode. Due to this characteristic, this paper put forwards a low voltage ride-through(LVRT) strategy which is based on VSG control. If the grid voltage sags, VSG can limit the output current amplitude and meanwhile inject reactive power to the grid. Once the voltage drop time exceeds the maximum ride-through time, VSG would smoothly switch to islanding mode and continues to carry loads independently. The proposed strategy achieves a control algorithm unity between normal operating and grid voltage sag status in grid-connected mode. Simulations can support the validity of the strategy.

1. Introduction

These guidelines, With the increasing penetration of distributed power, more and more power sources using inverters as interfaces are connected to the grid[1-2]. Traditional grid-connected inverter with fast response, low inertia and less damping, is difficult to participate in grid frequency and voltage regulation[3-4].To this end, scholars have proposed virtual synchronous generator(VSG) technology that can through controlling, the inverter possesses the same mechanism and external characteristics as what synchronous generator has through controlling, to provide inertia and damping for the grid[5-6]. The current researches about VSG are usually focusing on normal operation condition, but rarely involved in the aspects of fault ride-through[7-8]. Literature[7] classified the types of grid voltage drops in detail, and analyzed the worst operating conditions that each drop may cause to develop corresponding control strategies to compensate, with some reference value. Literature[8] used balanced current controller that transfer the stator terminal voltage reference to current reference value through the filter circuit equation. Thereby, current regulator was introduced to control negative sequence current directly in grid voltage imbalance condition.

On the basis of previous researches, considering that VSG can operate in islanding condition independently, when the grid voltage sag time exceeds the allowed limit value, VSG could smoothly switch to islanding mode from grid-connected LVRT mode and continue to carry loads. As for mode smooth switching algorithms, there are many references from both converter itself and the system level[9-10].

This paper analyzes the operation principles of synchronous generators(SG) in the first place, pointing that to make the VSG have the same external characteristics with the SG, the VSG should meet the operation equations through controlling. Then, according to the depth and the time of grid voltage sag during LVRT process, it is divided into different operation stages and the corresponding control strategies are designed in every stage and their switching process. Finally, simulations have verified the effectiveness of the LVRT strategy.

2. VSG control

2.1. Synchronous generator operation principles

In a synchronous generator[11], the operation satisfies the following equations (1) to (4):

$$E_A = K\phi\omega \tag{1}$$

$$V_\phi = E_A - jX_s I_A - R_A I_A \tag{2}$$

$$\begin{cases} P = \dfrac{3V_\phi E_A \sin\delta}{X_s} \\ Q = \dfrac{3V_\phi (E_A \cos\delta - V_\phi)}{X_s} \end{cases} \tag{3}$$

$$\begin{cases} T_m - T_e = J\dfrac{d\omega}{dt} \\ T_m = \dfrac{P_m}{\omega} \\ T_e = \dfrac{P_e}{\omega} \end{cases} \tag{4}$$

where E_A is the armature voltage, K is the proportional factor, ϕ is the magnetic flux, V_ϕ is the stator output voltage, $X_s$is the synchronous reactance, R_A is the synchronous resistance, δ is the phase angle between $E_A$and V_ϕ, Tm is the mechanical torque, T_e is the electromagnetic torque, P_m is the input mechanical power, P_e is the inductive electromagnetic power.

These four formulas completely describe the operation principle of a synchronous generator.

2.2. Analysis of VSG model

Virtual synchronous generator model should meet the operating principles of synchronous generator as expounded before and speed controller and excited controller are designed to satisfy different control demands. A speed controller is used to regulate the input mechanical power of the prime-mover to maintain the spinning speed whereas an excitation controller is utilized to regulate the armature voltage and the output reactive power.

If the DC side provides a stable input power to the inverter, VSG can operate in both grid-connected mode accessing to the grid and islanding mode with loads. Depending on the control requirements, VSG models can be constructed in various methods and the corresponding controllers will differ a lot.

3. LVRT Strategy

3.1. LVRT requirements analysis

At present, the standard of energy storage unit LVRT is not unified yet. Unlike PV and wind power, if the depth and duration time of the voltage drop do not meet the requirements of LVRT, inverters can switch to islanding mode and continue to operate.

Based on this, it can be divided into three operational stages: the grid-connected stage if the grid voltage is in normal range; the LVRT stage if the grid voltage meets the ride-through requirement and the islanding stage if the grid voltage exceeds the ride-through requirement.

3.2. LVRT algorithm

As mentioned above, the use of direct current control method can limit inductor current magnitude directly and control the reactive component of inductor current so as to satisfy the requirements of LVRT current limit and reactive power compensation.

The text of your paper should be formatted as follows:

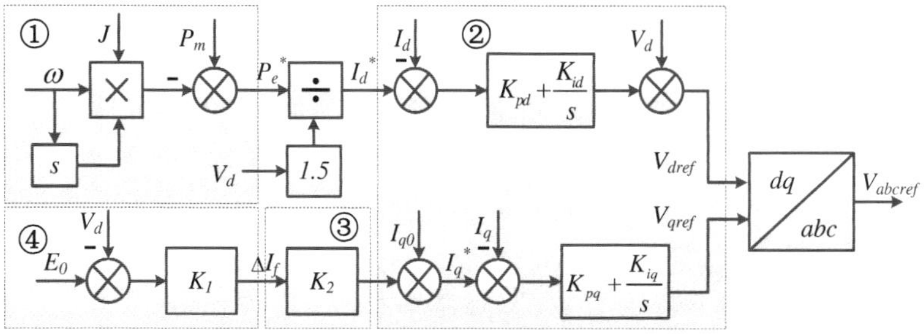

Figure 1. Grid-connected VSG model.

①is the rotor motion equation.

②is the terminal voltage equation.

③K2 characterized the V-curve which is obtained from simultaneous equations of (1) and (2) as reactive power control design basis.

④is the excitation controller.

In this algorithm model, ω is the angular velocity of the grid, V_d,V_q,I_d,I_q are the dq axis component of the grid voltage and inductor current respectively, E_0 is the rated grid voltage d-axis component, I_{q0} is the basis reactive current component,J is the rotational inertia. ω is achieved from phase locking loop that in grid-connected mode the speed controller is replaced by PLL block.

The reactive power instruction is regulated according to the grid voltage amplitude in real time.Without a separate LVRT algorithm to calculate the homologous current command based on the current grid voltage amplitude, this model is a unity of normal grid-connected operation model and LVRT algorithm. Selecting the appropriate parameters K_1 and K_2, the requirement of reactive power can be fully satisfied. When the voltage drop reaches the maximum depth, the VSG issues the maximum capacitive reactive power.

In order to facilitate islanding VSG algorithm design, take K_2=1, K_1=K. Active power instruction is set in accordance with the reactive power in equation (5):

$$P_m = \min\left(P_m, (\sqrt{S_{\max}^2 - Q^2}) \times \frac{V_d}{E_0}\right) \tag{5}$$

If the active current component keeps rated, the maximum reactive current component is calculated in equation (6) and the corresponding grid voltage drop is computed in equation (7). This means, if the grid voltage sags beyond 0.47V_N, the output active power should be restricted to ensure the reactive current component injection.

$$I_q = \sqrt{(1.1I_{Nm})^2 - I_{Nm}^2} = 0.46I_{Nm} \tag{6}$$

$$\Delta V = \frac{I_q}{K} = \frac{0.46 \times 1.13}{1.1} V_N = 0.47V_N \tag{7}$$

3.3. Islanding VSG algorithm

Close switches S_2 and S_3, then the islanding VSG model is derived as shown in Figure 2.Among them, ① to ④have the same meanings with those in section B and ⑤ is the speed controller.

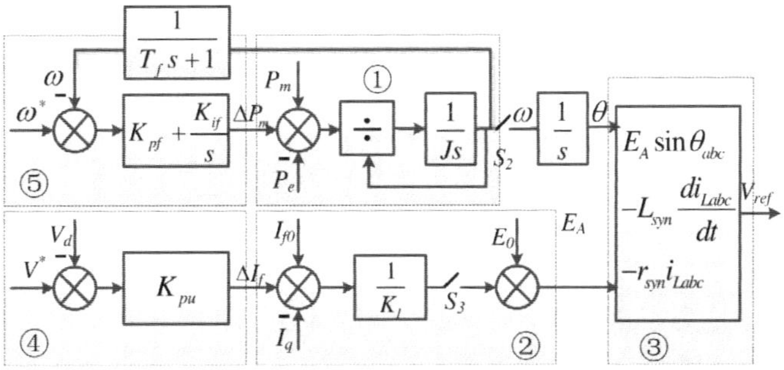

Figure 2. Islanding VSG model.

This algorithm ensures stable supports of frequency and voltage when off-grid, and normal operations of the loads can be guaranteed.

Since K_2=1, where I_f and I_q are equal in magnitude, that the field current feedback value I_f in armature equation (1) can be detected by the q-axis component of filter inductor current.

When the grid voltage drop depth and duration time exceed the LVRT requirements, the VSG should switch to islanding mode and operate continuously. The smooth switching algorithm is designed as Figure 3.

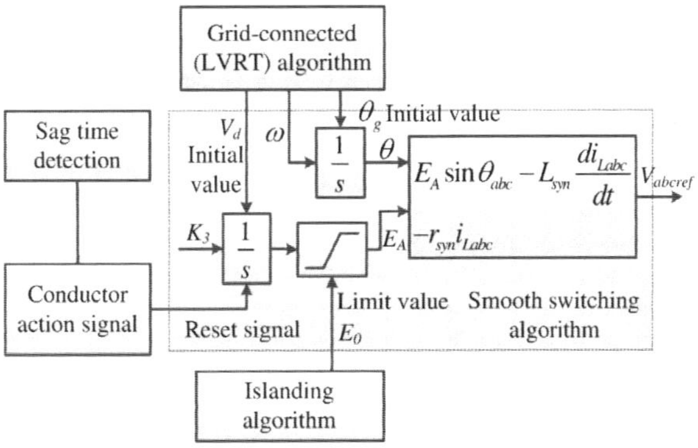

Figure 3. Smooth switching algorithm.

This algorithm takes the angular velocity of the PLL in grid-connected mode as the reference angle, the real-time grid voltage amplitude V_d as the initial value of the voltage integral block and the reset signal is provided by the contactor action signal. This control block guarantees the moment the contactor operates, the output of the integral block is consistent with the grid. The integral time constant K_3 is set based on the depth of voltage drop and the recovery time T_r whereas the limit of integrator is equal to the voltage reference value E_0 in islanding mode that ensures the output of integrator reach the armature voltage magnitude if saturated.

Considering equation (8), the setting of parameter T_r should take the regulation speeds of speed controller and excited controller into account to guarantee a stable output voltage and frequency in recovery process. Parameter K_3 can be optimized to adjust itself automatically according to the actual depth of voltage drop ΔV_N. The contactor opens after a time of T_r, the voltage restores, then removes the PLL and closes S_2 and S_3.

$$K_3 = \frac{\Delta V_N}{T_r} \tag{8}$$

4. Simulation Results

Several simulations are carried out in MATLAB/ Simulink environment to study the effectiveness of the proposed LVRT algorithm. Simulation parameters are shown in table 1.The LVRT process is shown in Figure 4. Three sub figures in Figure 4(a) are the inductor current during transient process.

Table 1. Simulation parameters.

DC voltage	700V	Rated power	50kVA
AC rms phase voltage	220V	Current limit value	1.1pu.
Filter inductance	0.5mH	Rotational inertia	0.5kg · m^2
Filter capacitance	50μF	Switching frequency	10kHz
Line impedance	0.05mH+0.05 Ω	Parameter T_r	0.2s

During 0-0.2s period, the grid voltage is in normal range that the inverter outputs 50kW rated active power with the active current component I_d=I_{Nm}=107A. At the moment of 0.2s, grid rms voltage sudden sags 0.3V_N with the corresponding d-axis component 0.42V_N which is less than 0.47V_N. Thus the active current component I_d keeps unchanged and the VSG issues capacitive reactive power to compensate the voltage drop. At 0.4s, the grid voltage drops further to 0.2V_N, leading the reactive current component I_q to the maximum value that the VSG outputs capacitive output power only. At 0.6s, the grid voltage recovers, and the VSG returns to work in rated state.

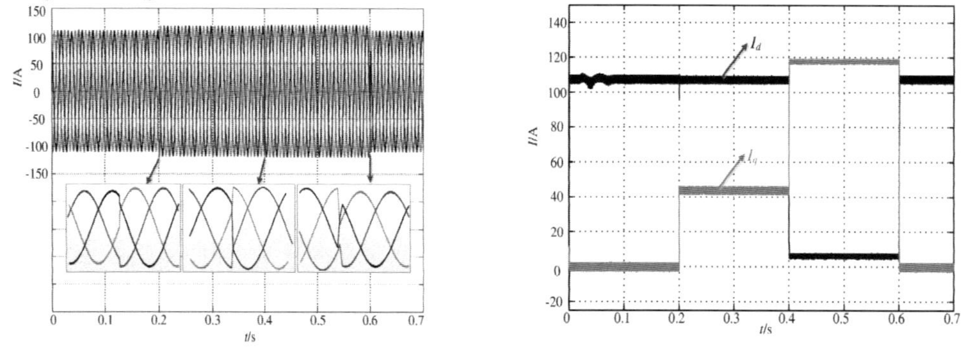

(a). Three-phase inductor current waveform. **(b).** Dq axis inductor current waveforms.

Figure 4. Simulation waveforms of LVRT process.

Figure 5 shows the smooth switching process waveforms when LVRT failure. At 0.2s, the grid voltage drops to 0.5V_N, setting the maximum duration time of 0.2s at this depth of sag, then the VSG will switch to islanding mode after the time of 0.4s since the grid voltage has not restored. Without a smooth switching algorithm shown in Figure 7(a), the moment the contactor opens, the inductor current jumps to eight times the rated value that far beyond the limit. But after the introduction of the smooth switching algorithm, the output voltage keeps unchanged and recovers to the rated value in 0.1s that matches the calculation result. After that, the VSG switches to islanding model and operates stably.

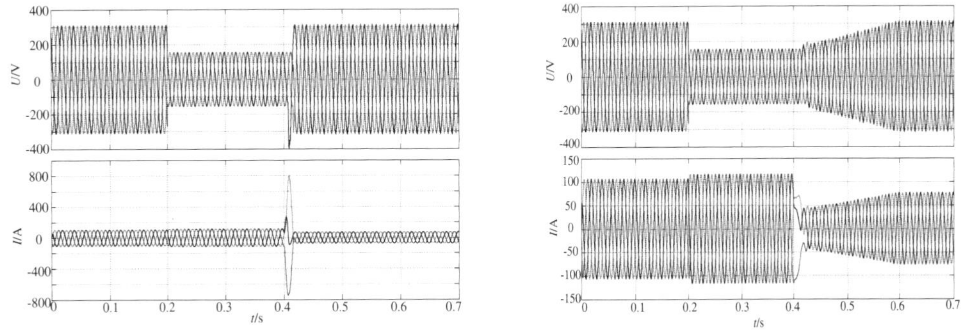

(a). Switch waveforms without a smooth algorithm.

(b). Switch waveforms with a smooth algorithm.

Figure 5. Simulation waveforms of switching process if LVRT failure.

5. Conclusion

This paper designed a LVRT strategy based on VSG control. This strategy can regulate reactive power based on synchronous generator characteristic in accordance with the grid voltage actual value in real time in grid-connected mode without a unique LVRT algorithm to calculate the compensate reactive power instruction. Moreover, considering that the VSG can be controlled as a voltage source, if the depth and duration time of sags beyond the LVRT requirements, this strategy will smooth switch to islanding mode and carry loads independently. Simulations and experimental results have valid that at the transient moments, this strategy will effectively restrict the output current and meet the demands of reactive compensate or voltage recovery.

References

[1] X.Yang, Y.Song, GWang, et al 2010 A comprehensive review on the development of sustainable energy strategy and implementation in China *IEEE Trans on Sustainable Energy* **1** 2 pp 57-65

[2] R H Lassetr 2007 Microgrids and distributed generation *Journal of Energy Engineering* **133** pp 144-9

[3] M T Fa, Z P Zang, A X Su, et al 2013 Enabling technologies for active distribution systems *Proceedings of the CSEE* **33** 22 pp12-8

[4] C L T Borges and V F Martins 2012 Multistage expansion planning for active distribution networks under demand and distributed generation uncertainties *Electrical Power and Energy Systems* **36** pp 107-16

[5] S D Arco and J A Suul 2015 A virtual synchronous machine implementation for distributed control of power converters in Smart Grids *Electric Power Systems Research* **122** pp 180-97

[6] Z P Lü, W X Sheng, Q C Zhong, et al 2014 Virtual synchronous generator and its applications in microgrid *Proceedings of the CSEE* **34** 16 pp 2591-603

[7] J Alipoor, Y Miura and T Ise 2014 Voltage sag ride-through performance of virtual synchronous generator *Power Elctronics Conference* (Hiroshima Japan) pp 3298-305

[8] T Y Chen, L J Chen, Y C Wang, et al 2016 Balanced current control of virtual synchronous generator *Power System Technology* **40** 3 pp 904-9

[9] J Wang, C P Nocilas, X W Feng, et al 2015 Design of a generalized control algorithm for parallel inverters for smooth microgrid transition operation *IEEE Transactions on Industrial Electronics* **62** 8 pp 4900-14

[10] C L Chen, Y B Wang, L Jason, et al 2010 Design of parallel inverters for smooth mode transfer microgrid applications *IEEE Transactions on Power Electronics* **25** 1 pp 6-15

[11] J C Stephen 2009 Synchronous Generators and Motors *Electric machinery fundamentals* (4th ed Beijing China) pp 166-84

IOP Conf. Series: Materials Science and Engineering **211** (2017) 012017 doi:10.1088/1757-899X/211/1/012017

The realization of the virtual synchronous generator based on the park model

L Yu*, J H Su, L Guo and J D Lai

School of Electrical Engineering and Automation, Hefei University of Technology, Hefei, China

*Email: m18247170205@163.com

Abstract. In this paper, a kind of virtual synchronous generator control algorithm is proposed that can accurately simulate the external characteristic of a synchronous generator. This algorithm based on the synchronous generator park mathematical model considers the motor salient pole effect and the influence of damper windings. Using two-step Adams method to solve the mathematical model of the synchronous motor for current of the motor's port, the converter is controlled to track the current so that the output status values have the same port characteristics with the actual motor. Simulation results verify that in many typical working conditions such as the synchronous motor starting, the output voltage mutating or the frequency changing, the converter controlled in the above algorithm can mimic a real synchronous generator.

1. Introduction

In the distributed generation and the micro grid technology, the grid-connected inverter has played an irreplaceable role as an important interface that connects the renewable energy unit and the grid. The control strategy of the inverter is always the key point of the study. However, because of the fast response speed, traditional control methods can hardly participate in the regulation of the grid and are unable to provide barely any inertia for the micro grid that relatively lacking stability or the necessary voltage and frequency support[1-4]. If a control strategy is able to make a grid-connected inverter has the same external characteristic like a synchronous generator, the traditional power system related control strategies can be easily transplanted in micro grid or other active distributed grid. This control strategy is so called the virtual synchronous generator (VSG) strategy.

Leuven University has put forward a kind of low-order VSG algorithm [5] that can simulate the characteristics of rotor inertia and the frequency modulation process. However, this algorithm only considers the motion equation of the motor which ignores the excitation adjustment mechanism. To further reflect the features, German Clausthal industrial university puts forward the VISMA scheme[6] that uses the third-order model of the motor considering the stator electric equation as well as the rotor mechanical equation. By controlling the filter inductance current, the inverter thus achieve the purpose of port characteristics simulations. Although this simplified model that neglects the salient pole effect and the influence of damper windings meets the demand for most steady circumstances, the dynamic process does differ a lot when the actual operating condition changes. Since the damper windings maintain the flux constant and suppress the rotor oscillation in dynamic variation, the effects should be taken into account. In order to mimic the damper effect, literatures[7-10] introduce the damper coefficient. Although this parameter is equivalent to damper windings in terms of impact, it can't match the actual motor coefficients. Thus, the transient characteristics between the VSG model and the actual motor are inconsistent.

To accurately simulate the transient characteristics of the actual synchronous generator, this paper proposes a virtual synchronous generator algorithm based on the synchronous generator full-order model. This method takes the salient-pole effect and the impacts of damper windings into account so that the output characteristics of the inverter match the actual synchronous generator in steady state as well as in transient process. Therefore, an inverter controlled in the VSG strategy is able to provide grid-connected interface in a micro grid and simulate the synchronous motor fault model.

2. synchronous generator modelling

In a synchronous generator, a dc current is applied to the rotor winding, which produces a rotor magnetic field. The rotor of the generator is then turned by a prime mover and produces a rotating magnetic field within the machine. This rotor magnetic field induces a three-phase set of voltages in the stator windings. The synchronous generator operation satisfies the following equations (1)-(3)[11].

$$E_A = K\psi\omega \tag{1}$$

$$V_\phi = E_A - jX_s I_A - R_A I_A \tag{2}$$

$$\left\{ T_m - T_e = J\frac{d\omega}{dt} \right. \tag{3}$$

where E_A is the armature voltage, K is the proportional factor, ψ is the magnetic flux, V_ϕ is the stator output voltage, X_s is the synchronous reactance, R_A is the synchronous resistance, δ is the phase angle between E_A and V_ϕ, T_m is the mechanical torque, T_e is the electromagnetic torque.

Commonly, synchronous generator model has two forms[12]: one is the park model based on the prototype circuit parameters and takes flux as the state variable; the other is the practical model based on the practical parameters and takes the electric potential as the state variable. This article chooses the park model.

2.1. *Flux equations*

The fluxes in stator and rotor can be governed by equation (4) where ψ_{dq0s} , ψ_{fq0r} , I_{dq0s} and I_{fdqr} refer to the stator flux, the excitation and the rotor flux, the stator current and the rotor current in damper windings in dq axis respectively.

$$\begin{bmatrix} \psi_{dq0s} \\ \psi_{fdqr} \end{bmatrix} = \begin{bmatrix} L_{11} & L_{12} \\ L_{12}^T & L_{22} \end{bmatrix} \begin{bmatrix} -I_{dq0s} \\ I_{fdqr} \end{bmatrix} \tag{4}$$

2.2. *Voltage equations*

Due to the presence of various types of power losses and voltage drops, the output voltage in stator can be expressed in equations (2). Through park transformation, the stator voltage equation can be described in equation (5) where U_{dq0s} is the stator voltage while U_{fdqr} refers to the excitation voltage and the rotor damper voltage sags in dq axis.

$$\begin{bmatrix} U_{dq0s} \\ U_{fdqr} \end{bmatrix} = \begin{bmatrix} R_s & 0 \\ 0 & R_r \end{bmatrix} \begin{bmatrix} -I_{dq0s} \\ I_{fdqr} \end{bmatrix} + \begin{bmatrix} \omega_r \Gamma & 0 \\ 0 & 0 \end{bmatrix} \begin{bmatrix} \psi_{dq0s} \\ \psi_{fdqr} \end{bmatrix} + p \begin{bmatrix} \psi_{dq0s} \\ \psi_{fdqr} \end{bmatrix} \tag{5}$$

Take the flux formula (4) into voltage equation (5), then equation (6) can be found.

$$p\begin{bmatrix} -I_{dq0s} \\ I_{fdqr} \end{bmatrix} = -\begin{bmatrix} L_{11} & L_{12} \\ L_{12}^T & L_{22} \end{bmatrix}^{-1} \begin{bmatrix} R_{11} & R_{12} \\ 0 & R_r \end{bmatrix} \begin{bmatrix} -I_{dq0s} \\ I_{fdqr} \end{bmatrix} + \begin{bmatrix} L_{11} & L_{12} \\ L_{12}^T & L_{22} \end{bmatrix}^{-1} \begin{bmatrix} U_{dq0s} \\ U_{fdqr} \end{bmatrix} \tag{6}$$

2.3. *Rotor motion equation*

The synchronous generator rotor motion equation has already shown in (3) and two more equation (7), (8) should be added:

$$\frac{\mathrm{d}\delta}{\mathrm{d}t} = \omega - 1 \tag{7}$$

$$T_{\mathrm{e}} = \psi_{\mathrm{ds}} i_{\mathrm{q}s} - \psi_{\mathrm{qs}} i_{\mathrm{d}s} \tag{8}$$

Thus, the rotor motion equation is reorganized in equation (9).

$$\begin{bmatrix} \dfrac{\mathrm{d}\omega}{\mathrm{d}t} \\ \dfrac{\mathrm{d}\delta}{\mathrm{d}t} \end{bmatrix} = \begin{bmatrix} 0 & 0 \\ 1 & 0 \end{bmatrix} * \begin{bmatrix} \omega \\ \delta \end{bmatrix} + \begin{bmatrix} \dfrac{T_{\mathrm{m}} - T_{\mathrm{e}}}{J} \\ -1 \end{bmatrix} \tag{9}$$

where T_m and T_e are the mechanical torque and electromagnetic torque per unit values respectively. ω is the actual generator mechanical speed p.u value, J refers to the inertia while δ is the phase difference angle between the excitation no-load EMF and the terminal voltage.

3. Realization of the synchronous generator mathematical model

Since the mathematical model has been determined by the equations above, it needs to be transformed into available numerical model for computer calculating. The key procedure is to select the appropriate state variables to equivalent convert the mathematical model into state equation form.

3.1. Establishment of the state equation

According to formula (6) and (9), an eight-order motor state equation model can be found using i-ω-δ as the state variables. Evidently, the synchronous generator is a high order nonlinear time-varying system with the terminal voltage u and the mechanical torque T_m as inputs. All the variables in the formulas are calculated in the form of the unitary value, and the time variable t and the motion inertia J need to be processed in non-dimension way.

3.2. Numerical algorithm selection

To obtain the solutions of the motor state equations, it is necessary to choose a proper numerical algorithm. The commonly used numerical algorithms are Euler method, trapezoidal method and Adams method[13]. Comparing the three methods that Euler method is quite simple but less precision, trapezoidal method has high calculation precision with a large number of computations. Relatively, as a typical linear multistep method, Adams method acquires a balance between precision and computations. Therefore, two step Adams explicit method is used in this paper. The two step Adams method is described as equation (10):

$$\begin{cases} x_{m+1} = x_m + h(3F_m - F_{m-1})/2 \\ F_m = f(t_m, x_m) \end{cases} \tag{10}$$

where x_m means the state variable at the moment t_m and F_m describes the differential value of x_m at time t_m. Parameter h represents the calculation step and in this paper it is selected to 100us. Take formula (10) into equation (6) and (9) ,and then fequation (11) is obtained.

$$\begin{cases} F_i^k = -L^{-1} * R_k * i_k + L^{-1} * u_k \\ i_{k+1} = i_k + 0.5h * (3F_i^k - F_i^{k-1}) \\ F_\omega{}^k = \dfrac{T_m - T_e}{J} \\ \omega_{k+1} = \omega_k + 0.5h * (3F_\omega{}^k - F_\omega{}^{k-1}) \\ F_\delta{}^k = \omega_{k+1}{}^k - 1 \\ \delta_{k+1} = \delta_k + 0.5h * (3F_\delta{}^k - F_\delta{}^{k-1}) \end{cases} \tag{11}$$

In the above equations, $F_{i\text{-}\omega\text{-}\delta}{}^{k}$ expresses the differential values of state variables at the moment k and L, R_k are the inductance and resistance matrix respectively.

4. Virtual synchronous generator control scheme

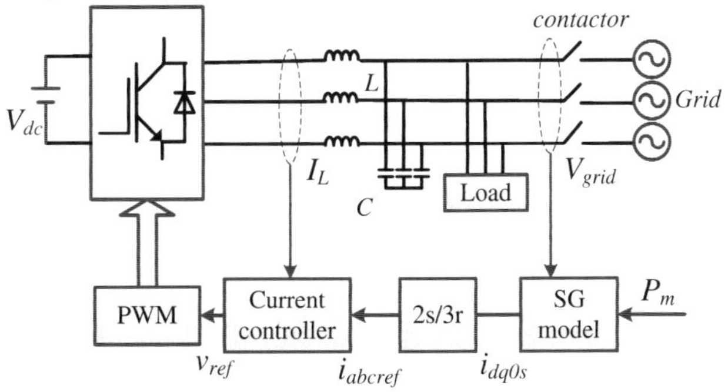

Figure 1. VSG scheme diagram.

Now that the synchronous generator mathematical model has established and the numerical solution has determined. The inverter VSG control scheme is then obtained in fig 1. The main circuit of the diagram is a three-phase bridge inverter circuit with L_f as the filter inductor. The port current reference values in dq-axis i_{dq0ref} are computed through motor mathematical model using grid voltage V_{grid} and the input power P_m. Then the reference values are transformed into abc static axis as the current controller instructions. Inductance current I_{Labc} are controlled by the current controller to track the instructions so the inverter finally has the same port characteristics as true synchronous generator does. Conventional generators are equipped with speed controller and excitation regulator to improve the stability and operation of the generator, therefore constrains mentioned above are introduced to make sure the EMF value is regulated when mechanical power changing.

5. Simulation results

To validate the feasibility of the proposed algorithm, simulations are executed in matlab/Simulink environment. The algorithm is operated in a s-function block and the topology of the main inverter circuit part is the same with the VSG scheme shown in Figure 1. For comparison, a 8.1kVA three-phase synchronous generator model of matlab own in PSB(powersystem blockset)block is also operating. The parameters of this model are exhibited in Table 1.

Table 1. The parameters of synchronous generator.

Parameters/p.u	values
stator resistance R_s	0.08201
excitation resistance R_{fd}	0.06117
damper winding resistance in d-axis R_{kd}	0.1591
damper winding resistance in q-axis R_{kq}	0.2416
armature reaction inductance in d-axis L_{md}	1.728
armature reaction inductance in q-axis L_{mq}	0.823
stator winding leakage inductance L_{ls}	0.072
excitation winding leakage inductance L_{lf}	0.1801
damper winding leakage inductance in d-axis L_{ld}	0.1166
damper winding leakage inductance in q-axis L_{lq}	0.1615

IOP Conf. Series: Materials Science and Engineering **211** (2017) 012017 doi:10.1088/1757-899X/211/1/012017

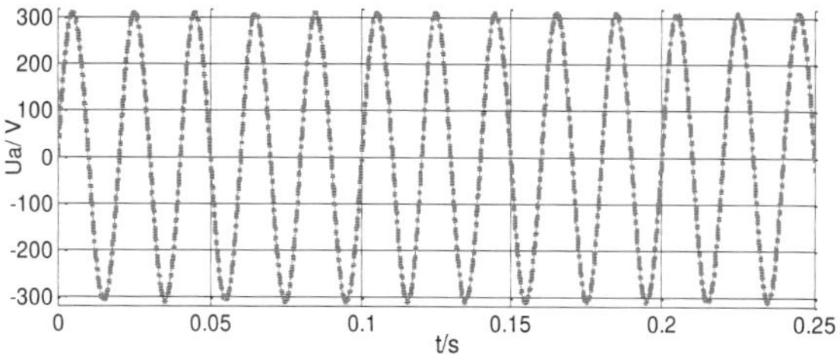

(a). Grid voltage in phase a.

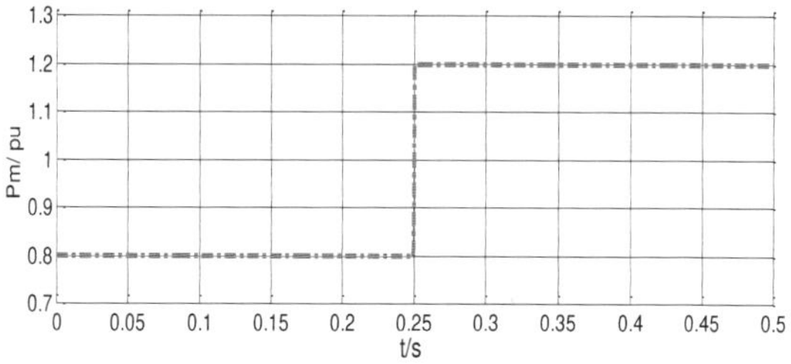

(b). Input mechanical power.

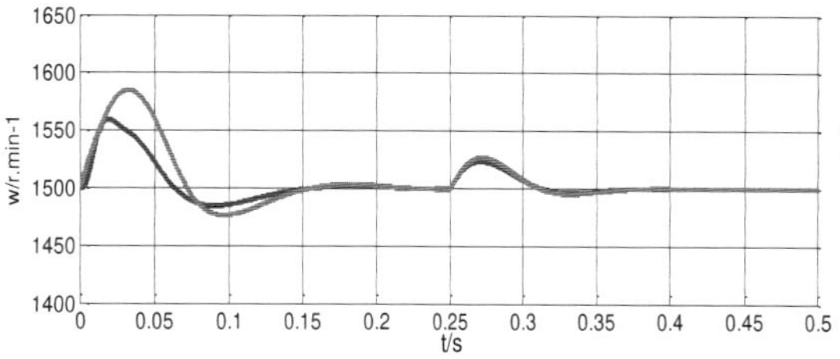

(c). The motor rotating speed.

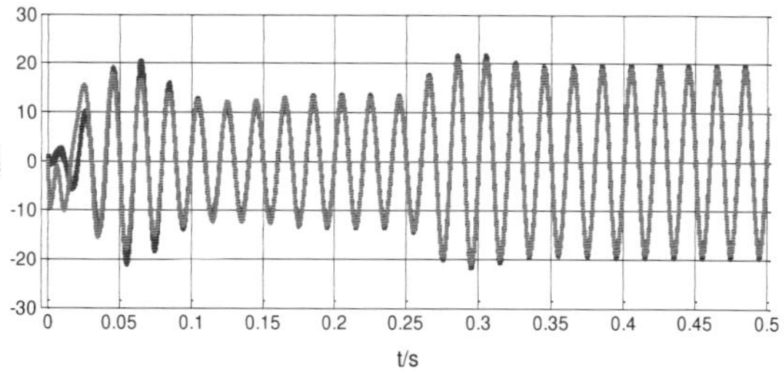

(d). The stator current in phase a.

Figure 2. Dynamic responses when Mechanical power mutation.

Under the circumstance that the VSG is connected to the grid, when the input mechanical power varies the motor rotating speed and the output stator current curves are exhibited in Figure 2. The red line is the matlab synchronous generator model while the blue line is the inverter controlled in the VSG algorithm.

Obviously during the whole mutation process, the output of the inverter controlled in VSG algorithm has the same response curve as the actual motor.

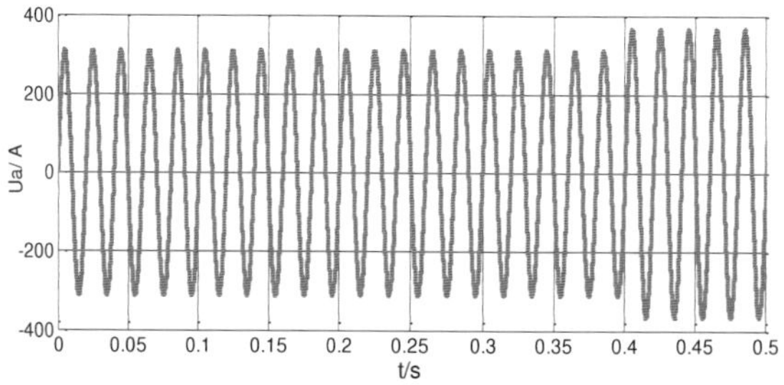

(a). The Grid voltage in phase a.

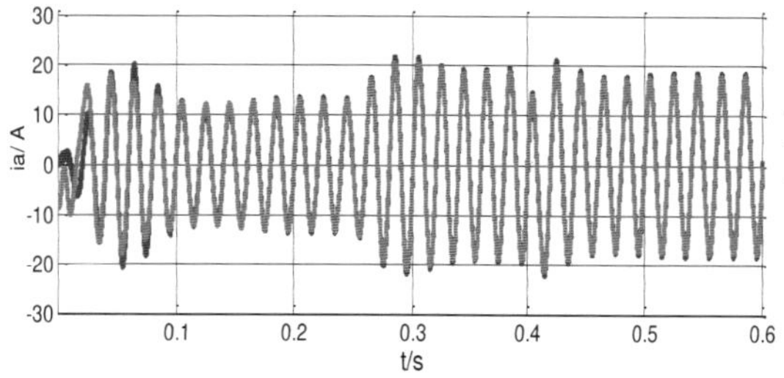

(b). The stator current in phase a.

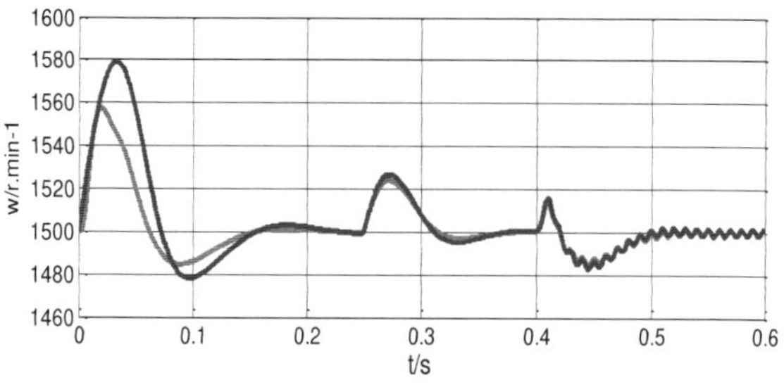

(c). The motor rotating speed.

Figure 3. dynamic responses when grid voltage mutation.

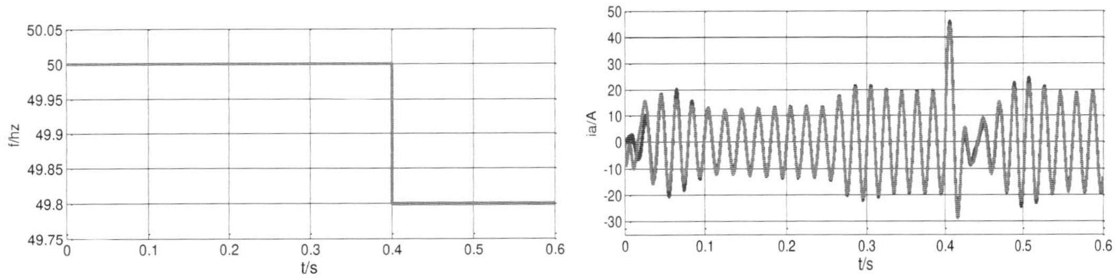

(a). The grid frequency. **(b).** The stator current in phase a.

Figure 4. dynamic responses when grid frequency mutation.

More simulations are executed to explore the port characteristics of the VSG if other variables changing. Figure.3 shows when the grid voltage mutates at the time 0.4s the output stator current and rotating speed also follow the motor model. And so does the grid frequency mutation shown in the figure 4.Compare these three Figures, it's evident that the VSG controlled inverter and the real synchronous generator model have the same characteristics of both the dynamic response and the stable value if under the same conditions. Thus, the practicability of the proposed VSG control algorithm has been verified.

6. Conclusion

This paper proposes a realization method of the virtual synchronous generator based on park model. The accurate model is achieved through the solution of the motor full-order equations. All the numerical variables are computed through initial values and port constraints. Simulations have proved the VSG dynamic responses are equal to the real synchronous generator with the same parameters. Although due to the deviation of the initial values, at the motor start period the two curves are not strictly coincidence, this algorithm does have some practical guiding significances.

References

[1] Fan Mingtian 2013 Enabling technologies for active distribution systems *Proceedings of the CSEE* **33(22)** pp 12-8

[2] Borges C L T and Martins V F 2012 Multistage expansion planning for active distribution networks under demand and distributed generation uncertainties *Electrical Power and Energy Systems* **36(1)** pp 107-16

[3] Zhang Jianhua 2011 Small-signal stability analysis of grid-connected microgrid with inverter-interfaced distributed resources *Automation of Electric Power Systems* **35(6)** pp 76-80

[4] Li Y W, Vilathgamuwa D M and Loh P C 2004 Design, analysis,and real-time testing of a controller for multibus microgrid system *IEEE Transactions on Power Electronics* **19(5)** pp 1195-203

[5] Tom L 2011 Participation of inverter-connected distributed energy resources in grid voltage control (Leuven:Katholieke Universiteit Leuven)

[6] Chen Y, Hesse R, Turschner D, et al 2011 Improving the grid power quality using virtual synchronous machines *International Conference on Power Engineering, Energy and Electrical Drives* pp 1-6

[7] Li Tan, Yang Honggeng, Wen Ke, et al 2015 Design of adaptive control strategy for a low voltage micro-grid mode *Electrical Measurement& Instrumentation* **52(17)** pp 28-32

[8] Li Qing, Zhang Bide, et al 2015 Control strategies of inverters based on virtual synchronous generator in a micro-grid *Electrical Measurement& Instrumentation* **52(13)** pp 11-15

[9] Wang Jialei, He Jianping, et al 2015 research on sinusoid-locked technology based on virtual synchronous generator *Electrical Measurement& Instrumentation* **52(16)** pp 17-21

[10] Qing-Chang Zhong and George Weiss 2011 Synchronverters: Inverters That Mimic Synchronous Generators *IEEE Transactions on Industrial Electronics* **58(4)** pp 1259-67

[11] Ni Yixin, Chen Shousun, Zhang Baolin, et al 2012 *Theory and Analysis of Dynamic Power System* (Beijing: tsinghua university press)

[12] Gu Chenglin 1998 *Electromechanical power system analysis* (Wuhan: Huazhong University of science and Technology Press)
[13] Hu Jianhua, Chen Xinhua, et al 2008 *Numerical Methods* (Jiangsu: China University of Mining and Technology Press)

A maximum power point prediction method for group control of photovoltaic water pumping systems based on parameter identification

B Chen*, **J H Su, L Guo and J Chen**

School of Electrical Engineering and Automation, Hefei University of Technology, Hefei, China

*Email: 2276224426@qq.com

Abstract. This paper puts forward a maximum power estimation method based on the photovoltaic array (PVA) model to solve the optimization problems about group control of the PV water pumping systems (PVWPS) at the maximum power point (MPP). This method uses the improved genetic algorithm (GA) for model parameters estimation and identification in view of multi P-V characteristic curves of a PVA model, and then corrects the identification results through least square method. On this basis, the irradiation level and operating temperature under any condition are able to estimate so an accurate PVA model is established and the MPP none-disturbance estimation is achieved. The simulation adopts the proposed GA to determine parameters, and the results verify the accuracy and practicability of the methods.

1. Introduction

PV pump, directly driven by solar energy has many advantages such as environmentally friendly, low maintenance cost and simple operation [1-3]. In group control of PVWPS, the output power amplitude of the PVA is scheduled according to the P-V curve. In the process of the group system operation, if the pump output frequency has not reached the maximum limit yet, the output power of the converter is MPP of the PVA. However, if the irradiation gradually increasing, the pump output frequency achieves the maximum limit, thus the array operating voltage deviates from the maximum power point voltage. Under this circumstance, the actual maximum power is not able to be measured directly.

To solve this contradiction, in recent years, domestic and foreign scholars have proposed MPP estimation method based on PVA mathematical model. The main idea is to identify the uncertain parameters of a PVA and get the complete P-V characteristic curve on the basis of mathematical model. Literatures [4-5] present a five parameter mathematical model based on the equivalent circuit model which take the influences of parallel resistance and the photogenic current into account. This method has a relatively high precision, but quite complicated that is not suitable for engineering calculation. Literatures [6-7] simplified the model under the premise of ensuring the accuracy but it is also needed to determine the slope value of the open circuit voltage. Besides, literature [8-9] neglect the series resistance and the shunt resistance in the model and provide a conversation relationship between the different solar irradiation level and operating temperature which have a lot of practical engineering values.

In this paper, a new MPP estimation method is proposed, in which influencing parameter, solar irradiation level and operating temperature are regarded as unknown parameters. The improved GA is used to identify and predict PV curves, and the least squares method is used to correct multiple sets of parameters. On this basis, solar irradiation level and operating temperature under any working condition are predicted, and the P-V curve is obtained, and finally the estimation of MPP is realized. In this paper, the accuracy and practicability of the proposed method are verified by comparing the power predicted value and the measured value under three kinds of operating conditions.

2. Group control of PVWPS

During the operation of group control of PVWPS, when the pump speed does not reach the saturation speed and the system has been in MPPT state, the PVA will operate at its MPP (A) as shown in Figure 1 As shown in curve 1. When the array power increases, the array operating point will deviate from point A ', and the pump speed does not reach the critical saturation state, thus the system runs in MPPT mode. At this point the array still works at its MPP (B), the array of working conditions as shown in curve 2, when the pump has reached its upper speed limit; As the illumination continues to increase and the array power continues to rise to curve 3, the array operating point will deviate from point B ', and MPP can't be further tracked due to the upper pump speed, Since the array operating voltage point deviates from MPP, MPP of the array can't be sampled directly, so MPP in curve 3 need to be estimated.

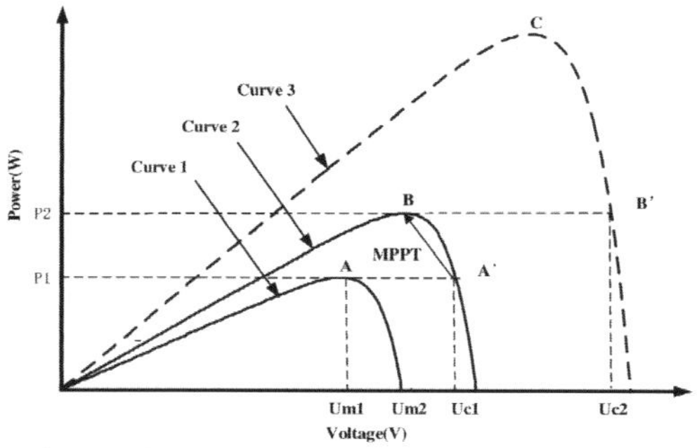

Figure 1. PV curves of illumination changed.

3. PV modeling

In practical application, considering the internal defects of semiconductor crystal and effect of series and parallel, we can think the contact resistance and material body resistance are equivalent to the internal series resistance and the effect of the leakage current in the component is equivalent to a parallel resistance in the PV equivalent circuit shown in the Figure 2.

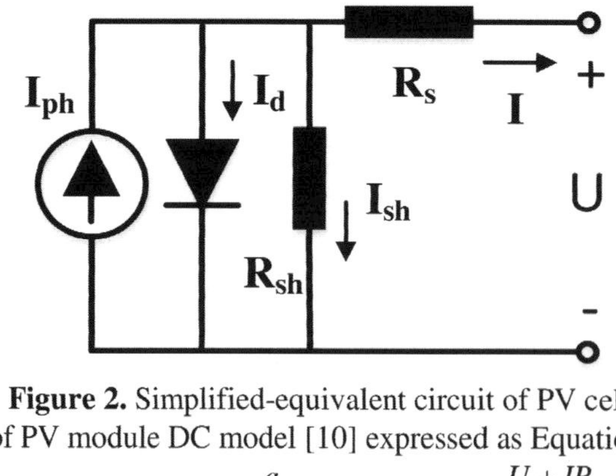

Figure 2. Simplified-equivalent circuit of PV cell.

The current equation of PV module DC model [10] expressed as Equation (1):

$$I = I_{ph} - I_0\{\exp[\frac{q}{nKT_c}(U + IR_s)] - 1\} - \frac{U + IR_s}{R_{sh}} \tag{1}$$

where I is cell output current, A; I_{ph} is photocurrent, function of irradiation level and junction temperature, A; I_0 is reverse saturation current of diode, A; q is electron charge ($1.6\times10^{-19}C$); K is Boltzmann constant ($1.38\times10^{-28}J/K$); T_c is reference cell operating temperature, K; n is diode factor; R_s is series resistance of cell, ψ; R_{sh} is shunt resistance of cell, ψ; U is cell output voltage. Under the standard test conditions, the equivalent circuit model parameters are denoted by the subscript *ref*, which is different from other operating conditions.

In ensuring a certain accuracy, solar cell model for engineering uses the following two approximations.

1) $(U+IR_s)/R_{sh}$ is very small and the effect on the battery current is very small, which can be ignored.
2) R_s is much less than the diode forward resistance, we can consider that $I_{ph}= I_{sc}$.

Under the assumption above to further reduce the parameters of I-V equation, we can use as Equation (2) to fit the I-V characteristic curve of silicon solar cells:

$$I = I_{sc}\{1-C_1[\exp(U/C_2)-1]\} \tag{2}$$

We can derive as Equation (3) with the equations: $U=U_m$ and $I=I_m$ when reaching the maximum power point, $I=0, U=U_{oc}$ when in open state.

$$\begin{cases} I_m = I_{sc}\{1-C_1[\exp(U_m/C_2)-1]\} \\ 0 = I_{sc}\{1-C_1[\exp(U_{oc}/C_2)-1]\} \end{cases} \tag{3}$$

On the condition of room temperature, we can think $exp(U/C_2)>>1$ and simplify the Equation (3) to Equation (4):

$$\begin{cases} C_1 = (1-I_m/I_{sc})\exp(-U_m/C_2) \\ C_2 = (U_{oc}-U_m)[\ln I_{sc}/(I_{sc}-I_m)]^{-1} \end{cases} \tag{4}$$

Thus, we can obtain a two-parameter model of PV consisting of only 2 unknown parameters C_1 *and* C_2 .According to the Equation (3), we can get the value of the unknown parameters of C_1 and C_2 when the parameters of the solar cell in different operating conditions the short-circuit current I_{sc}, the open-circuit voltage $V_{oc,}$ the maximum power point current $I_{m,}$ the maximum power point voltage V_m are obtained. As the characteristics of photovoltaic cells is relevant to the solar irradiation level S and operating temperature T_c, voltage and current will change with T_c and S changes, and thus C_1 *and* C_2 will change together. When the ambient temperature T_a and S change, T_c also change. The influence factor of T_a changes on the output voltage and current of the PV array is expressed by C_{TV} and C_{TI}, as Equation (5) respectively. The change of S will promote the photovoltaic current and operating temperature changes resulting in changes in output voltage. The influence factor of irradiation levels changes on the output voltage and current can be expressed by C_{SV} and C_{SI} [11]:

$$\begin{cases} C_{TV} = 1+\beta(T_a-T_{ref}) \\ C_{TI} = 1+\lambda(T_a-T_{ref})/S_{ref} \\ C_{SV} = 1+\alpha(S-S_{ref}) \\ C_{SI} = 1+(S-S_{ref})/S_{ref} = S/S_{ref} \end{cases} \tag{5}$$

where α means the influence of solar irradiation level of the change on the output voltage the photovoltaic cell, β,λ mean the influence of the ambient temperature on the output voltage and current of the photovoltaic cell. T_{ref} and S_{ref} are the ambient temperature and the irradiation level under standard test conditions (STC). Considering the effect of T_a and S change on the output voltage and current, the current output voltage and current can be obtained by using the correction factors C_{TV}, C_{TI}, C_{SV} and C_{SI} at the new T_a and S as Equation (6) :

$$\begin{cases} U' = U_{ref}C_{TV}C_{SV} \\ I' = I_{ref}C_{TI}C_{SI} \end{cases} \tag{6}$$

Because the α,β,λ parameters is related to battery performance, and can be effected by the photovoltaic cell technology or series and parallel, the experience factor is not precise enough. At the same time, the current T_a and S of photovoltaic cells can't be accurately obtained in the project.

4. Application of GA in group control of PVWPS

4.1. Genetic algorithm

Genetic algorithm is an efficient parallel global search algorithm based on the natural selection and genetic principle [12]. In this paper, we propose an improved GA, that is, the optimal individual preservation method, which preserves the optimal individuals of the parent each time, which improves the convergence and does not lose the overall situation. The steps includes initialization control parameters, population initialization, fitness evaluation, selection operation, cross operation, termination condition judgment and so on. The novel genetic algorithm flow chart is shown in Figure 3:

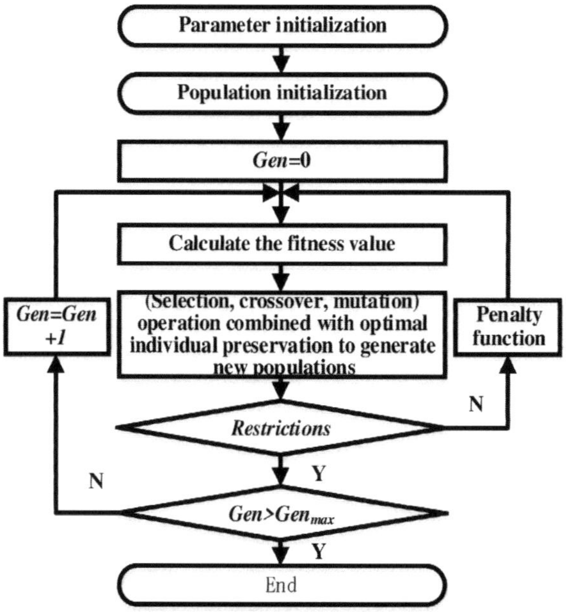

Figure 3. The flowchart of improved GA.

4.2. Application of genetic algorithm in parameter recognition

Objective function: The problem of parameter identification of photovoltaic array model is transformed into the minimum value of objective function as Equation (7):

$$minC = \sum_{j=1}^{k}(P_j - P_j^*)^2 + \Delta C \tag{7}$$

P_j^* is the predicted power, I_{scref}, U_{ocref}, I_{mref} and U_{mref} of a PV module under STC provided by manufacturers with identify parameter values, combining the Equation (1), Equation (3) and Equation (5); P_j is the measurement power of the photovoltaic array; k is the number of measured data; ΔC is the penalty function that does not satisfy the constraint condition.

Constraint condition: As Equation (8)

$$\begin{cases} |U^*_m - U_m| \le \delta_1 \\ |I^*_m - I_m| \le \delta_2 \\ |P^*_m - P_m| \le \delta_3 \end{cases} \tag{8}$$

U_m, I_m, P_m are the actual maximum power voltage, current, power collected by photovoltaic water pump MPPT under different operating conditions. U_m^*, I_m^*, P_m^* are maximum power point of photovoltaic array model based on GA parameter identification under different operating conditions. $\delta_1, \delta_2, \delta_3$ are the identification range and the actual error range.

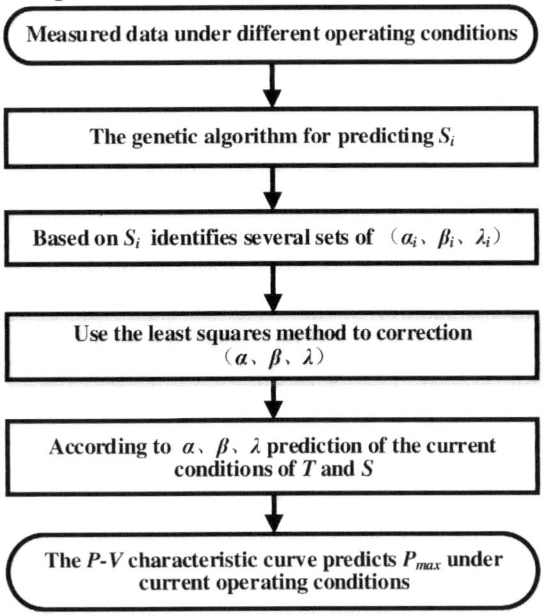

Figure 4. The diagram of parameter identification.

Parameter identification process shown in Figure4, according to different conditions under MPP of the PV pump to capture the collected data, The k group data are extracted separately from the curves 1 and 2 in Figure 1. Using the improved GA for parameter identification and estimation, $\alpha_i, \beta_i, \lambda_i$, T_i and S_i under different conditions were obtained, and the least squares method was used to modify α, β and λ. Using the data collected by curve 3 and combining α, β, λ to predict T and S, we can obtain the P-V curve 3 under the current condition. However, it is found that the α, β and λ is not high enough. Thus, in response to this problem, this paper uses a step-by-step method to improve the accuracy of parameter identification parameters. GA is used to estimate S under different conditions. Then, the estimated parameters of S_i are taken as the known parameters and then identification of $\alpha_i, \beta_i, \lambda_i$ and T_a are carried out according to the improved GA. The multi-group identification parameters are modified by the least squares method. Finally, using the modified parameters and the measured data to estimate the current T_a and S, combined with PV model to predict the current MPP of the PVA.

5. Simulation experiment and result analysis

According to the empirical data, the range of PV model parameters: $0.0008 \leq \alpha_i \leq 0.001$; $0.0035 \leq \beta_i \leq 0.0045$; $0.05 \leq \lambda_i \leq 0.07$; $0 \leq S_i \leq 2000$; $0 \leq T_i \leq 40$. In genetic algorithm, the relevant parameter values are set as shown in Table 1. In the table, N is the population size; Gen_{max} is the maximum evolutionary algebra; P_c is the crossover probability; P_m is the mutation probability.

Table 1. Parameter setting of genetic algorithm.

Parameters	Traditional program	Improve program
N	200	200
Gen_{max}	100	500
P_c	0.85	0.85
P_m	0.08	0.08

Track the collected data when the PV pump is at the maximum power point, then run the genetic algorithm to get the convergence curve as shown in Figure 5:

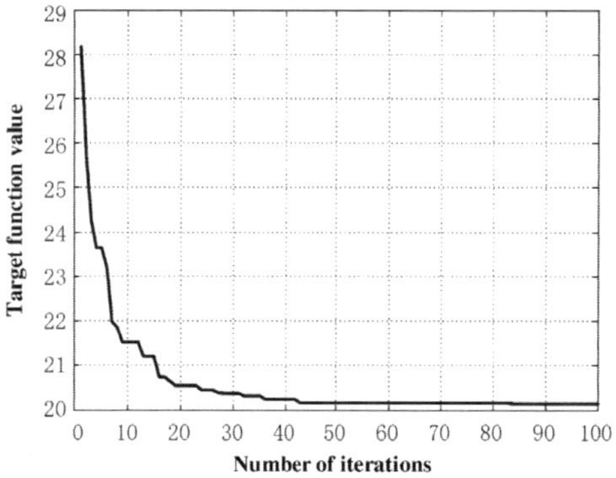

(a). Traditional method.

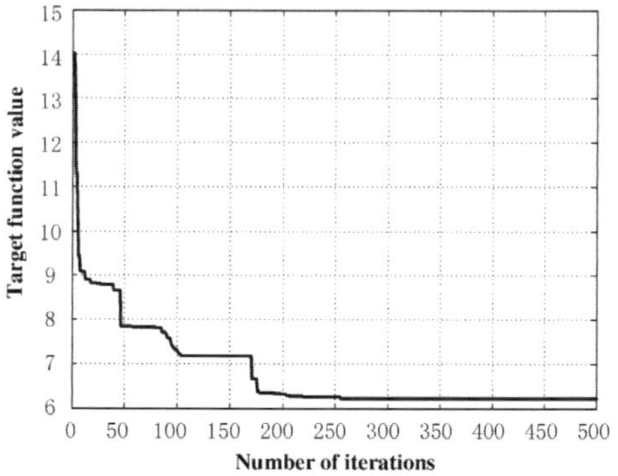

(b). Improved method.

Figure 5. The relationship between objective function and evolution generations of genetic algorithm.

It can be seen from Figure 5 in the graph (b) the objective function is smaller than the graph (a), and parameter identification result is more accurate.

In this paper, we use four groups of photovoltaic in the state of water MPP to capture the collected data, and identify the model parameters. The PV cell parameters are identified in Table 2.

Table 2. Identification of parameter values under different operating conditions.

Parameters	Curve 1	Curve 2	Curve 3	Curve 4	Correction parameters
$\alpha/10^{-3}$	0.8007	0.8010	0.8001	0.8039	0.8012
β	0.0039	0.0041	0.0041	0.0040	0.0040
λ	0.0602	0.0605	0.0602	0.0601	0.0602
T_a	20.0117	22.1575	24.9999	25.8682	\
S	599.5	799.6	1000	1199.9	\

Based on α, β and λ three sets of measured data were used to estimate the current T_a and S by GA. Figure 6 shows the comparison chart of the model simulation results and the measured value, we can see that the model based on the identification of the model output is better matched to the measured value and meet the requirements of photovoltaic modules engineering modelling better.

According to the above analysis, it can be used to accurately estimate the operating temperature and irradiation level based on the parameter values identified by the PVA parameter identification method, and then we can draw out a much more accurate P-V curve to estimate MPP which proves the validity and practical value of the method.

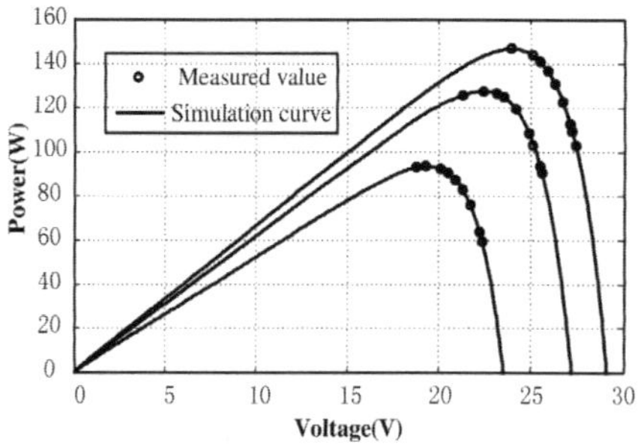

Figure 6. Comparison of simulation results with measured values.

6. Summary

In this paper, a new method combined with parameter identification of PVA, prediction of temperature and irradiation level, is proposed. Using the PVA measured data, GA can be used to identify parameters and predict T_a and S. The identification accuracy of parameter identification is not enough, and improve the accuracy of parameter identification by improved method. The actual example shows that the parameter value based on the method proposed in this paper is more accurate, and then an accurate PVA model can be established to predict the MPP of the PVA.

References

[1] Su Jianhui 2003 A Research on Photovoltaic Water Pumping System and its Control Strategy *Hefei University of Technology*

[2] Xu Zheng, Zeng Pijiang, Zhang Fei and He Shaoqiang 2015 Research and Development of PV Reverse Osmosis Desalination Systems *Acta Energiae Solaris Sinica* **36(4)** pp 886-92

[3] Liu Peng, Su Jianhui and Wu Wenjin 2016 Research on Control Strategy and Stability of Common DC Bus Solar Pumper Cluster System *Power Electronics* **50(10)** pp 67-9

[4] Kennerud K L 1969 Analysis of Performance Degradation in CdS Solar Cells *Aerospace and Electronic Systems* **5(9)** pp 912-7

[5] Jian Xianzhong, Wei Kai and Guo Qiang 2015 Artificial Bee Swarm Algorithm in the Application of Photovoltaic Cell Five-parameter Double-diode Model *Acta Photonica Sinica* **44(1)** pp 168-72

[6] Chan D S H, Phillips J R and Phang J C H 1986 Acomparative Study of Eextraction Methods for Solar Cell Model Parameters *Solid-State Electronics* **29(3)** pp 329-37

[7] Xu Xiaobing, Wang Jianping and Zhang Chongwei 2009 Analytical Model and Parameter Revising Method of Silicon Solar Cells *Journal of Hefei University Of Technology* **32(2)** pp 166-9

[8] Su Jianhui, Yu Shijie and Zhao Wei 2001 Investigation on Engineering Analytical Model of Silicon Solar Cells *Acta Energiae Solaris Sinica* **22(4)** pp 409-12

[9] Qiu Chun, Cai Tao, Duan Shanxu, Dai Qian, Shen Yanbo and Chen Zhenghong 2013 Extraction of Solar Cell Model Parameters under Arbitary Irradiance and Cell Temperature *Acta Eenergiae Ssolaris Sinica* **34(9)** pp 1626-32

[10] Rauschenbach H S 1980 *Solar cell and its application* (USA: Litton Educational Publishing Inc)

[11] Altas I H, Sharaf A M 2007 A Photovoltaic Array Simulation Model for Matlab-Simulink GUI Environment *International Conference on Clean Electrical Power IEEE* pp 341-5

[12] Cao Daoyou 2010 Research and Applications Based on Improved Genetic Algorithm *Anhui University*

Power train and emission control: allocation procedure by OBD-II system for automotive technology

Porag Kalita

Automobile Engineering Department,Vocational Education (+2), M R S Higher Secondary School, Govt. of India, Titabor, Jorhat, ASSAM, India.

Email:poragkalita@rediffmail.com

Abstract. OBD-II, systems were designed to maintain low emissions of in use vehicles, including light and medium duty vehicles. In 1989, the California code of Regulations (CCR) known as OBD – II was adopted by the California Air Resource Board (CARB) and the objective to reduce hydrocarbon (HC) emission caused by malfunction of the vehicles emission control systems. OBD-II provides additional information to engineer for diagnosis and repair of emissions related problems. OBD-II, standardizes on the amount of memory (Freeze Frame) it uses to store the readings of the vehicle sensor when it logs on emission related Intermittent Trouble code (IT).The intent of OBD-II, systems is to detect most vehicle malfunctions when performance of a power train component or system deteriorates to the point that the vehicle's HC emission exceed standard. The vehicle operator is notified at the time when the vehicle begins to marginally exceed emission standards, by illuminating the Malfunctions Indicator Light (MIL).

1. Introduction

The biggest question today before the metropolitan population of our country like India, is "pollution" is snuffing us out and will pollution ever be controlled? The reason for this type of thinking may be because of the fact that in spite of several legislation from 1964 onwards in India, not much of progress can be visualized in this area, even through; we are much more apparently concerned and aware of the environmental degradation today than a few decades earlier . In our country like India, after spending a lot of money and also enacting several statuary laws, the country could not make much headway because of several factors i.e. we could not make this problem an electoral-issue like that of USA.

The first major clean air act was adopted by the congress in 1970 in USA. Congress established the Environmental Protection Agency (EPA), with the overall responsibility of regulating motor vehicle pollution to the atmosphere. Congress also identified the inspection and Maintenance for improving the air quality.

The power train components towards emission controls are:

- Throttle and Manifold,
- Exhaust and Fuel system,
- Combustion and Rotational dynamic,
- Automatic Transmission, etc.

Throttle & intake manifold, the throttle body assemble is an air valve. It regulates the air flow into the engine and thereby contributes to the control of engine and power. Idle air control valve; bypass the throttle to provide additional air to compensate for the loads during closed throttle.

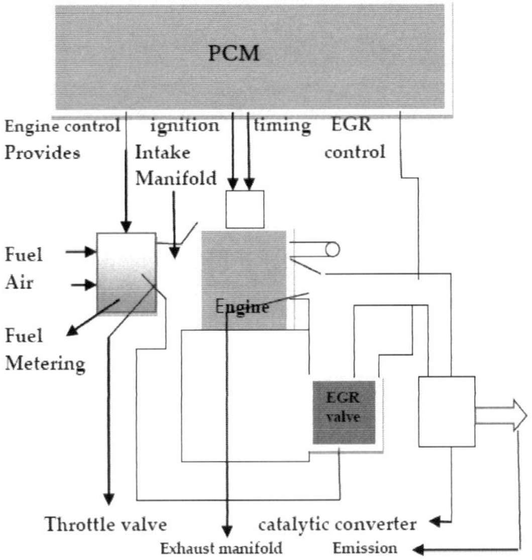

Figure 1. Major controller out-put to engine.

2 Present engaged (in India)

In June 1999, the honourable supreme court of India ruled that vehicular emission had to be reduced at a much quicker pace than planned so far, and therefore, engine manufactures are facing the challenging engineering task to quickly find solution by Euro-norms and remarkable change in the concept of I.C. engines designs, MPFI, catalytic converter and incorporation of EGR.

Catalytic converter converts harmful gases to harmless gases. As India prepares to make the use of catalytic converter compulsory for drastically reduce automobile exhaust emission and however, EPA of USA has observed that catalytic converter has become a significant and growing cause of Global warming. EGR is a small amount of exhaust gases, which are feed back into intake system by EGR valve. However, as per study on Indian condition, while the vehicle speed is increasing the peak combustion temperature also increasing, as a result less time available for heat transfer in the combustion chamber, because the maximum bsfc value increases by the increasing of speed. At the higher speeds there is a need to supply more fuel to overcome greater frictional losses and due to heat losses, will increase turbulence and consequent improper combustion. Therefore, OBD –II regulations in the USA for light and medium duty vehicle for I.C. Engines are introduced to implement the air quality standard and need of implementing on Indian condition.

On-line diagnosis of internal combustion engines in passenger vehicles is mandated due to the strict environmental regulations in the U.S.A. and in some European Countries (e.g. , the EFTA (European Free Trade Agency) Partners') to control Hydro carbon emissions from the exhaust.

3. Methodology

The power train functions are described to show how the PCM controls the emissions while delivering the torque to the vehicle requested to the driver.

3.1. Throttle and intake manifold

The throttle body assembly in an air valve, it regulates the air flow into the engine and thereby contribute to the control of engine speed and power. IACV (Idle Air Control Valve) provides additional air flow during starting of the engine and during idle. IACV by pass the throttle to provide additional air to compensate for the load during closed throttle. EGR provides exhaust gases to the intake manifold. This has the effect of reducing oxygen content in the engine.

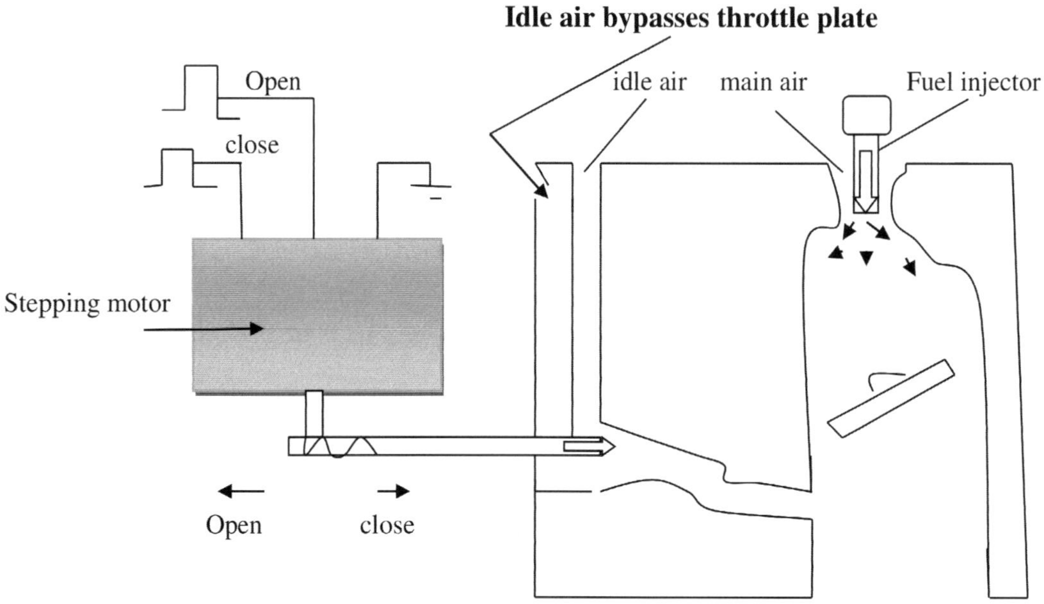

Figure 2. Idle air control.

3.2. Exhaust & fuel system:

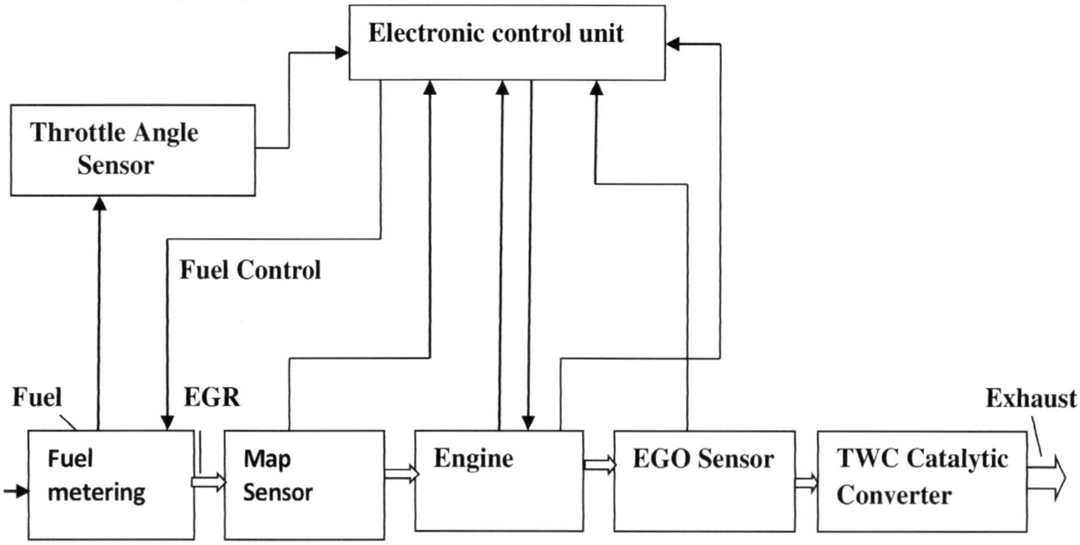

Figure 3. Simplified block diagram of fuel control systems.

Exhaust valves of the engine cylinders purge the exhaust through the Exhaust Gas line which then passes through the catalytic converters in which most of the HC and CO are oxidized to CO_2 and Water. The extra oxygen required for this oxidation is supplied by adding air to exhaust stream from an engine driven air pump. This air called secondary air is normally introduced into the exhaust manifold. This has a considerable effect in reducing emissions. The Fuel Pump supplies metered fuel which is electronically injected through nozzles operated by solenoids under control of the PCM. The

IOP Conf. Series: Materials Science and Engineering **211** (2017) 012019 doi:10.1088/1757-899X/211/1/012019

fuel in the fuel tank is filtered. The Fuel Level Sensor measures the inlet vacuum which is a measure of fuel pump suction which affects pump priming.

4. Result & Discussion

The oxygen sensor is used to monitor the residual oxygen (after catalysts in the converter) in the exhaust gases, because catalytic converter is most efficient in eliminating all pollutants by oxidizing HC to CO_2 and reducing NOx to N_2 when the exhaust gases indicate Stochimetric (14.7) air /fuel ratio, indicated by the Exhaust Gas Oxygen sensor. Catalytic converter is a three way catalyst which will oxidize the Hydro carbons including CO and CO_2 and reduce the NOx to N_2 in the exhaust gases simultaneously thus removing.

Automatic Transmission: The Automatic Transmission uses a hydraulic or fluids coupling to transmit engine power to the wheels, (*Fluid Coupling: A device in the power train consisting of two rotating members. It transmits power through a fluid from the engine to the remainder of the power train).* Efficient transmission of engine output to the automatic transmission input shaft is performed through a transmission lockup clutch similar to a standard pressure plate clutch placed inside the torque converter.

In order to smoothly engage the lockup clutch the hydraulic pressure is adjusted by controlling the output current applied to the lockup solenoid valve. *(Solenoid is a type of electro-magneto, often used to operate the starter motor switch).*

Automatic transmission is controlled by inputs from the vehicle speed sensor and throttle position sensor which senses the vehicle load. The automatic gear shift points, the point at which the lockup clutch is activated, and the clutch's hydraulic pressure level are controlled by the Power Control Module (PCM). The optimal shifts and lockup operations are carried out using a solenoid valve to open and close the hydraulic circuit, primed by the hydraulic pump.

The transmission's input-shaft speed is monitored during shifting by the speed sensor after the ON/OFF signal is output from the shift solenoid valves. The shifting process is adjusted by the hydraulic pressure of the clutch so that the clutch is smoothly engaged. The engine torque is controlled in synchronism with the shift to reduce impact due to shift. During cruise, the lockup clutch is engaged and is disengaged during shifts, which improves fuel economy and emissions.

Resistance to the Motion of the vehicle: A thrust known as tractive effort or force is provided by the power unit of a vehicle at the driving road wheels. Varying at different engine speeds and gear positions, it is mainly required to overcome the force of opposing motion of the vehicle. The resistance to the motion of the vehicle is known as tractive resistance.

4.1 Rolling resistance

It is the force necessary to maintain constant speed on a level road and the rolling resistance generally varies with the type of the road surface, load on each tyre, inflation pressure and type of tyre trend.

The value rolling resistance can be calculated from the formula:

$R_r = k_r . W$

where, W = Total weight of the vehicle in kg.

K_r = constant of rolling resistance.

<u>Gradient resistance:</u> It is the force opposing forward motion of a vehicle up a gradient. This resistance does not depend upon the speed of the vehicle. It is expressed as,

$R_g = W \sin\theta$,

where, W= Total weight of the vehicle kg.

θ = inclination of the slope to the horizontal,

When expressed as a percentage, it is percent gradient,

$= 100 \times \tan\theta = 100 \sin\theta$

IOP Conf. Series: Materials Science and Engineering **211** (2017) 012019 doi:10.1088/1757-899X/211/1/012019

4.2 Wind or air resistance

Wind or air resistance is dependent upon speed, the shape of the vehicle body and wind velocity and it is given by:

$$R_a = K_a . A.V^2$$

where, A= projected frontal area in m^2

V = speed of the vehicle in km/hr.

K_a = Co-efficient of air resistance.

= 0.00235 for best stream lined cars,

= 0.0032 for average cars,

= 0.0046 for trucks and Lorries.

5. Type of data

Power train subsystem consist of the engine and transmission including the exhaust emission control apparatus which needs to be continuously monitored by the engine controller i.e. computer for potential defects leading to decreased effectiveness in emission control system. Exhaust & Fuel system consists of the following components : Exhaust valves: Exhaust Gas line; Fuel pump; Fuel Level Sensor; Vacuum Sensor ; Canister vent; Fuel feed and metering ; Fuel injection nozzles; Oxygen sensor; catalytic converter etc.

For example, a petrol engine consumes 6.35 kg of petrol per hour. The specific gravity of petrol is 0.7. The fuel air ratio is 0.066. The diameter of the single jet of the carburetor 1.27 mm and its top is 3 mm above the petrol level. Average condition of air is 15.5 C and 1,027 kg/cm². The values discharge coefficient for fuel and air respecting are 0.6 and 0.8. Find the critical air velocity and effective throat diameter of the venturi. What is the drop of pressure in the venture, expressed in cm of water.

Solution:

Density of air at given conditions:

$$P\gamma = \frac{1.027 \text{ x } (10)4}{29.27 \text{ x } 2888.5}$$

$= 1.215 \text{ kg/m}^3$

Neglecting the compressibility effect on air, critical air velocity:

$$V_a \text{ (critical)} = K_a\sqrt{2g \times \frac{\text{pf}}{\text{pa}}}$$

$$= 0.8\sqrt{\frac{2 \times 9.81 \times 3 \times 700}{1000 \times 1.215}}$$

= 4.65 m/sec

Petrol flow per second,

$$W_f = A_f . K_{f.}\sqrt{2g .pf\ (\Delta\, Pa - x\, .Pf}$$

$$\text{Or, } = \frac{6.35}{60 \times 60}$$

$$= \frac{\pi\ (1.27)2 \times 0.6}{4 \quad (10)4}\sqrt{(2 \times 9.31 \times 700)}\sqrt{(\Delta\, Pa - \frac{3}{1000} \times 700)}$$

$$\text{Or, } \sqrt{1.4000 \times 9.81\ (\Delta\, Pa - 2.1}$$

$= 2.310 \qquad \therefore \Delta\, Pa = 390.1\ kg/cm^2$

Using the relation, p = wh, h = $\frac{\text{p}}{\text{w}}$

$\Delta\, Pa = (\frac{390.1}{1000})$meters of water.

= .39 meters of water,

= 39 cm of water.

Air flow per second:

$W_a = K_a.A_a\sqrt{2g.Pa.\Delta Pa}$

$\frac{6.35}{0.066 \times 3.600} =$

$0.8 \times Aa\sqrt{2 \times 9.81 \times 1.215 \times 390.1}$

$\therefore 0.8 \times A_a\sqrt{9.320}$

$= 2.7 \times (10)^{-2}$

$\therefore \frac{\pi}{4} d_a^2 = \frac{2.67 \times (10)2}{0.8 \times \sqrt{9.320}}$

$\therefore d_a^2 = 4.41 \times (10) - 4$

$d_a = 2.1 \times 10^2$ meters.

$\therefore$Throat diameter venturi

= 2.1 cm

6. Conclusion

Combustion and Rotational dynamics: The engine provides the mechanical power to the vehicle. The engine cylinders perform the combustion of air/fuel mixture at stoichiometric ratio (14.7). The Crankshaft assembly and flywheel house the crank angle which senses the position of the Top Dead Centre (TDC) of the cylinder and provides the necessary ignition spark at the correct crank angle between the reference point on the flywheel and the horizontal centerline of crank shaft. The amount of fuel needed for the combustion in the engine cylinder is a direct function of the throttle position and the mass of air through the intake manifold which is controlled by the drivers accelerator pedal. This mass of air is measured with the Mass Air Flow (MAF) sensor. The throttle plate is mechanically linked to the accelerator pedal which is operated by the driver. When the pedal is pressed the throttle plate rotates and allows more air to pass through the intake manifold. During engine off condition, the fuel stored in the fuel system tends to evaporate into the atmosphere. To reduce these HC emissions, they are collected by a charcoal filter in a canister. The collected fuel is released into fuel intake through a purge solenoid valve controlled by the PCM periodically.

References

[1] 2000 *Proc. of SAE International Total Life Cycle Conference & Exposition in Michigan*, USA,.
[2] Dr. Porag Kalita 2000 SAE paper no. 2000-01-1477 M R S Higher Secondary School, Titabor, Assam, India.
[3] Prof. S. Ganeshan 2000 *Seminar On Board Diagnostics – II*, Oakland University, USA.
[4] www.ijcert.org
[5] Porag Kalita *Automotive Service Business,* Vol. **I**, a Book.
[6] P.Kalita 2002 paper presented at the SAE International Body Engineering Conference & Exposition, Paris, & France, dated July 9th to 11th, Paper I/D No. 02 IBECC-3.
[7] P. Kalita 2015 paper presented at Dr. K.V.Subha Reddy Institute of Technology, Kurnool, A.P.,Dated 25th April, paper I.D. 1505001, www.ieee.org

Acknowledgement

I would like to thank Mr. Alakesh Kalita, Student of B. Tech in Engineering Physics (2nd Semester-Dual) in Indian Institute of Space Science & Technology, Government of Space, under ISRO, Kerala for the valuable cooperation .

IOP Conf. Series: Materials Science and Engineering **211** (2017) 012020 doi:10.1088/1757-899X/211/1/012020

Unbalanced voltage control of virtual synchronous generator in isolated micro-grid

Y Z Cao*, H N Wang and B Chen

School of Electrical Engineering and Automation, Hefei University of Technology, Hefei, China

*Email: 814390474@qq.com

Abstract. Virtual synchronous generator (VSG) control is recommended to stabilize the voltage and frequency in isolated micro-grid. However, common VSG control is challenged by widely used unbalance loads, and the linked unbalance voltage problem worsens the power quality of the micro-grid. In this paper, the mathematical model of VSG was presented. Based on the analysis of positive- and negative-sequence equivalent circuit of VSG, an approach was proposed to eliminate the negative-sequence voltage of VSG with unbalance loads. Delay cancellation method and PI controller were utilized to identify and suppress the negative-sequence voltages. Simulation results verify the feasibility of proposed control strategy.

1. Introduction

Micro-grid that using wind, solar and other renewable energy as main source is an effective local power supply, which has broad application prospects in the mountain area and island [1]. However, isolated micro-grid often has no synchronous power supply to support its voltage and frequency. Therefore, the virtual synchronous generator(VSG) control is usually adopted in the isolated micro-grid inverter [2-3]. The existing VSG works as voltage source, and the output voltage is usually open-loop controlled. When unbalanced loads connect to the isolated micro-grid, the unbalance degree of output voltage is usually higher than national standard—2%.

The circuit topology of the neutral-splitting circuit or the three-phase four-leg circuit has solved the problem brought by unbalance loads. But the applications are not common in existing micro-grid inverter products, as this type of topology increases the cost of the system and introduces the problem of neutral point voltage balance control. For the control of unbalance voltage in isolated micro-grid, the existing research mainly focused on the compensation of negative-sequence voltage [4-5]. Reference [6] derived universal transformation from synchronous to stationary reference frame, then proposed the concept of virtual composite impedance to control the positive- and negative-sequence voltage respectively. References [7-8] pointed that the negative-sequence voltage caused by unbalance loads can be controlled indirectly by controlling negative-sequence admittance. Reference [9] proposed a virtual negative impedance control method to compensate the voltage drop on the inverter output impedance, which effectively reduced the voltage unbalance degree. However, the existing relevant papers had no analysis of the generation mechanism of VSG output voltage imbalance, and the unbalanced voltage control method remains to be further studied.

This paper aims at solving the electrical power quality problem of isolated micro-grid inverter operating with VSG control strategy under unbalanced loads situation. On the basis of the positive-

and negative-sequence equivalent circuit of VSG and negative-sequence voltage depressing methods, an improved VSG voltage control algorithm has been proposed. Simulation results have verified that using this control strategy can promote the ability of carrying unbalanced load and improve the power supply quality.

2. Basic principle of VSG

The typical main circuit topology of three-phase three-wire inverter is shown in Figure 1.

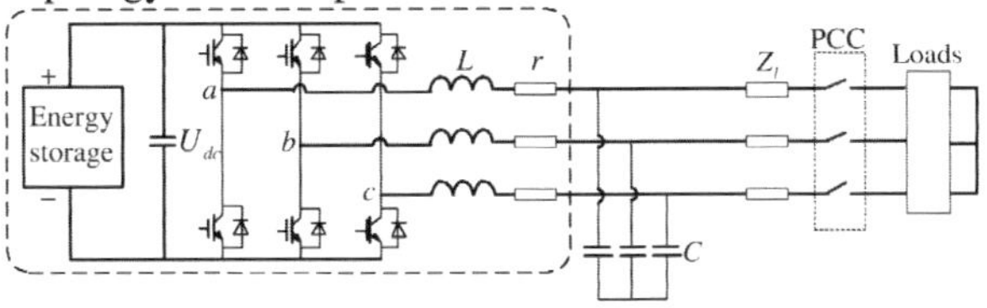

Figure 1. Topology of VSG main circuit.

2.1. VSG ontology model

The motion equation of the rotor and the electrical equation of the stator are respectively shown in equation (1), which using the mathematical language to describe the rotary synchronous generator electromagnetic changes and mechanical motion.

$$\begin{cases} T_m - T_e = \dfrac{P_m}{\omega} - \dfrac{P_e}{\omega} = J\dfrac{d\omega}{dt} - D(\omega - \omega_g) \\ u_{abc} = E_0 \sin\theta_{abc} - L_{SG}\dfrac{di_{abc}}{dt} - r_{SG}i_{abc} \end{cases} \tag{1}$$

In equation (1), ω_g is the synchronous angular velocity of the AC bus; E_0 is the open-circuit voltage amplitude of the synchronous generator; L_{SG} is the synchronous reactance, and r_{SG} is the stator resistance.

The excitation equation of the synchronous generator is shown in equation (2), where i_{f0} is the excitation current required to synchronize the generator to establish the rated no-load electromotive force when the rotor is running at rated speed.

$$i_f = i_{f0} + \Delta i_f \tag{2}$$

But in VSG, the excitation current is not an intuitive controllable variable, which is not easy to set the parameters. In the case of neglecting the saturation of the magnetic circuit, the open-circuit voltage E_0 and the excitation current i_f is approximately proportional relationship [4]. Therefore, we can see that the no-load EMF E_{s0} is: $E_{s0}=\omega M_f i_{f0}$, M_f is the inter-rotor mutual inductance. By multiplying both sides of equation (2) by the constant term $\omega_0 M_f$, we can obtain equation (3):

$$\omega_0 M_f i_f = \omega_0 M_f i_{f0} + \omega_0 M_f \Delta i_f \tag{3}$$

That is the same expression as equation (4):

$$E_0 = \frac{E_{s0} + \Delta E_s}{\omega_0} \times \omega = \frac{E_s}{\omega_0} \times \omega \tag{4}$$

Thus, the adjustment of the excitation current i_f can be converted into a direct adjustment to the open-circuit voltage E_0. The excitation process of real synchronous generator has a certain hysteresis, so join the first-order inertia link to make sure that the excitation system of VSG is closer to the excitation system of real synchronous generator. Since the realization of the synchronous reactance is behind the combined of EMF, it is placed outside the VSG ontology model. Together with virtual speed regulator and virtual excitation regulator, the VSG ontology model can be shown in Figure 2.

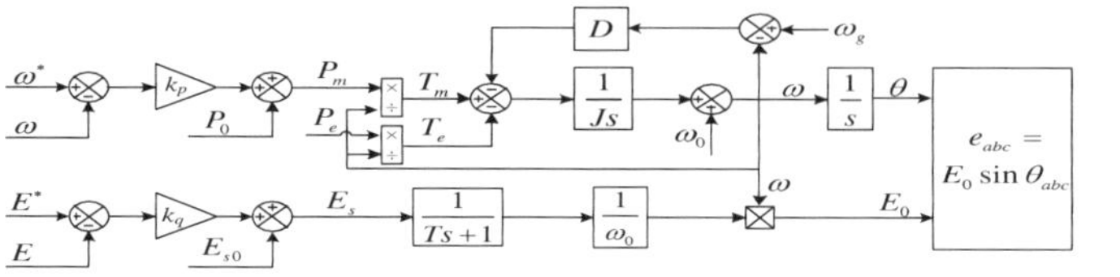

Figure 2. Structure diagram of VSG ontology model.

2.2. Realization of synchronous reactance

The synchronous reactance is an important parameter in synchronous generator. In order to simulate the synchronous generator accurately, the synchronous reactance must be introduced as a control parameters of VSG. If the equation in (1) is rewritten as a current integral form, as shown in equation (5), the output can the stator current of synchronous generator.

$$i_{abc} = 1/L_{SG}\int(e_{abc} - u_{abc} - r_{SG}i_{abc})dt \tag{5}$$

By using the stator current as the reference current of VSG, and introducing current closed-loop, the output current of VSG can be consistent with the output current of synchronous generator ,which ensuring the VSG can accurately simulate the running characteristics of synchronous generator.

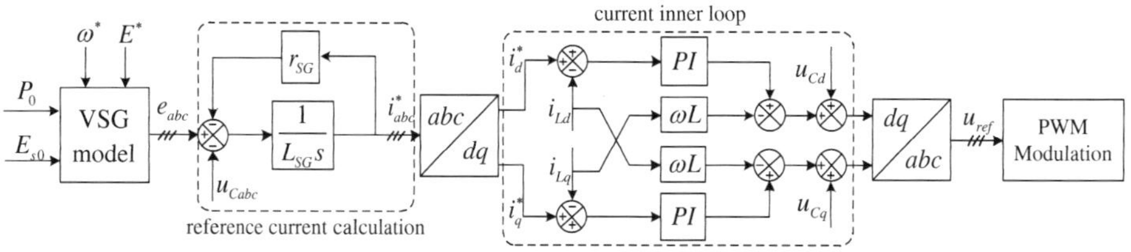

Figure 3. The realization of VSG synchronous reactance.

3. Negative-sequence voltage control of VSG

There is a large amount of single-phase loads in micro-grid. The load current presents asymmetry when common VSG connects to these loads in island mode, which leads to output voltage unbalanced.

3.1. Analysis of VSG under unbalance loads

Based on symmetric component method, we can know that three-phase unbalanced phasors can be decomposed into three sets of symmetric phasors: positive-sequence, negative-sequence and zero-sequence component. For the three-phase inverter without neutral line, the zero-sequence is usually neglected.

According to the deduction and analysis in reference [10], we can obtain positive- and negative-sequence equivalent circuit of VSG in island mode, as shown in Figure 4.

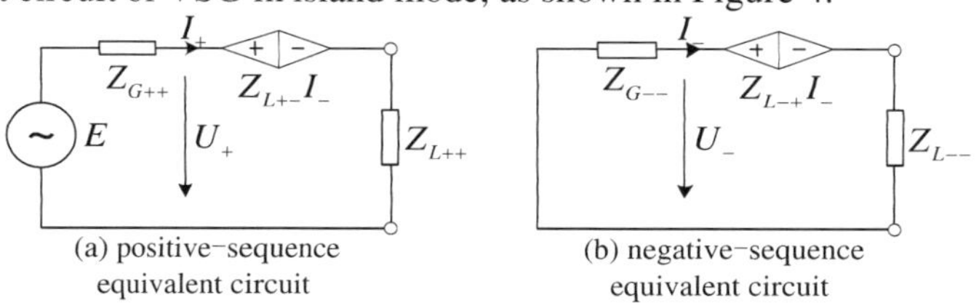

Figure 4. Positive- and negative-sequence equivalent circuit of isolated VSG.

If the negative-sequence voltage in VSG can be controlled to zero, only leaving the positive-sequence component, it can be ensure that the output terminal voltage is three-phase balanced. which will improve the voltage quality of power supply in isolated micro-grid.

3.2. The extraction of negative-sequence component

In order to suppress the negative-sequence component, we need to extract it from unbalanced voltages first. Three-phase asymmetric voltage can be decomposed according to equation (6).

$$\begin{bmatrix} u_a \\ u_b \\ u_c \end{bmatrix} = U_{m+} \begin{bmatrix} \cos(\omega t + \varphi_+) \\ \cos(\omega t + \varphi_+ - 120^\circ) \\ \cos(\omega t + \varphi_+ + 120^\circ) \end{bmatrix} + U_{m-} \begin{bmatrix} \cos(\omega t + \varphi_-) \\ \cos(\omega t + \varphi_- + 120^\circ) \\ \cos(\omega t + \varphi_- - 120^\circ) \end{bmatrix} \quad (6)$$

where U_{m+} ,U_{m-} respectively means the positive- and negative-sequence peak value of fundamental voltage. And φ_+,φ_- respectively means the initial phase angle of the positive- and negative-sequence of fundamental wave voltage. Transform the positive- and negative-sequence components into α-β axis, then equation (7) and (8) are obtained.

$$\begin{bmatrix} u_{\alpha+} \\ u_{\beta+} \end{bmatrix} = \frac{2}{3} \begin{bmatrix} 1 & -\frac{1}{2} & -\frac{1}{2} \\ 0 & \frac{\sqrt{3}}{2} & -\frac{\sqrt{3}}{2} \end{bmatrix} \begin{bmatrix} u_{a+} \\ u_{b+} \\ u_{c+} \end{bmatrix} = U_{m+} \begin{bmatrix} \cos(\omega t + \varphi_+) \\ \sin(\omega t + \varphi_+) \end{bmatrix} \quad (7)$$

$$\begin{bmatrix} u_{\alpha-} \\ u_{\beta-} \end{bmatrix} = \frac{2}{3} \begin{bmatrix} 1 & -\frac{1}{2} & -\frac{1}{2} \\ 0 & \frac{\sqrt{3}}{2} & -\frac{\sqrt{3}}{2} \end{bmatrix} \begin{bmatrix} u_{a-} \\ u_{b-} \\ u_{c-} \end{bmatrix} = U_{m-} \begin{bmatrix} \cos(-\omega t + \varphi_-) \\ -\sin(-\omega t + \varphi_-) \end{bmatrix} \quad (8)$$

We can find that $u_{\alpha+}$ and $u_{\beta+}$ have the cosine-sine function relationship with the same amplitude and frequency. Thus, the delay cancellation method[11] can be used to acquire the $u_{\alpha-}$,$u_{\alpha+}$ and $u_{\beta-}$,$u_{\beta+}$ component, where they are still AC values in the α-β axis. In this paper, the positive- and negative-sequence components are transformed into the d-q axis rotating coordinate to control. The control diagram of negative-sequence component extraction method is shown in Figure 5.

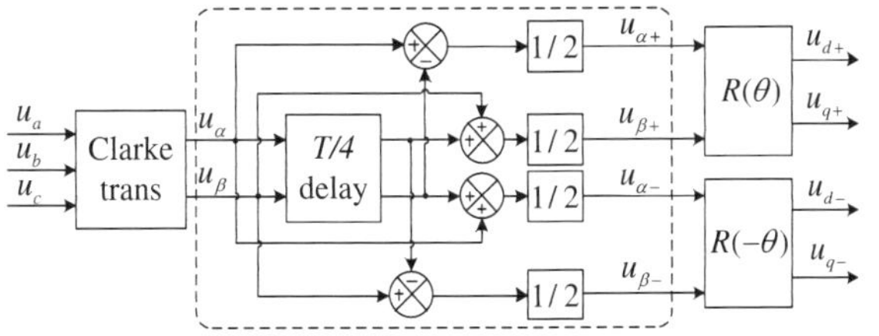

Figure 5. Control diagram of negative-sequence component extraction method.

In the above figure, θ is the angle between d-axis and a-axis, $R(\theta)$ represents the transformation matrix of positive-sequence from α-β axis to d-q axis ,while $R(-\theta)$ represents the negative one.

In order to achieve 90° phase shift, the SOGI phase shift system is adopted, and its transfer function can be reorganized as equation (9):

$$\begin{cases} D(s) = \dfrac{u'_\alpha}{u_\alpha} = \dfrac{k\omega s}{s^2 + k\omega s + \omega^2} \\ Q(s) = \dfrac{ju'_\alpha}{u_\alpha} = \dfrac{k\omega^2}{s^2 + k\omega s + \omega^2} \end{cases} \quad (9)$$

When k =1, ω =100π, $D(s)$ has a gain of 1dB at 50Hz and the phase offset is 0°, while $Q(s)$ has a gain of 1dB and the phase offset is 90° at the same frequency.

3.3. Unbalance voltage control strategy

In order to suppress the negative-sequence voltage and increase the unbalance load capacity of VSG, the unbalance voltage control strategy for VSG is designed (shown in Figure 6).

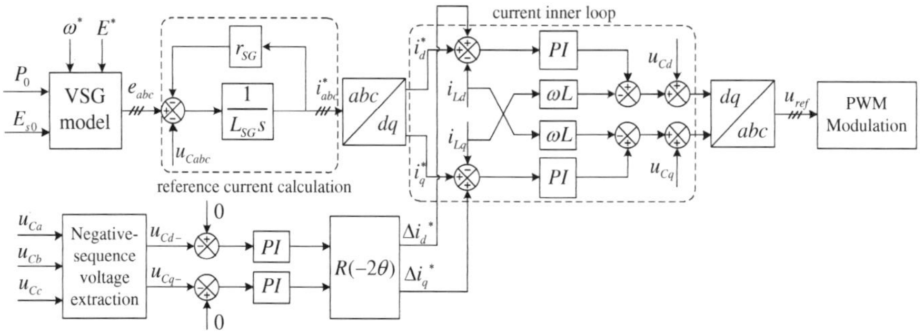

Figure 6. Unbalance voltage control structure diagram of isolated VSG.

The measured output terminal voltage u_{Cabc} are calculated through delay cancellation method and thus the positive- and negative-sequence components in the α-β axis are obtained. The negative-sequence component $u_{C\alpha-}$ and $u_{C\beta-}$ are transformed into d-q axis to achieve the direct-current negative-sequence components u_{Cd-} and u_{Cq-}. According to the classical control theory, in the synchronous rotating coordinate system, the negative sequence output voltage can track the reference quantity without static error using PI controller. The reference DC negative-sequence components u_{Cd-} and u_{Cq-} are set to 0 and the PI controller output values are added on the current loop reference instructions i_d^* and i_q^* as compensations. This improved VSG control can suppress the negative-sequence voltages to zero.

4. Simulation results

The simulation was performed in Matlab/simulink environment to verify the feasibility of proposed unbalance voltage control strategy. Simulation parameters are shown in Table1.

Table 1. Simulation parameters.

Block name	Parameter value
Inverter	U_{dc}=600V; LC Filter: L=0.35mH,C=50μF; Switching frequency: f=6000Hz; Rated capacity: P_N=100kW
Current loop	PI regulator:k_p=1.2,k_i=60
NSV control	PI regulator:k_p=2.4,k_i=120
Simulation duration	1s（intercept displaying 0.5~0.6s）
Load	P_l=90kW,Q_l=30kvar
Unbalance condition	（1）Inserting 5Ω load between A and C phase （2）B phase load out of running

Unbalance degree is usually adopted to indicate how bad the unbalance condition is. It is defined as the amplitude ratio of negative-sequence and positive-sequence component. Unbalance degree is expected to keep as small as possible.

The output voltage and load current waveform under unbalance condition(1) can be seen in Figure 7. When inserting 5Ω load between A and C phase, the output voltage of common VSG become unbalanced distinctly. The unbalance degree of output voltage comes to 7.5%, which is higher than relevant standard. In contrast, the improved VSG outputs three-phase balanced voltage, whose unbalance degree is 0.17%.

IOP Conf. Series: Materials Science and Engineering **211** (2017) 012020 doi:10.1088/1757-899X/211/1/012020

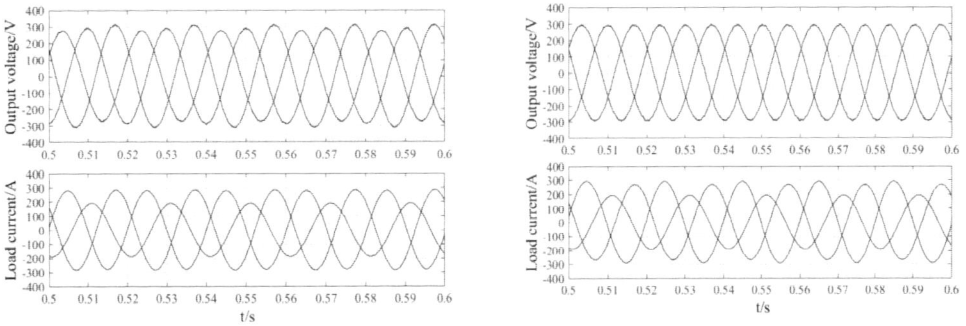

(a). Common VSG under unbalance condition(1).

(b). Improved VSG under unbalance condition(1).

Figure 7. Comparison of simulation results under unbalance condition(1).

Figure 8 shows the output voltage and load current waveform under unbalance condition(2) . The unbalance degree of output voltage comes to 12.4% when B phase load is out of running, which is too much higher than 2%. In contrast, the improved VSG outputs three-phase balanced voltage, whose unbalance degree is 0.21%.

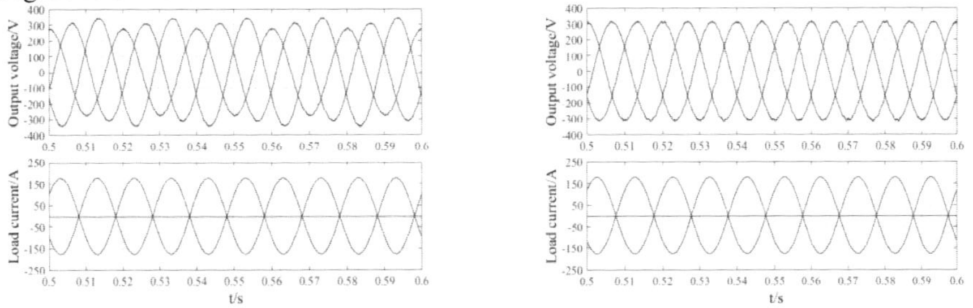

(a). Common VSG under unbalance condition(2).

(b). Improved VSG under unbalance condition(2).

Figure 8. Comparison of simulation results under unbalance condition(2).

By comparing the simulation results, we can draw a conclusion that the proposed unbalance voltage control strategy can make VSG adapt to various kind of unbalance loads, even the worst condition: one phase load out of running. The unbalance degree of output voltage can be kept under 0.5%.

5. Conclusion

In isolated micro-grid, the output voltage unbalance degree of common VSG is difficult to fit the relevant technical standards because of the existence of a large number of asymmetric loads. In this paper, an improved VSG voltage control algorithm has been proposed based on the positive- and negative-sequence equivalent circuit of VSG and negative-sequence voltage depressing methods. The simulation results verify the feasibility and effectiveness of the proposed VSG unbalance voltage control strategy. However, the problem of voltage unbalance in multiple VSG co-governed island mode is still to be further studied.

References

[1] Robert H Lasseter 2007 Microgrids and Distributed Generation *Intelligent Automation & Soft Computing* **133(3)** pp 225-34

[2] Sakimoto K and Miura Y 2011 Stabilization of a power system with a distributed generator by a Virtual Synchronous Generator function *Power Electronics and ECCE Asia (ICPE & ECCE), 2011 IEEE 8th International Conference on IEEE* pp 1498-1505

[3] Zhong Q C and Weiss G 2011 Synchronverters: Inverters That Mimic Synchronous Generators *IEEE Transactions on Industrial Electronics* **58(4)** pp 1259-67
[4] Liu J and Yushi M 2016 Parallel operation of a synchronous generator and a virtual synchronous generator under unbalanced loading condition in microgrids *IEEE 8th International Power Electronics and Motion Control Conference (IPEMC-ECCE Asia)* pp 3741-8
[5] Liu Q W and Tao Y 2014 Voltage unbalance and harmonics compensation for islanded microgrid inverters *IET Power Electronics* **7(5)** pp 1055-63
[6] Wang Y C and Luo A 2015 An unbalanced voltage compensation method for parallel inverters in microgrids *Proceedings of the CSEE* **35(19)** pp 4956-64(in Chinese)
[7] Alipoor J and Miura Y 2014 Voltage sag ride-through performance of virtual synchronous generator *Power Elctronics Conference* pp 3298-305
[8] Zhao X and Wu X H 2015 A direct voltage unbalance compensation strategy for islanded microgrids *IEEE Applied Power Electronics Conference and Exposition (APEC)* pp 3252-9
[9] Savaghebi M and Jalilian A 2013 Autonomous voltage unbalance compensation in an islanded droop-controlled microgrid *IEEE Transactions on Industrial Electronics* **60(4)** pp 1390-1402
[10] Zeng Z and Shao W H 2017 Unbalanced Voltage Control of Virtual Synchronous Generator in Islanded Micro-grid *Proceedings of the CSEE* **37(2)** pp 372-80
[11] Karthikeyan S S and Kumar R S 2012 Design and analysis of controller for three-phase UPS system *Electrical,Electronics and Computer Science (SCEECS)* pp 1-4

IOP Conf. Series: Materials Science and Engineering **211** (2017) 012021 doi:10.1088/1757-899X/211/1/012021

Evaluation of acoustical comfort in mosque

M S Prawirasasra* and S Mubarok

Engineering Physics Department, School of Electrical Engineering, Telkom University, Bandung, Indonesia.

*Email: bibinprawirasasra@telkomuniversity.ac.id

Abstract. There are various worship activities held at mosque, such as reciting, prayers and Friday sermon and acoustically, it is categorized as speech. Syamsul Ulum Mosque (MSU) located at Telkom University, Indonesia was taken as an object for research since there were no information nor historical data about actual acoustic condition and acoustical design. Main hall, considered as primary venue, is taken as analysis area. Measurement covers only speech-related objective parameters; reverberation time (RT), Definition (D50), Sound Strength (G) and Noise Criteria (NC). Furthermore, identification of proper area-for-speech activity and uniformity of distribution energy of impulse response parameters is illustrated by map contour. Since there is some value at certain areas does not have value within recommendation range, the main hall of MSU is inappropriate venue to accommodate worship activities.

1. Introduction

Mosque is foremost place of worship for Muslims and various activities are using it as a venue. Basically, there are three routine worship acoustic-related activities held at mosque: audibility of prayer order, listening & understanding speech during sermon and listening and/or reciting verses from Holy Quran [1]. All these activities require a high level of speech audibility and intelligibility because it is directly related to devoutness of followers.

According to previous research, acoustic requirements to be fulfilled are sufficient loudness, prevalent sound distribution, optimum reverberation time, low background noise, and high speech intelligibility [2]. Those conditions are quantified into objective parameters such as sound pressure level (L_p), Reverberation Time (RT), Sound Strength (G), Noise Criteria (NC) and Definition (D50). The objective of this research is to measure objective parameters at mosque and compared it with recommendation value for speech activities hence actual acoustic condition of object study is identified.

2. The object study

Object of this research is Syamsul Ulum Mosque (MSU) which is located at Telkom University, Indonesia. MSU consist of two floor but area of 2nd floor is lesser than 1st floor so there are void inside mosque. Focus of this research only at 1st floor because main activities are located at it. 1st floor has square shape with dimension 24x24 m and height around 9 m. There is jutted area in front of hall for prayer leader.

Boundary of mosque dominantly covered by wood material. All floor is made from solid hardwood and so ceiling is. Sun light penetrates into main hall through glasses with wooden frame from right, left and back side of object. The perimeter of front wall is semi ellipsoidal form made of smooth concrete. Interior of main hall is illustrated at **Figure 1**.

(a) (b) (c)

Figure 1. Main hall of MSU. (a). rear view, (b). right side view and (c). front view.

3. Measurement

Data is taken by using microphone, DAQ, omnidirectional speaker and personal computer. Speaker was located at front of main hall while position of microphone represents audience position. During Friday sermon, main hall is fully occupied by around 200 persons so measurement must to cover all observer area. It was accommodated by took sample from 56 different position with certain configuration. Measurement point has distance around 3 m between point and 2 m from point to the nearest wall. The height of receiving sensor was set to 75 cm or height of human ear in seating position.

Sound source is fed by signal impulse and located at front of main hall. The decaying sound that emitted from source is captured as impulse response. Each parameter is calculated from impulse response with one-octave filter from 125-4000 Hz [3]. Reducing noise exposure to sensor is very critical, as consequence, measurement was taken at night while there were nearly no activities in nearby environment.

4. Result

As basis of reference for speech activity, the object is expected to have result within recommendation value as presented at **Table 1** Measurement results are presented in two type a) statistical value (mean, maximum, minimum and standard deviation) and b) contour map. By presenting both, the variation value and its position is visualized. For contour map, area for leader prayer is excluded.

Table 1. Recommendation value [4][5].

Parameter	Value
RT	0.6-0.7 s
NC	25-30
G	>0 dB
D50	>0.86

4.1. Background noise

Background noise measurement was taken at several position in order to get the largest value of sound level pressure for each middle frequency plotted at NC curve. Since there were no significant activities around object and no active mechanic & electronic devices inside mosque, practically energy level of noise was low and it shown by value of NC is 25.

4.2. Reverberation time

The definition of reverberation Time (RT) is time it takes for a room to diminish sound energy by 60 decibels [6]. This parameter is very important and being used as baseline for designing room whether

IOP Conf. Series: Materials Science and Engineering **211** (2017) 012021 doi:10.1088/1757-899X/211/1/012021

for speech or music. In subjective perspective, RT is perceived as 'dead' or 'live' room. Room with long reverberation time is perceive as 'live' while short time 'dead'. The measurement result of RT is plotted at Table 2 and the energy distribution pattern from RT parameters can be seen on contour map at Figure 2.

Table 2. Measurement result of reverberation time.

Category	Value (s)
Average	0.51
Max	0.82
Min	0.33
Standard deviation	0.10

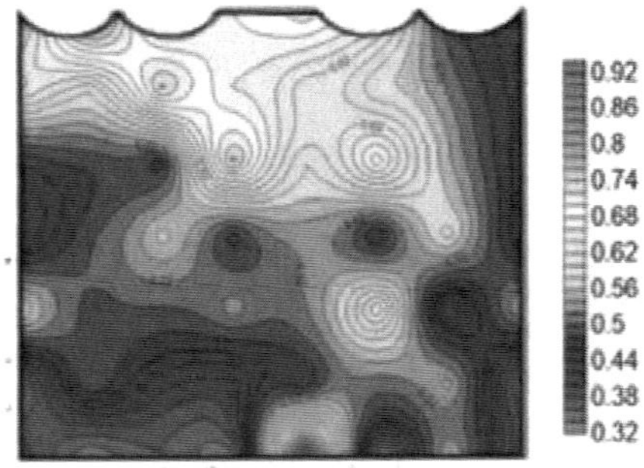

Figure 2. Contour map of reverberation time.

Comparing measurement result at **Table 2** with recommendation value, most area is categorized as 'dead room' and acoustically does not suitable condition for speech activities. According to Figure 2, the best positions for listening activity are at front near prayer leader, middle and rear area while the rest is inappropriate for listening activities.

4.3. Sound strength (G)

By definition, sound strength (G) is the comparison of received energy to energy of direct sound at 10 m from sound source. Whenever energy of direct sound is larger than total energy, G < 0 dB, it means acoustical characteristics tend to absorb energy. That is unexpected condition and every measurement points have to be larger than 0 dB [7]. Material surfaces of MSU is quite reflective, it is demonstrated by value of G larger than 0 dB. Illustration of G contour is at Figure 3.

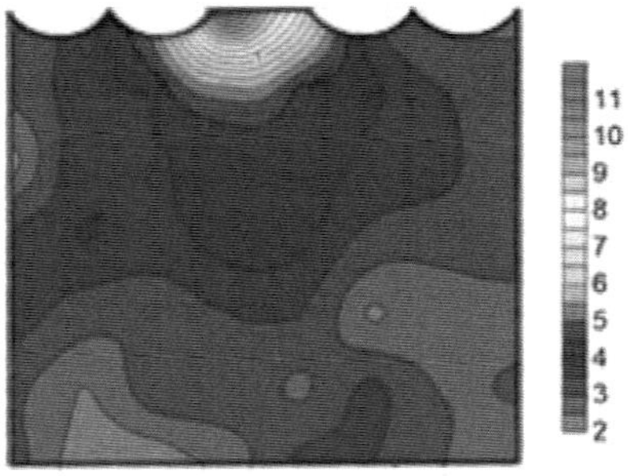

Figure 3. Contour map of sound strength.

4.4. Clarity (D50)

Clarity or D50 is defined as ratio of early reverberant energy after direct sound to total energy received [8]. For speech, energy of early reverberant in first 50 ms has significant impact to intelligibility. Table 3 shows the value of D50 parameter, representing statistical data from all measured points. There are positions with value larger than 75% but as most data is scattered around 19.97% from 24.24 %, it indicates that most area of main hall has 'good' category of speech intelligibility. The position of each value could be accessed at Figure 4.

Table 3. Measurement result of clarity.

Category	Value (%)
Average	24.24
Max	90.76
Min	5.32
Standard deviation	19.97

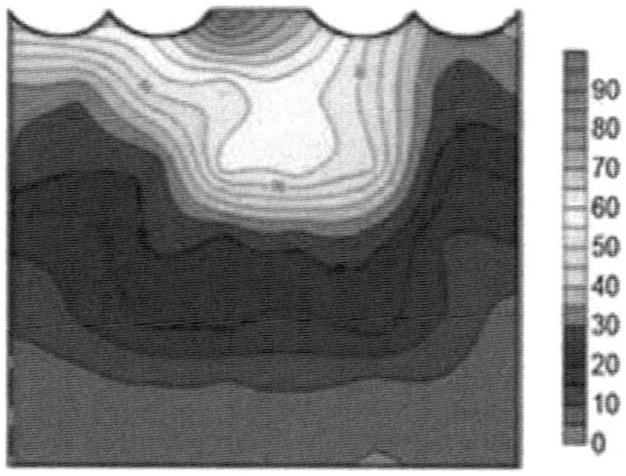

Figure 4. Contour map of clarity.

5. Conclusion

The conclusion of this research have found that MSU have improper acoustic condition to perform speech activities. The result of acoustic parameters measurement shows the average value of RT is 0.51 s, 24.24 % of D50 and G is larger than 0 dB in all receiver position. According to recommendation value, impulse response based parameters (RT and D50) do not fulfil the minimum threshold. Only certain position meets the requirements, especially at front row (near podium). However, there is good thing about acoustic performance at MSU is the boundary has sufficient characteristic to reflect acceptable energy across main hall. Also, background noise is very low so the change of sound masking is nearly zero.

References

[1] Abdou A A 2003 Measurement of acostical characteristics of mosques in Saudi Arabia *Acoustical Society of America* **113(3)** pp 1505-17
[2] Soegijanto 2001 *Penelitian Kinerja Akustik Masjid di Indonesia: Laporan Hasil Hibah Bersaing Perguruan Tinggi IX* Bandung
[3] ISO 3382-1 *Acoustic Measurement of room acoustic parameters*
[4] Brandly J S 2010 Review of objective room acoustics measures and future needs *Proceedings of the International Symposium on Room Acoustics, ISRA*
[5] Kuttruff H 1991 *Room Acoustic* (London: Elsevier)

IOP Conf. Series: Materials Science and Engineering **211** (2017) 012021 doi:10.1088/1757-899X/211/1/012021

[6] Kinsler L E 1982 *Fundamental of Acoustic* (New York: Palatino)
[7] Barron M 1981 *Spatial Impression due to Early Lateral Reflection in Concert Halls: The Derivation of a Physical Measure* J Sound Vib **77** pp 211-232
[8] Kasim H, Putra A, Nor M J M 2015 The Acoustical Characteristics Of The Sayyidina Abu Bakar Mosque, UTeM *Journal of Engineering Science and Technology* **10**, No. 1

Effect of air velocity and direction for indirect evaporative cooling in tropical area

Tri Ayodha Ajiwiguna*, Fadhlin Nugraha Rismi and Mukhammad Ramdlan Kirom

Engineering Physics Department, Telkom University, Jalan Telekomunikasi, Bandung, 40252, Indonesia.

*Email: triayodha@telkomuniversity.ac.id

Abstract. In this research, experimental study of heat absorption rate caused by indirect evaporative cooling is performed by varying the velocity and direction of air. The ambient is at average temperature and relative humidity of 28.7 °C and 78% respectively. The experiment is conducted by attaching wet medium on the top of material reference plate with the dimension of 14 x 8 cm with 5 mm thickness. To get evaporative cooling effect, the air flow is directed to the wet medium with velocity from 1.6 m/s to 3.4 m/s with the increment of 0.2 m/s. The direction of air is set 0° (parallel), 45° (inclined), and 90° (perpendicular) to the wet medium surface. While the experiment is being performed, the air temperature, top and bottom of plate temperature are measured simultaneously after steady state condition is established. Based on the measurement result, heat absorption is calculated by analysing the heat conduction on the material reference. The result shows that the heat absorption rate is increased by higher velocity. Perpendicular direction of air flow results the highest cooling capacity compared with other direction. The maximum heat absorption rate is achieved at 13.9 Watt with 3.4 m/s velocity and perpendicular direction of air.

1. Introduction

The largest energy consumption in building is on air conditioner sector which is estimated 40 to 50% of total energy consumption [1]. In tropical area, the air conditioner is mainly for cooling the spaces thus the thermal comfort is achieved. Currently, the most common technology for air conditioner is vapor compression refrigeration heat pump since the COP is very high and flexible to use. However, it still consumes large of electrical energy for compressor. One of the alternative cooling technologies is evaporative cooling system which has very large COP at 15-20 [2].

Evaporative cooling is the phenomena of simultaneous heat and mass transfer [3]. The concentration difference between water and air causes the evaporation rate. The heat transfer rate is required for evaporation so that temperature difference must exists and resulting the cooling effect. In tropical area, the average relative humidity is about 70 to 80%. To avoid the increasing of humidity, indirect evaporative cooling is more suitable in this condition.

The development of indirect evaporative cooling technologies and studies have been performed by researchers and engineers. I Maclaine-Cross *et al* studied the general theory of wet surface and use it for indirect evaporative cooling [4]. J. F. J Alonso *et al* developed the simulation model for indirect evaporative cooler [5]. N. J. Stoitchkov *et al* studied the effectiveness of cross flow plate heat exchanger for indirect evaporative cooling [6]. X. C. Guo *et al* analyze the parameter affecting the indirect evaporative coolers [7].

The aim of this study is to investigate the effect of velocity and direction of air on indirect evaporative cooling phenomena. The study is performed at unconditioned ambient tropical area. Heat absorption is then calculated from measured data of experiment. Comparison between experimental conditions are then compared.

2. Experimental setup and procedure

Figure 1 shows the scheme of experimental setup. The experiment was conducted in tropical area at 78% average relative humidity and 28.7 °C ambient temperature. The direction of air were varied from 0° (parallel), 45° (inclined), and 90° (perpendicular) to the wet medium surface. The experiment was set so that heat absorption equals to heat conduction at the material reference.

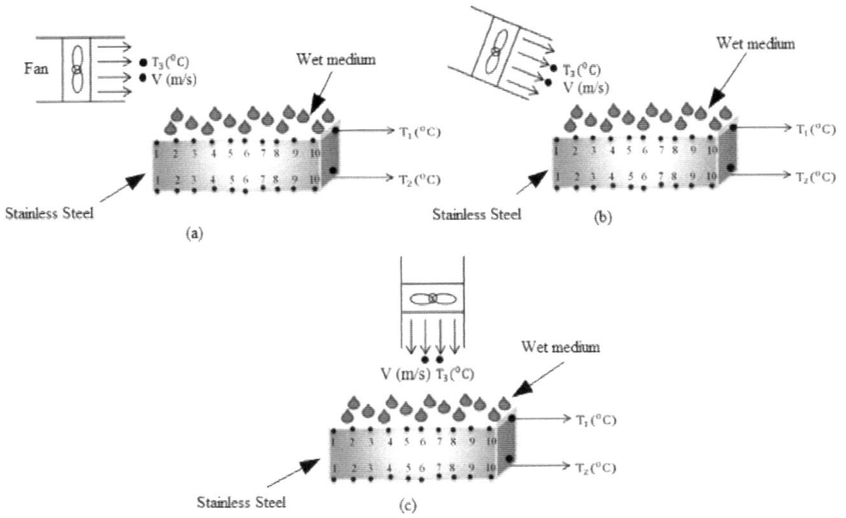

Figure 1. Experimental setup: (a) parallel flow (0°), (b) inclined flow (45°), (c) perpendicular flow (90°).

Stainless steel as material reference was used as base plate since its thermal conductivity is well known. The dimension of material reference is 14 x 8 cm with 5 mm thickness. Wet medium was attached on the top of base plate. The air was flown on the top of wet medium by using electrical fan to get the evaporative cooling effect on the surface. The voltage of fan was adjust so that air had velocity of 1.6 m/s to 3.4 m/s with the increment of 0.2 m/s. After the steady state condition had been established, T-type thermocouples were used to measure the temperature of top base plate (T_1), bottom base plate (T_2), ambient (T_3). To get the average temperature difference, T_2 and T_3 were measured at ten different locations (no. 1 to 10 at stainless steel) along the length of base plate.

Heat is absorbed from the bottom to the top of base by evaporative cooling. Since the thickness of base plate is very thin, so it is assumed that heat absorption equals to conduction heat transfer rate from the bottom to the top of base plate. Therefore, the heat absorption rate is calculated by equation (1):

$$Q_{abs} = \frac{kA(T_2 - T_1)}{L} \quad (1)$$

where *k*, *A*, and *L* are thermal conductivity, surface area, and the thickness of base plate respectively.

3. Result and discussion

After the steady state condition, the result shows that the top and the bottom of base plate temperature are lower than ambient temperature, shown in Figure 2. When the air flows on the top of wet medium, water contained in wet medium is evaporated and causes evaporative cooling effect [3]. The curve temperature of top and bottom base plate are relatively the same with the curve of ambient temperature with the offset of temperature difference. It means that ambient temperature is strongly effect the temperature resulted by evaporative cooling effect.

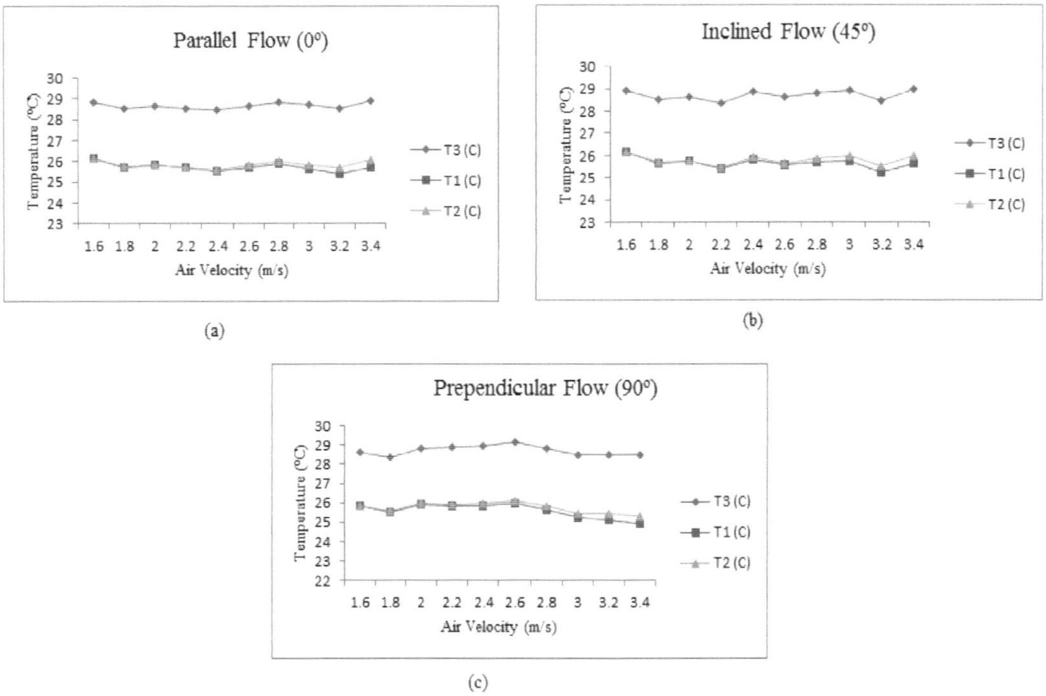

Figure 2. Temperature difference between ambient, top and bottom of hot plate for: (a) parallel, (b) inclined (45°), (c) and perpendicular flow.

The effect of air velocity and direction to the heat absorption of evaporative cooling is shown in Figure 3. Horizontal axis is related to the air velocity and vertical axis is related to heat absorption of evaporative cooling. Blue, red, and green lines show the direction of air for 0°, 45°, and 90°. The result shows that the higher velocity causes higher heat absorption. Meanwhile, the perpendicular direction causes the highest heat absorption compared with other direction. The combination condition of highest velocity and perpendicular direction results the highest heat absorption at 13.9 Watt.

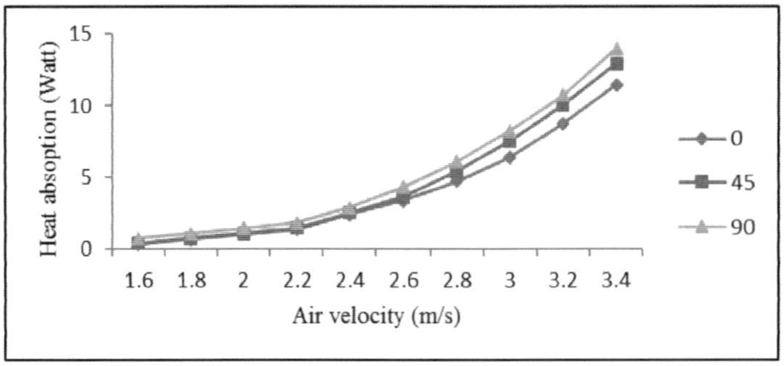

Figure 3. Effect of air velocity and direction to heat absorption.

The heat absorption is proportional to the evaporation rate of water contained in wet medium. Higher velocity with perpendicular direction causes the highest evaporation rate, therefore the heat absorption is higher as well.

4. Conclusion

The apparatus and experiment to investigate the effect of velocity and direction of air to the evaporative cooling effect in tropical area has been constructed and analysed in this study. The result shows that the higher cooling capacity is achieved at higher velocity with perpendicular direction. The velocity and direction of air affect the evaporation rate, so that evaporative cooling heat absorption is also affected. The maximum cooling capacity from this experiment is 14 W which is resulted by 3.4 m/s of air velocity with perpendicular direction to the wet medium surface. The same study using another method is recommended for further study.

References

[1] V Vakiloroaya, B Samali, A Fakhar, and K Pishghadam 2014 *A review of different strategies for HVAC energy saving Energy Conversion and Management* **77** pp 738-54

[2] Z Duan et al Indirect evaporative cooling *Past, present and future potentials Renewable and Sustainable Energy Reviews* **16** pp 6823-50

[3] Yunus A Cengel 2002 *Heat Transfer: A Practical Approach 2nd Edition Mcgraw Hill*

[4] I Maclaine-Cross and P Banks 1981 *A general theory of wet surface heat exchangers and its application to regenerative evaporative cooling* Journal of Heat Transfer **103** pp 579-85

[5] J F J Alonso, F J Martínez, E V Gómez, and M A Plasencia 1998 *Simulation model of an indirect evaporative cooler* Energy and Buildings **29** pp 23-27

[6] N J Stoitchkov and G I Dimitrov 1998 *Effectiveness of cross flow plate heat exchanger for indirect evaporative cooling* International Journal of Refrigeration **21** pp 463-71

[7] X C Guo and T S Zhao 1998 *A parametric study of an indirect evaporative air cooler* International Communications in Heat and Mass Transfer **25** pp 217-26

IOP Conf. Series: Materials Science and Engineering **211** (2017) 012023 doi:10.1088/1757-899X/211/1/012023

Heterogeneous carbon nano-tube window layer with higher sheet resistance improve the solar cell performance

Khomdram Jolson Singh[1]*, Thokchom Jayenta Singh[1], Dhanu Chettri[1] and Subir kumar Sarkar[2]

[1]Dept. of Electronics and Communication Engineering, Manipur Institute of Technology ,Imphal-795004,India
[2]Dept. of Electronics and Telecommunication Engineering, Jadavpur University ,Kolkata 700032,India

*Email:jolly4u2@rediffmail.com

Abstract. The heterogeneous Carbon Nano-Tube (CNT) layers deposited on the window surface of solar cells that allow better charge carrier collection was numerically analyzed and studied in modern TCAD tools. The quantum efficiency(EQE) as well as power conversion efficiency(η) were found to be improved significantly based on the light transmission capabilities of the CNT layer. Two CNT network models using experimental sheet resistance values of 75 Ω/□ and 128 Ω/□ as a top conducting layer in a GaAs solar cell were compared.It is found that the CNT networks allow for a greater area of charge collection as well as serve as a lower resistance path for charge carriers with minimum voltage loss to travel to the top contact. This model thus significantly improve the η up to 30% under AM0 with more than 90% EQE. The effect of cell width varying starting from 200 μm to 4000 μm with CNT top layer on J_{sc} , V_{oc} and P_{max} parameters were studied and found that these parameters remain almost constant irrespective of cell width. This work thus shown that a thin CNT top layers of lower sheet resistance with a higher light transmission can greatly improve the efficiency of solar cells.

1. Introduction

Carbon nanotubes (CNT) possess a wide range of direct band gaps matching the solar spectrum, strong photo absorption, from infrared to ultraviolet, and high carrier mobility, reduced carrier transport scattering, making them ideal photovoltaic material[1].Photovoltaic effect can be achieved in ideal single wall carbon nanotube (SWCNT) diodes. Individual SWCNTs can form ideal p-n junction diodes [2]. This discovery has influenced a wide research in the solar cell community regarding the use of CNTs as transparent conducting layers.This work focus on heterogeneous networks of SWCNTs as transparent conductors, which, when grown, produce a variety of semiconducting and metallic nanotubes[3]. It has been found that due to the somewhat random distribution of CNTs formed in heterogeneous networks, one third of the tubes are metallic, and the others semiconducting[4].All these discovery has influenced a wide research in the solar cell community regarding the use of CNTs as transparent conducting layers. The aim of this work was to model a heterogeneous CNT network in Silvaco ATLAS software and use the model to show the possible benefits of the use of CNT networks as a surface layer in all solar cells.

2. Numerical modelling method

It is well known that the metallic nanotubes are highly conductive, therefore they dominate the electrical characteristics of the layer of CNTs. In recent research, heterogeneous networks of CNTs have been simulated to find the conductance and sheet resistance using the Stick Percolation Model and again verified by experimental data in [5],[6].A thin layer of CNTs on the top layer of a solar cell has the same random orientation of conductors that have a net conductance based on the density of conducting tubes in the area. Higher densities of conducting tubes allow for higher conductance values. The model created in this design was based on parameters from widely accepted recently published experimental data on heterogeneous CNT networks

2.1 Modeling the material Properties of CNT

A heterogeneous CNT network would be modeled as a metal with the optical properties and resistivity of the CNT network. However, due to the limitations of ATLAS software, the CNT network in this work is modeled as a low bandgap semiconductor. We found that the ATLAS software has problems obtaining solutions when a metal makes contact with an electrode in some specified structure. Since the main aim of this work is to study the flow of charge carriers through a semi- transparent material, the use of a CNT layer as a semiconductor is optimal because the ATLAS software has no problem in obtaining solutions for all of the charge carriers moving towards the electrodes in semiconductors. Though a heterogeneous CNT network does not exhibit semiconductor properties, the material statements in the ATLAS code can be utilized to create a semiconductor with metallic properties. Here, 4-H Silicon Carbide (4H-SiC) was chosen as the semiconductor to alter based on the work of Brunton in the modeling of carbon nano-fiber interconnects in Silvaco Atlas software[7].The parameters that were changed in the 4H-SiC to model a CNT network are given in Table I.The electron and hole mobilities were modified to match the experimental sheet resistances of 75 Ω/□ and 128 Ω/□ of the material given in Table I. The resistivity of a semiconductor material and the sheet resistance is given by equation (1) as follows

$$\rho = \frac{1}{q(\mu_n n + \mu_p p)}, \qquad R_s = \frac{\rho}{t} \tag{1}$$

where q is the equivalent charge of one electron or hole, μ_n and μ_p are the electron and hole mobilities, respectively, μ_n and μ_p are the electron and hole concentrations in the material, and t is the thickness of the material

Table 1. The material parameters that were changed to give 4H-SiC metallic properties in Silvaco ATLAS.

Critical parameters	**ATLAS Identifier**	**Top CNT**	
		4H-SiC	
		128Ω/□	*75Ω/□*
Layer band gap E_g (eV)	EG	0.0	0.0
Electron affinity X_e (eV)	Affinity	5.8	5.8
Relative permittivity ε_r (F cm^{-1})	Permittivity	5.4	5.4
Electron mobility μ_n (cm^2/V s)	MUN	8138.02	13889
Hole mobility μ_p (cm^2/V s)	MUP	8138.02	13889
Conduction band effective density of states Nc (cm^{-3})	NC300	3×10^{17}	3×10^{17}
Valence band effective density of states Nv (cm^{-3})	NV300	3×10^{17}	3×10^{17}
Recombination	Taun0,taup0 Augn, Augp, Etrap,Copt etc.	0.0	0.0

2.2 Modeling the Optical Properties of CNT

Without altering the optical properties of 4H-SiC material model, firstly it was made to be completely transparent. Then,the input solar power AM0 in simulations involving a CNT network was altered to model the resultant light transmitting through the material while leaving the CNT material completely transparent. To achieve this, the transmission data from the Institute for Microstructural Science of the National Research Council of Canada[8] as shown in Figure 1 was used to apply a series of scale factors to alter the AM0 spectrum at different wavelengths[8].The AM0 spectrum compared with the input spectrums of associated with heterogeneous CNT layers of 75Ω/□ and 128Ω/□ of sheet resistance are shown in Figure 2

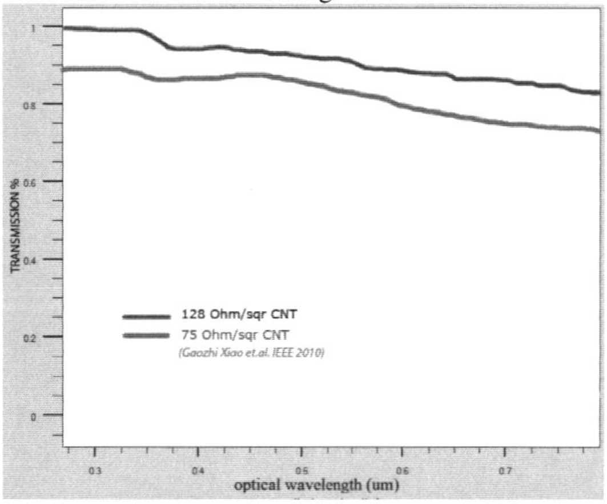

Figure 1. The experimental light transmission data of a single-dip coated and doubledipcoated CNT network show an inverse relationship between sheetresistance and light transmission. From [1].

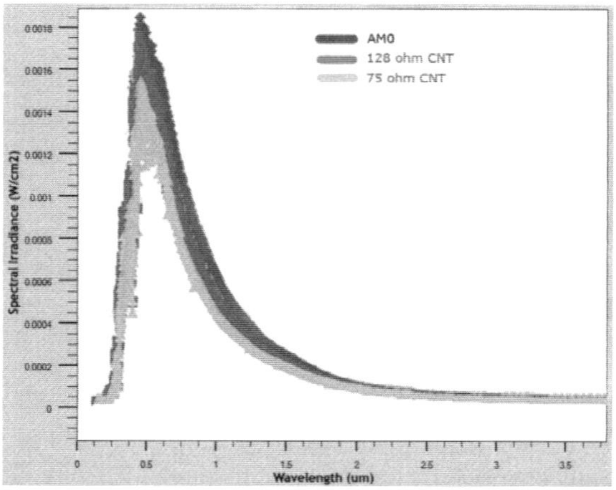

Figure 2. The AM0 spectrum is plotted against the spectrums used inin simulation of CNT networks with 128Ω/□ and 75Ω/□ of sheet resistance.

3. Modelling CNT based solar cell structure

For the CNT/GaAs heterostructure system, the drift-diffusion equation and current continuity equation (2) are shown as follows [9]:

$$\left.\begin{aligned} J_n &= -qn\mu_n \nabla V + qD_n \nabla n \\ J_p &= -qn\mu_p \nabla V - qDp\nabla p \\ &\frac{1}{q}\nabla . J_n - R_n + G_n = 0 \\ &-\frac{1}{q}\nabla . J_p - R_p + G_p = 0 \end{aligned}\right\} \quad (2)$$

J_n, μ_n , D_n are the electron current density, the electron mobility and the electron diffusion coefficient in GaAs respectively. J_p, μ_p, D_p are the hole current density, the hole mobility and the hole diffusion coefficient in GaAs respectively. G_n, G_p are the electron and hole generation rate due to light illumination, R_n and R_p are the electron and hole recombination rate respectively. The above equations are solved self-consistently by finite difference method until converged results are obtained. ATLAS permits the user to identify a variety of physics models for calculating carrier mobility and recombination. In our design, we used the following models for our analysis. The Concentration-Dependent Low Field Mobility model (CONMOB) was used to model the doping-dependent low-field mobilities of electrons and holes in AlGaAs/GaAs cell at 300K.The recombination models utilized were the Optical Recombination (OPTR) and the Shockley-Read-Hall (SRH) recombination models. SRH recombination model takes into account the electrons being emitted or captured by donor and acceptor- like traps. The OPTR determines the possibility that a photon is generated when an electronand hole recombine. Green has shown that the OTPR model increase the accuracy of the solar cell simulation [10].Thus we found that ATLAS solves the Poisson, the continuity and current density equations including SRH and AUGER (Auger recombination) mechanisms. The simulation further takes into account the bandgap narrowing effect (BGN), temperature and doping dependent mobility models for the light propagation and photogeneration use the TMM approach. We incorporated SRH FERMI CONMOB OPTR AUGER BGN model in our design shown in Figure 3 and Figure 4.

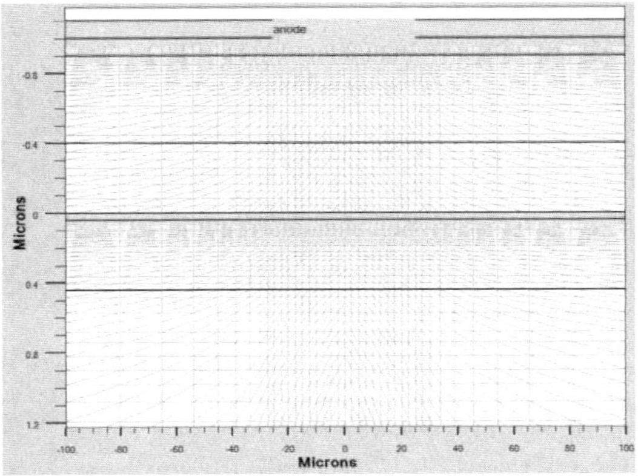

Figure 3. Generated finer mesh structure of the model for accurate realization.

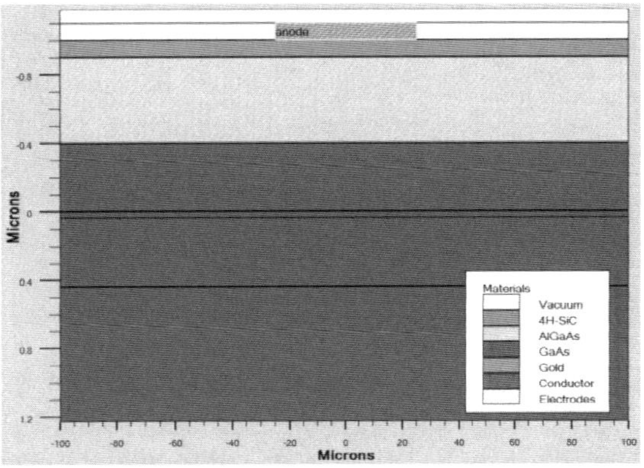

Figure 4. Final realization of the single junction GaAs cell with top CNT Layer.

4. Result Analysis and discussion

It is necessary to determine the different performance parameter of the design in order to optimize it. The I-V curve of a solar cell is the superposition of the IV curve of the solar cell diode in the dark with the light-generated current [11].The following equations(3),(4),(5),(6) and (7) are used to calculate Isc, Voc, Pmax, FF and Efficiency respectively as follows

$$I_{sc} = I_0\left[\exp\left(\frac{qV}{nKT}\right) - 1\right] \tag{3}$$

$$V_{oc} = \frac{nKT}{q}\ln\left(\frac{I_L}{I_0} + 1\right) \tag{4}$$

$$P_{\max} = V_{OC} I_{SC} FF \tag{5}$$

$$FF = \frac{V_{OC} - Ln(V_{OC} + 0.72)}{V_{OC} + 1} = \frac{I_m V_m}{I_{sc} V_{oc}} \tag{6}$$

$$\eta = \frac{V_{OC} I_{SC} FF}{P_{in}} = \frac{P_{\max}[W] \times 100}{1000[Wm^{-2}] \times CellArea[m^2]} \tag{7}$$

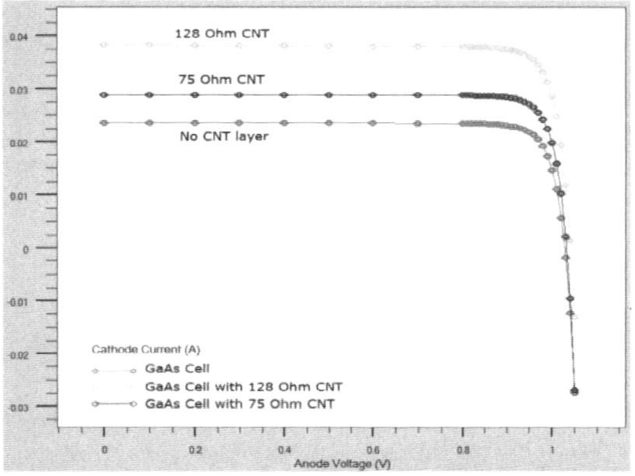

Figure 5. The I-V curves for the standard GaAs solar cell and GaAs solar cells.

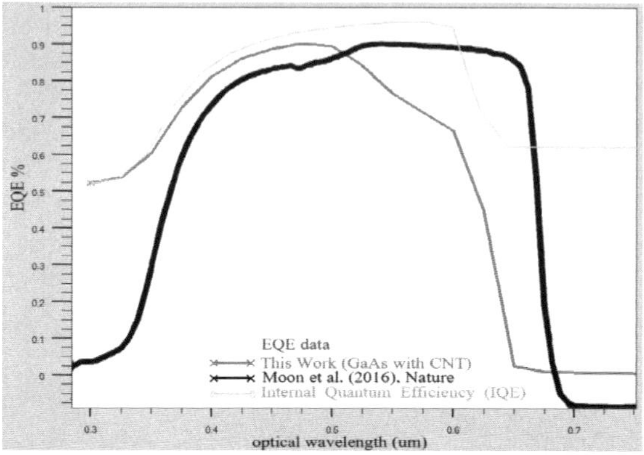

Figure 6. The EQE curves for the GaAs solar cells having top CNT layers of 128 Ω sheet resistance compared with other published experimental data.

The I-V curves for the normal GaAs solar cell and a cell including a 128 Ω/□ and 75 Ω/□ sheet resistance CNT layer are compared as shown in Figure 5. As expected with an optimal cell width, the cells with the CNT layers have higher current values than the standard GaAs cell due to the greater charge collection of the CNT layers. A comparison of solar cell performance parametersis given in Table 2 including from Experimental Data[12]. The numerical External Quantum Efficiency(EQE) given in Figure 6 imitates well the EQE of a single junction GaAs solar cell with 90% efficiency as compared with other experimental data[12]

Table 2. Comparison of cell performance parameter with and without CNT.

Cell	Isc	Voc	Vm	Im	Pm	FF	Eff
128 Ω/□ CNT	33.23	1.044	0.94	32.46	34.12	86.43	**29.18**
75 Ω/□ CNT	28.45	1.032	0.93	27.37	33.05	87.02	26.04
GaAs only	27.31	1.012	0.90	25.42	32.01	84.91	23.32
Moon *et al.* (2016) [12]	27.06	0.981	-	-	-	83.35	22.08

Again,the cell width was varied from 200 to 4000 microns, and the contact width was set at 50 microns constant for all simulations. The maximum power values for each solar cell were then obtained and compared for each cell width and are shown in Figure 8. Each cell reached a maximum P_{max} at a certain width and then showed reductions in P_{max} for greater widths. The most pronounced decline in P_{max} for greater cell widths was seen in the GaAs solar cell without a CNT layer. This sharp decline was due to the voltage losses from resistance in the AlGaAs and GaAs layers as charge carriers had to travel further to the contact. Since the CNT layers provided charge carries in the solar cell with CNT charge collector layer with lower resistance paths to the contact, the reduction in P_{max} for greater cell widths was very less pronounced as shown in Fig.8.The current density in a GaAs solar cell with a top CNT layer of 128 Ω/□ of sheet resistance is shown in Figure 7. The current converges in the top CNT layer to travel to the top contact for charge collection. The charge carriers in this region see less net resistance in the more highly conducting CNT layer with greater generation of electron current density.

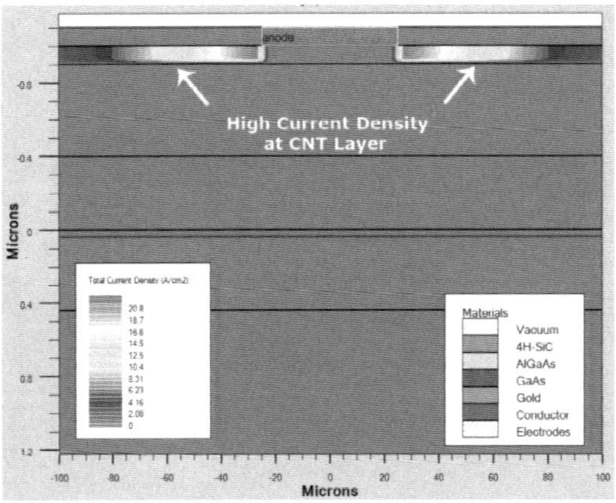

Figure 7. The high current density in CNT layer of a GaAs solar cell with a 128 Ω/□ sheet resistance layer of CNT on the surface.

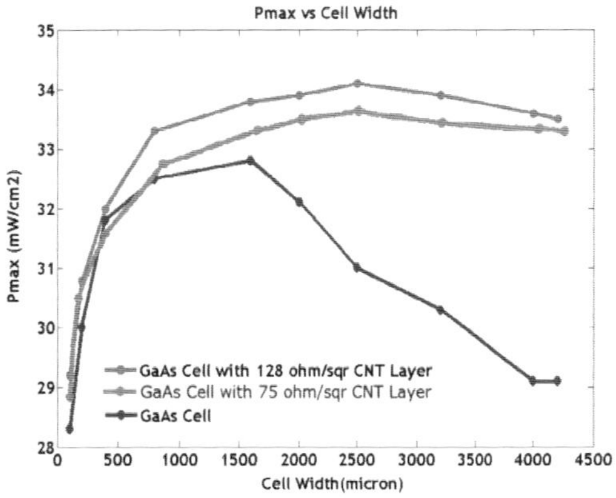

Figure 8. The maximum power obtained at each cell width simulated is plotted for a GaAs solar cell with 128 Ω/□ sheet resistance CNT layer and GaAs solar cell.

5. Conclusion

The unique properties of CNT layers on the surface of solar cells allow better charge carrier collection while only losing a portion of light input power based on the light transmission of the CNT layer. Although our cell is p-on-n structure, the addition of this CNT layer effectively increases the efficiency from the other reported similar cell without CNT. The validation of our numerically modelled and simulated cell with other good experimental data can conclude the reliability and accuracy of our method used in the design.Thus, this work can conclude that CNT layer can act as semi-transparent charge collector for better efficiency in solar cell. The method can be used for future development and research in the field of CNT based solar cell.

References

[1] J Guo 2004 *Carbon Nanotube Electronics, Modeling, Physics, and Applications* Doctoral Dissertation Purdue University

[2] L Ji Ung 2003 Photovoltaic effect in ideal carbon nanotube diodes, Appl. Phys J Lett **87(3)** 073101

[3] A Alam, M Chowalla and N Pimparkar 2008 *Device Optimization for OrganicPhotovoltaics with CNT Networks as Transparent Electode* Proc 33rd IEEE Photovoltaic Specialists Conference pp1–4, 11–16

[4] M S Dresselhaus, G Dresselhaus and R Saito 1995 *Physics of Carbon Nanotubes* Carbon vol 33 no 7 pp 883–91

[5] A Alam, M Chowalla and N Pimparkar 2008 *Device Optimization for OrganicPhotovoltaics with CNT Networks as Transparent Electode* Proc 33rd IEEE Photovoltaic Specialists Conference pp1–4, 11–16

[6] D Stauffer and A Aharony 1992 *Introduction to percolation Theory* Taylor and Francis: London

[7] J Brunton 2011*TCAD Analysis of Heating and Maximum Current Density inCarbon Nanofiber Interconnects* Master's Thesis Naval Postgraduate School

[8] G Xiao, Y Tao, J Lu and Z. Zhang 2010 *Highly transparent and conductive carbon nanotube coatings deposited on flexible polymer substrate by solution method* Proc 3rd International Nanoelectronics Conference (INEC) pp 208–9, 3–8

[9] Singh, Khomdram Jolson, N Basanta Singh, and Subir Kumar Sarkar 2015 *Textured window with DLAR coating design for an effective minimization of electrical and optical losses in an efficient III–V solar cell* Journal of Computational Electronics **14.1**

[10] SILVACO 2013 Data Systems Inc: Silvaco ATLAS User's Manual

[11] Lindholm, FA, Fossum, J G, Burgess and E L 1979 *Application of the superposition principle to solar-cell analysis* IEEE Trans Electron Devices **26(3)** pp 165–171
[12] Moon, Sunghyun, et al 2016 *Highly efficient single-junction GaAs thin-film solar cell on flexible substrate* Scientific Reports 6

IOP Conf. Series: Materials Science and Engineering **211** (2017) 012024 doi:10.1088/1757-899X/211/1/012024

A performance optimization and analysis of graphene based schottky barrier GaAs solar cell

Khomdram Jolson Singh[1]*, Dhanu Chettri[1], Thokchom Jayenta Singh[1], Terirama Thingujam[2] and Subir kumar Sarkar[3]

[1]Dept. of Electronics and Communication Engineering, Manipur Institute of Technology, Imphal-795004, India
[2]Dept. of Electronics and Communication Engineering, NIT, Manipur, India
[3]Dept. of ETC, Jadavpur University, Kolkata 700032, India

*E-mail: jolly4u2@rediffmail.com

Abstract. Performance optimization of Graphene-GaAs schottky barrier solar cell have been performed by considering variables such as substrate thickness, Graphene thickness, dependence between graphene work function and transmittance. The optimized parameter was extensively used to numerically model the design using TCAD Atlas. The results show the enhanced performance of the design with the optimized thickness of Graphene (0.3μm) and GaAs (10μm), resulting in significant increase in power conversion efficiency from 0.732% to 2.581% and reasonable fill factor up to 70%. It was further analysed that maximum potential was developed in the vicinity of the anode, which results in better charge collection hence improving the overall performance of the solar cell. The results are validated with the reported experimental work.

1. Introduction

The widely used Indium tin oxide (ITO) and fluorine tin oxide (FTO) as window electrodes in optoelectronic devices appear to be increasingly problematic mainly due to the limited availability of indium. Thus, a new unique material with a single atomic layer of carbon hexagon, Graphene have attracted lots of research in the last few years. This transparent single atomic layer has great potential application to replace high cost ITO and FTO as a window layer in solar cell [1]. The flexibility, robustness, environmental stability and optical transparency of Graphene with high mobility and near zero bandgap [2] is believed to have a great potential in optoelectronics and photonics application. Few layer graphene and single layer graphene have been produced by Chemical vapor deposition (CVD) [3]. Graphene sheets have also been produced by mechanical exfoliation via repeated peeling of highly ordered pyrolytic graphite (HOPG) or by chemical oxidation of graphite [4]. Graphene based solar cells have been recently fabricated on various substrate such as Si [5], CdS [6] and GaAs [7]. Wenjing et al. successfully transferred single and bilayer graphene on n-type GaAs to form Schottky junction yielding a power conversion efficiency of 1.95% [8]. GaAs in comparison with most commonly used Si Substrate, GaAs being highly resistant to radiation [9] and having direct band gap [10] makes it suitable for highly efficient solar cell both for terrestrial and space application, but it is necessary to study the band parameter and optimize various thickness of the structure of solar cell to make it more efficient. Hence, in this paper we have optimized the thickness of the GaAs substrate with graphene layer in SILVACO TCAD, and the result obtained have been validated with published experimental data.

2. Optimization of GaAs and graphene thickness

The illuminated AM 1.5 spectrum is plotted using Silvaco Atlas in Figure 1, which shows the wavelength corresponding to the peak solar spectrum is 0.473 um, this value has been precisely extracted using HP4151 Tonyplot tool in Silvaco, hence the absorption coefficient of GaAs corresponding to the wavelength is approximately calculated as 152086.11014706021 (cm^{-1}).

The Beer-Lambert's equation $\frac{dI_x}{I_x} = -\sigma N dx$, where I_x intensity of the incoming wave in x direction, σ is the lattice vibration/absorption per unit area, which gives the equation (1)

$$I = I_0 e^{-\alpha t} \tag{1}$$

where I is the intensity of the transmitted wave and I_0 is the intensity of the incoming light, and t is the thickness of the material in which the light is illuminated, obtaining a fitness function using equation (1), the different optimized GaAs thickness is plotted in Figure 2 for different transmittance of graphene. Kostya S. et al [11] gives experimentally obtained the transmittance and thickness for different number of graphene layer, using this experimental value and curve fitting, Figure 3 gives the plot of different transmittance of graphene layer with corresponding graphene thickness.

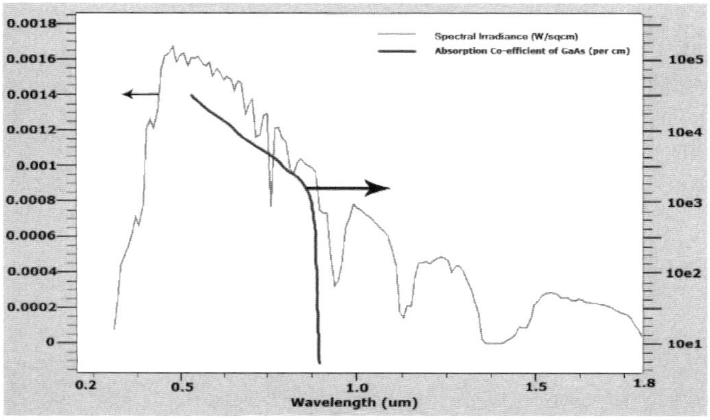

Figure 1. The illuminated AM 1.5 spectral irradiance varying with the wavelength, absorption coefficient of GaAs and wavelength.

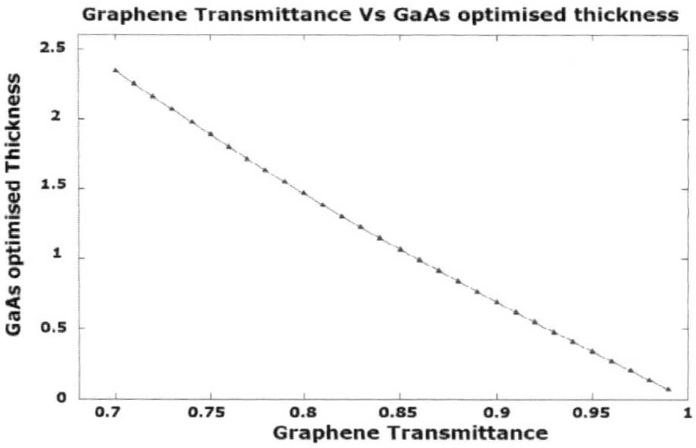

Figure 2. Graphene transmittance vs GaAs thickness.

Using this values from published experimental data in Figure 3 and the calculated optimized GaAs thickness in Figure 2, we obtain a noble optimized thickness of the GaAs corresponding to different thickness of graphene which is given in Figure 4, which can be used for modeling of optimized structure of Graphene-GaAs solar cell.

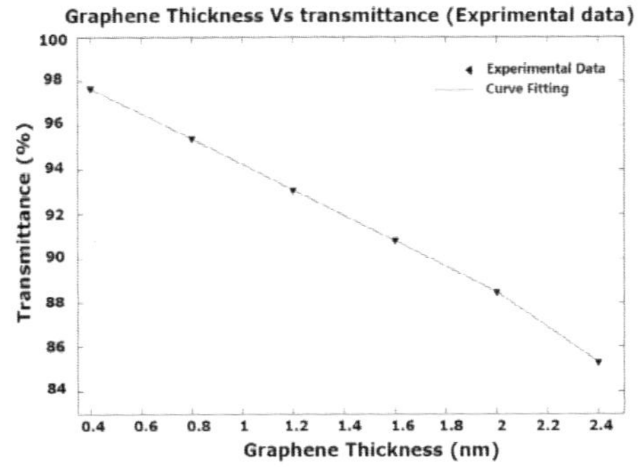

Figure 3. Plot of different transmittance of graphene for different thickness of graphene.

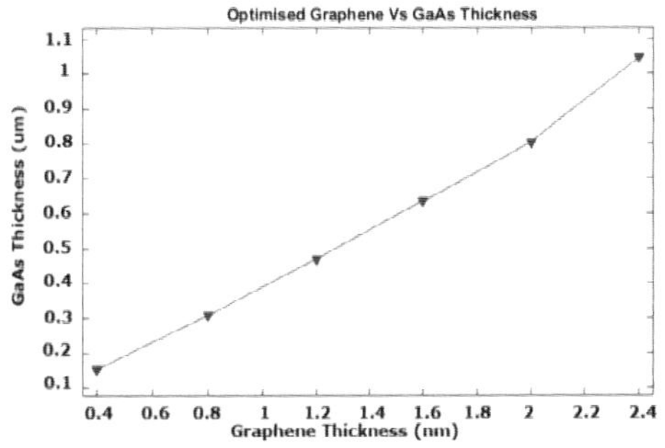

Figure 4. Optimised GaAs thickness for different layer of graphene.

3. Device structure and modelling

Figure 5 shows the proposed structure of the graphene, the device consists of three regions. The top area is covered with multilayer graphene followed by SiO_2 window layer which have been grown upon GaAs substrate. The first step in the modelling of device in any TCAD is the definition of mesh, the mesh must be defined in such a manner that it doesn't not take much unnecessary simulation time at the same time it must give high accuracy, thus Figure 6 shows the meshing of the modelled structure, it can be seen that the density of mesh is high in the top portion of the device as it consists of Graphene GaAs junction and also Graphene, SiO_2 junction. Any semiconductor could form schottky junction with certain metal provided that the semiconductor has a moderate carrier density and the difference of workfunction between the metal and the semiconductor is big enough which would give rise to a schottky barrier.

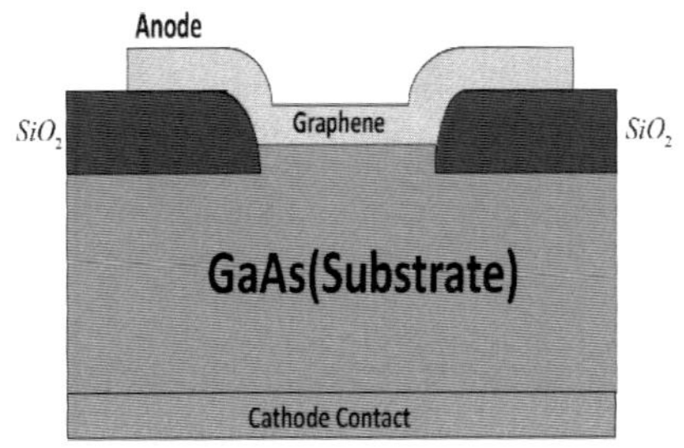

Figure 5. Structure of the proposed schottky barrier graphene GaAs solar cell.

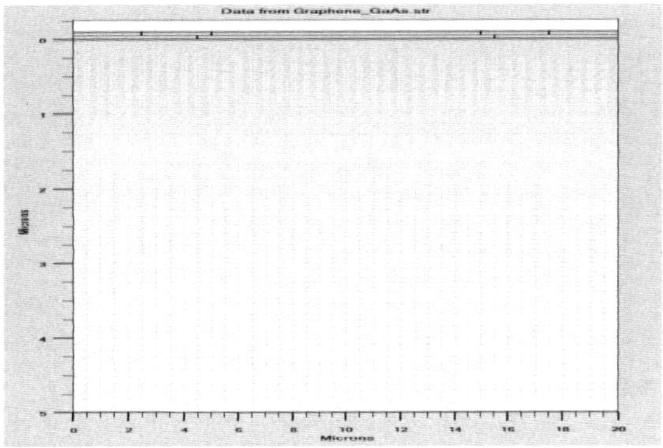

Figure 6. Generated finer mesh model of the design.

The I-V characteristics of the schottky-barrier solar cell is given by [12] which is expressed in equation (2)

$$I_o = \alpha A**T^2 \exp(-e\phi_{Bn}/kT) \quad (2)$$

where α is the cell area, A** the effective Richardson constant (which depends on the current –transport mechanism considered), and ϕ_{Bn} the metal-semiconductor (taken here to be n-type) barrier height as expressed in equation (3). T is absolute temperature and B is the Boltzmann's constant. And barrier height for p-type semiconductor ϕ_{Bp} is expressed in equation (4)

$$\phi_{Bn} = \phi_G - \chi \text{, for an n-type semiconductor} \quad (3)$$

$$\text{And, } \phi_{Bp} = E_G - \phi_G + \chi \text{, for p-type semiconductor} \quad (4)$$

where, ϕ_G is the workfunction of graphene, χ is the electron affinity of the semiconductor and E_G is the semiconductor energy gap. The major parameters used in the modelling and simulation in the Graphene GaAs are given in Table 1, which have been obtained from various literature reviews, when sunlight is illuminated in the Graphene GaAs solar cell, electron hole pair are formed this electron hole pairs are separated by the built in electric field at the schottky junction, thus electrons are collected at the electrode, Figure 7 shows the energy band diagram and the photovoltaic action of Graphene n-GaAs solar cell.

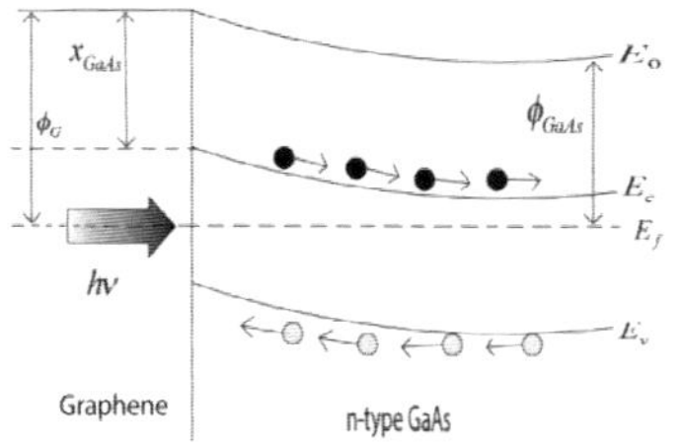

Figure 7. Energy band diagram of graphene n-GaAs solar cell.

Table 1. Major parameter for Graphene GaAs solar cell.

Materials	GaAs
Band gap Eg (eV) at 300 K	1.42
Lattice constant α (Å)	5.653
Affinity (eV)	4.07
e-mobility, MUN ($cm^2/V \times s$)	8000
h+ mobility, MUP ($cm^2/V \times s$)	400
Conduction band effective Density of state, N_c (cm^{-3})	4.35×10^{17}
Valance band effective Density of state, N_v (cm^{-3})	8.16×10^{18}
Nature of Energy Gap	Direct

Since Graphene is a new material, it is currently unavailable in the material library of SILVACO Atlas, thus in order to create an acceptable and accurate model of Graphene layer, initially Graphene layer have been defined with material as 4H-SiC and then the material properties of 4H-SiC were modified to give the material metallic characteristics while also matching experimental sheet resistance values [13]. Without altering the optical properties of 4H-SiC material model, firstly it was made to be completely transparent, than the transmittance of the Graphene is input by constructing an .nk file for graphene using the experimental value from [14]. Thus, graphene is modeled as a semimetal, with carrier mobility of 15,000 cm^2/Vs. The carrier densities were calculated by Fermi distribution and by adjusting the values of effective masses, a thickness of 10 nm, and band gap such that the carrier densities agree with experimental results.

4. Results and discussion

4.1. Effects of GaAs substrate thickness

The modelled device was illuminated with AM1.5G solar spectrum to simulate the global terrestrial sunlight, the LUMINIOUS 3D takes into account for the simulation of sunlight in SILVACO Atlas. The GaAs crystal substrate is acts as the absorption area hence it can be seen in that the photogeneration is maximum on GaAs as shown by the contours in Figure 3, on the other hand Graphene layer acts as a transparent electrode, for the intensity of photogenerated carriers and the workfunction difference between Graphene-GaAs junction is the built in Schottky junction. Figures 8 and 9 show the photogeneration rates of GaAs solar cell under AM1.5 illumination with depth for absorption layer were maintained at 5μm, 10μm respectively. The light could be absorbed in barrier layer and inside the semiconductor. The photogeneration rate is given by $G = \eta_0 \dfrac{P\lambda}{hc} \alpha e^{-\alpha y}$, where P is the cumulative effect of reflection, transmission and loos due to the absorption over the ray path, y the relative distance for the given ray, h the planck's constant, λ the wavelength, c the speed of light and α the absorption coefficient calculated for each set of (n,k) values, η_0 internal quantum efficiency, which represents number of carrier generated per photon per pair. Both results show the most effective absorption area located in the surface attachment about 0.1 μm. For a graphene thickness of 0.5 μm the optimized GaAs thickness is approximately 10 μm, hence Figure 9, which is the optimized structure shows batter photogeneration rate which validates our optimization data.

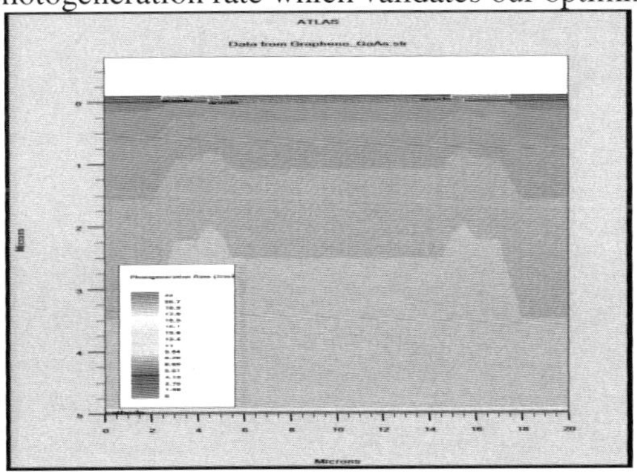

Figure 8. Photogeneration rates in GaAs cell of 5μm.

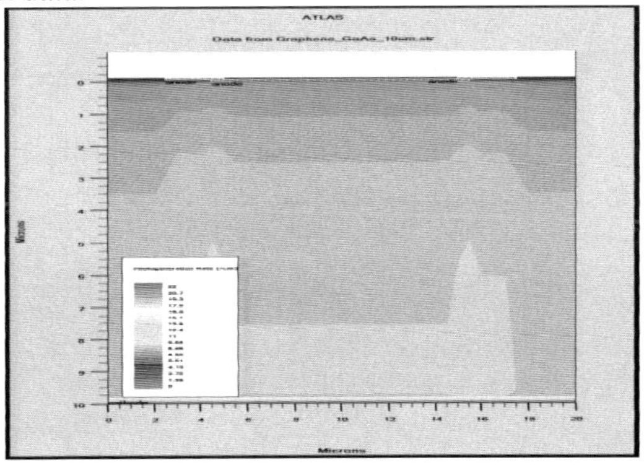

Figure 9. Photogeneration rates in GaAs cell of 10 μm.

4.2. Analysis of I_{SC}, V_{OC}

Short circuit current (I_{sc}) & open circuit voltage (V_{oc}) are very important parameter in the analysis of the performance of the solar cell as the output power of the solar cell depends on these two factors, I_{sc} is the maximum current that can flow, whereas the maximum voltage in open circuit is called the V_{oc}. The expression for open circuit voltage is given equation (5)

$$V_{oc} = \frac{kT}{q} \ln\left[\frac{q r_g \phi_B + J_0(1 + r_R \beta)}{J_0(1 + r_R \beta)}\right] \quad (5)$$

where r_g is the generation enhancement ratio, and the field factor can be calculated using equation (6) as

$$FF = \frac{V_{OC} - Ln(V_{OC} + 0.72)}{V_{OC} + 1} \tag{6}$$

Table 2 gives the output result of the modelled solar cell for different GaAs substrate thickness, it must be noted that for the substrate thickness of 10 um, maximum Jsc, Voc and FF is obtained, giving maximum efficiency of 2.581%. Figure 10 shows the overlay of I-V curve of various GaAs thickness for 0.3 μm graphene thickness. The potential developed is shown in Figure 11, which shows that maximum potential is developed at the anode vicinity, which implies in better charge collection.

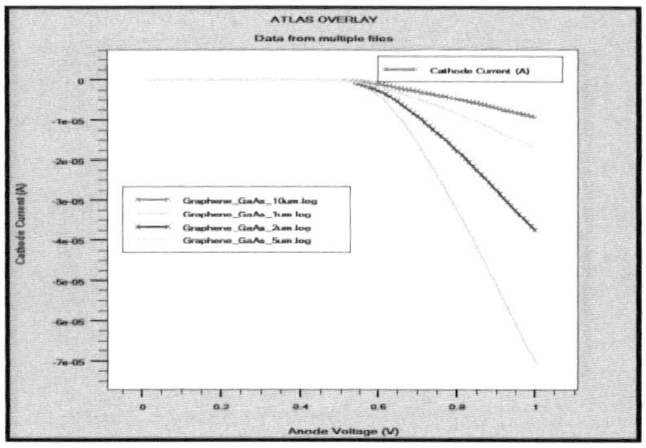

Figure 10. Overlay of I-V plot of modelled Graphene solar cell with different thickness.

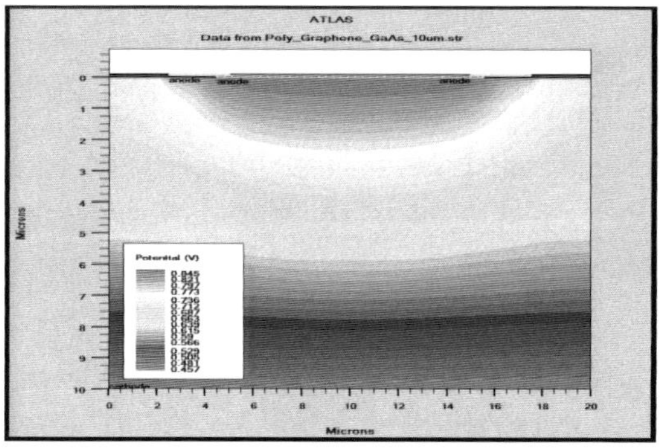

Figure 11. Potential developed at the optimized modelled Graphene-GaAs solar cell.

Table 2. Performance output from various GaAs substrate thickness cell.

GaAs substrate thickness	Short circuit current, J_{SC} (mA/cm^2)	Open circuit Voltage, V_{OC}(V)	Field Factor, *FF (%)*
1 um	1.555e-009	0.337566	63.9252
2 um	1.917e-009	0.351637	65.9488
5 um	2.091e-009	0.356200	69.0916
10 um	2.141e-009	0.357314	69.7376

5. Conclusion

A novel optimization has been performed using result from published experimental data and generating fitness function for the thickness of GaAs and Graphene in schottky barrier graphene GaAs solar cell. These optimised values have used for numerically modelling of the solar cell in SILVACO TCAD Atlas, and it was verified that the performance of the solar cell increases significantly with the obtained optimised parameter. It was further analysed that maximum potential was developed in the vicinity of the anode, which results in better charge collection hence improving the overall performance of the solar cell. This remarkable optimization technique as well as data can be used in the study, design and development of different structure of graphene based solar cell.

References

[1] Wang Xuan, Linjie Zhi and Klaus Müllen 2008 *Transparent, conductive graphene electrodes for dye-sensitized solar cells* Nano letters **8.1** pp 323-7

[2] Bonaccorso, Francesco, et al 2010 *Graphene photonics and optoelectronics* Nature photonics **4.9** pp 611-22
[3] Reina, Alfonso, et al 2008 *Large area, few-layer graphene films on arbitrary substrates by chemical vapor deposition* Nano letters **9.1** pp 30-5
[4] Schlatmann A R, Floet D W, Hilberer A, Garten F, Smulders P J M, Klapwijk T M and Hadziioannou G 1996 *Indium contamination from the indium–tin–oxide electrode in polymer light-emitting diodes* Applied physics letters **69(12)** pp 1764-6
[5] Miao, Xiaochang, et al 2012 *High efficiency graphene solar cells by chemical doping* Nano letters **12.6** pp 2745-50
[6] Lei Yun, et al 2014 *Hydrothermal synthesis of graphene–CdS composites with improved photoelectric characteristics* Journal of Materials Science: Materials in Electronics **25.7**
[7] Luo, Lin-Bao, et al 2014 *Near-Infrared Light Photovoltaic Detector Based on GaAs Nanocone Array/Monolayer Graphene Schottky Junction* Advanced Functional Materials **24.19**
[8] Jie W, Zheng F and Hao J 2013 *Graphene/gallium arsenide-based Schottky junction solar cells* Applied Physics Letters 103(23):233111
[9] Aukerman L W, et al 1963 *Radiation effects in GaAs* Journal of Applied Physics **34.12**
[10] Döhler, G H, et al 1981 *Observation of tunable band gap and two-dimensional subbands in a novel GaAs superlattice* Physical Review Letters **47.12** p 864
[11] Novoselov, Kostya S, et al 2004 *Electric field effect in atomically thin carbon films* science **306.5696** pp 666-9
[12] Pulfrey, D L, and R F McOuat 1974 *Schottky-barrier solar-cell calculations* Applied Physics Letters **24.4** pp 167-9
[13] Nair, Rahul Raveendran, et al 2008 *Fine structure constant defines visual transparency of graphene* Science **320.5881** p 1308
[14] Weber, J W, Calado, V E, Van de Sanden and M C M 2010 *Optical constants of graphene measured by spectroscopic ellipsometry* Applied Physics Letters **97 (9)**

Influence of oxygen flow rate on structural, optical and electrical properties of copper oxide thin films prepared by reactive magnetron sputtering

Thitinai Gaewdang* and Ngamnit Wongcharoen

Physics Department, King Mongkut's Institute of Technology Ladkrabang, Bangkok 10520, Thailand

*Email: thitinai.ga@kmitl.ac.th

Abstract. In this research, copper oxide thin films were prepared by reactive dc magnetron sputtering method on glass substrates with oxygen flow rate in the range of 1-10 sccm. From XRD patterns, formation of Cu_2O cubic structure or CuO monoclinic structure was controlled by adjusting oxygen flow rate. Nanocrystallite size of the as-grown films was observed by AFM. From transmittance spectra, direct energy gap varied between 1.97 and 2.55 eV. Electrical conductivity and Hall effect measurements were performed on the films with van der Pauw configuration. The positive sign of the Hall coefficient confirmed the p-type conductivity in all studied films. Important electrical parameters of films as a function of oxygen flow rate were observed. With low resistivity and high mobility values, the films prepared at oxygen flow rate of 8 sccm are identified as suitable candidates for fabrication as absorber layer in solar cell devices.

1. Introduction

Oxides of copper are well-known to show p-type conductivity and have many advantages for catalysts, chemical sensors and solar energy conversion application. Non-toxic, economic, abundant availability and relatively simple formation of oxide make copper oxide as an interesting material [1]. Two common forms of copper oxide are cuprous oxide (Cu_2O) and cupric oxide (CuO). A meta-stable copper oxide, Cu_4O_3, which is an intermediate compound between the previous two, has also reported [1]. Cu_2O and CuO belong to cubic and monoclinic structures with band gap of 2.1-2.6 eV and 1.9-2.1 eV, respectively [2]. It has been demonstrated by varying the deposition conditions, films phase such as Cu_2O, CuO and Cu_4O_3 can be obtained and this also leads to different stoichiometry for a given phase. The optical transmittance and electrical resistivity of the films are two critical parameters in evaluating the performance of photovoltaic cells and electrochromic devices. A variety of deposition techniques have been employed for the growth copper oxide films. Among these thin films deposition techniques, reactive magnetron sputtering is one of the industrially practiced techniques for preparing of uniform films on large area substrates with required chemical composition. In this paper, copper oxide thin films were obtained through reactive dc magnetron sputtering on glass substrates with different oxygen flow rates. The physical properties of the films were studied in order to find the possibility to use them in p-n junction low cost thin film solar cells.

2. Materials and methods

The copper oxide films were deposited by reactive dc magnetron sputtering from a high purity Cu target of 99.99% with diameter of 50 mm and thickness 3 mm. Oxygen and argon were used a reactive gas and working gas, respectively. In order to study the effect of oxygen flow rate and eliminate the action of sputtering pressure, the sputtering pressure was kept at 0.73 Pa, with a variation of oxygen flow rate in the range of 1-10 sccm and an argon flow rate in the range of 12-32 sccm. A diffusion pump was used to achieve a base pressure of 6.90×10^{-3} Pa. The substrate temperature during deposition was not intentional controlled. Target to substrate distance was kept constant at 100 mm. Sputtering time was 30 min. The crystal structure of these films was checked by X-ray diffraction technique with a Brucker D 8 diffractometer using CuK_{α} radiation. Surface morphology was examined by Park XE-100 AFM. The EDS spectra were recorded with a Zeiss EVO MA 10 Scanning Electron Microscopy. Optical transmission measurements were performed with UV-VIS double beam spectrophotometer in the wavelength range of 300-1,000 nm. FTIR spectra were performed by Thermo Scientific Nicollet 6700 in the range between 450 and 4000 cm^{-1}. Electrical properties of the films were evaluated by Hall effect and resistivity measurements in the van der Pauw configuration at room temperature.

3. Results and discussion

3.1 *Structural properties*

Figure 1 shows XRD patterns of Cu-O films grown on glass substrates as a function of oxygen flow rate $R(O_2)$. For $R(O_2)$ at 1 sccm, the as-grown films showed amorphous characteristics. In contrast, the XRD spectra of the films grown at 2 and 4 sccm showed the well-resolved two diffraction peaks. These peaks corresponded to the reflection of (111) and (200) planes of standard JCPDS data card of Cu_2O phase (JCPDS No. 74-1230). For $R(O_2)$ at 6 sccm, two diffraction peaks corresponded to the (-111) and (111) of CuO phase (JCPDS No. 80-76). When $R(O_2)$ increased to 8 sccm intensity of (-111) and (111) diffraction peaks decreased and then disappeared at $R(O_2)$ of 10 sccm attributed to amorphous phase. Inadequate or excessive oxygen atoms leads the films difficulty crystallise [3]. That is because oxygen atoms are reflected from the target and collided the substrate with appreciable energy, thereby causing nucleation. However, too much oxygen content will decrease the presence of Cu atoms in the vapor phase, thereby hindering nucleation. Simillar behavior was observed in TiO_2 films [3], that too much oxygen content also make the films become amorphous phase.

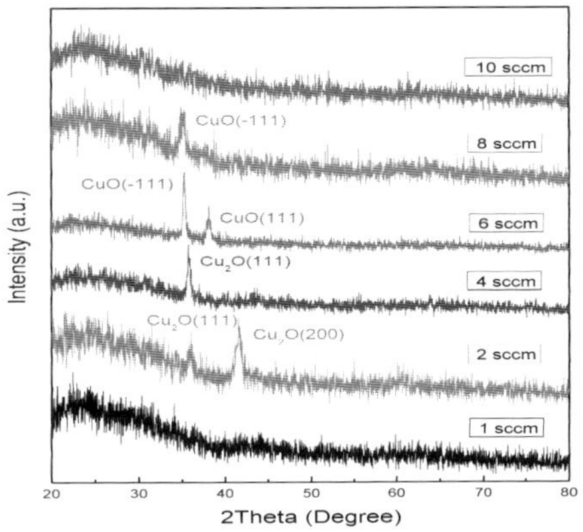

Figure 1. XRD patterns of Cu-O films as a function of $R(O_2)$.

AFM images of the films prepared as a function of $R(O_2)$ are shown in figure 2. The films formed at $R(O_2)$ of 1 sccm showed isolated particles with different sizes due to a low deposition rate and deficient O_2 content for reacting with Cu precusor atoms. At $R(O_2)$ of 2 sccm, AFM image showed that the films crystallised in pillar shape with average roughness and particle size of 3.82 and of 31.25 nm, respectively. With increasing the $R(O_2)$ from 4 to 6 sccm, the surface roughness of the films with pillar shape decreased from 2.84 to 2.29 nm but the particle size increased from 33.20 to 36.88 nm. However, the pyramidal shape of particles at film surface was noticed at $R(O_2)$ of 8 sccm with average roughness and particle size of to 1.75 and of 37.10 nm. At $R(O_2)$ of 10 sccm, flattened surface with different particle sizes in irregular shape was observed.

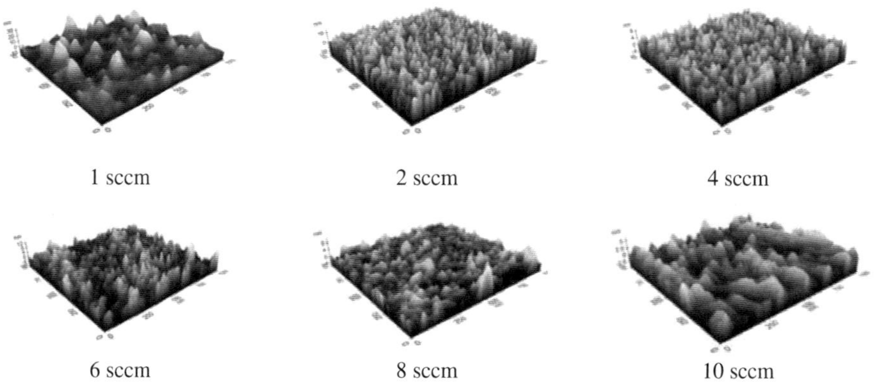

Figure 2. AFM images of Cu-O films as a function of $R(O_2)$.

3.2 Optical properties

Figure 3 shows transmittance spectra of Cu-O films obtained from various oxygen flow rates. From the transmittance spectra, the absorption coefficient (α) was calculated using relationship of Sunds $\alpha = (1/d)\ln(1/T)$ where T is transmittance and d is film thickness. The absorption coefficient (α) and the incident photon energy (hv) according to Tauc relationship $(\alpha hv)^2 = A(hv - E_g)$ where A and E_g are constant and energy gap values, respectively. The E_g can be determined by extrapolation of the linear portion of the curve to the photon energy axis. Figure 4 shows the curve of $(\alpha hv)^2$ vs. photon energy (hv) of the films obtained from various oxygen flow rates. The dependence of E_g on $R(O_2)$ of the films in the range of 1.97- 2.55 eV is shown in figure 5. Figure 6 shows the FTIR spectra of Cu-O films as a function of $R(O_2)$. The nature of peaks are presented in table 2. The peaks with wave numbers of ~630 and ~770 cm^{-1} are characteristics of the stretching vibration of Cu(I)-O and Cu(II)-O bonds, respectively [4,5].

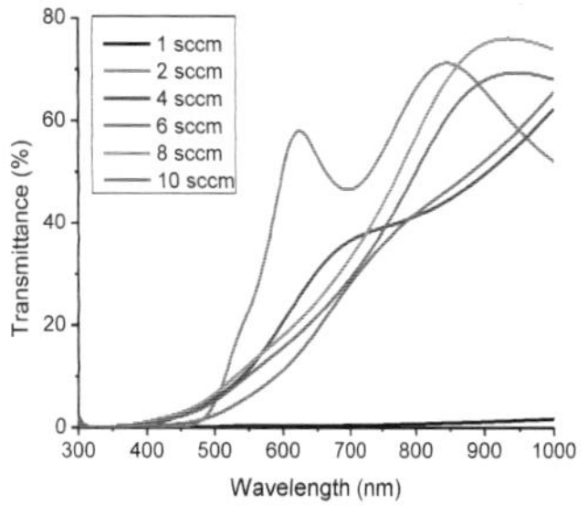

Figure 3. Transmittance of Cu-O films as a function of $R(O_2)$.

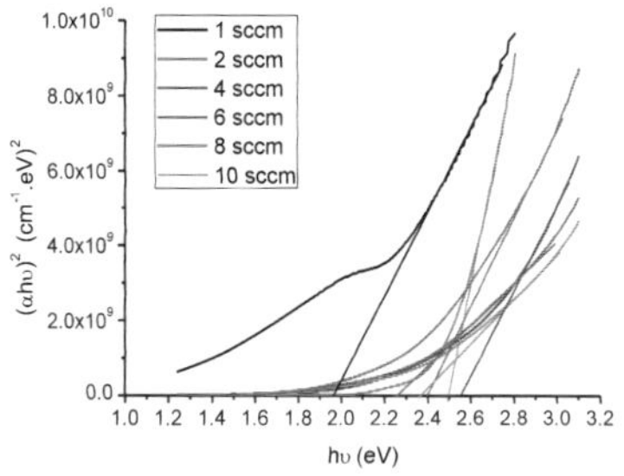

Figure 4. $(\alpha hv)^2$ vs. hv plot of Cu-O films as a function of $R(O_2)$.

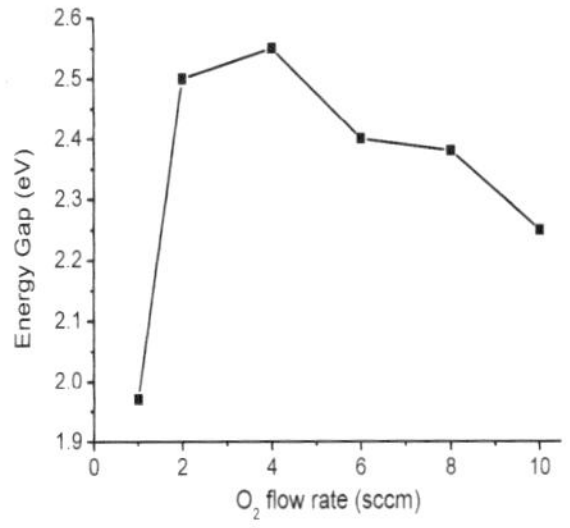

Figure 5. Variation of E_g of Cu-O films as a function of $R(O_2)$.

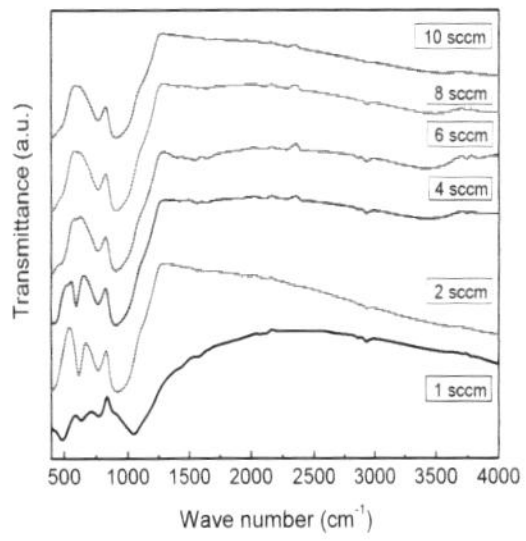

Figure 6. FTIR spectra of Cu-O films as a function of $R(O_2)$.

Table 1. FTIR spectra of Cu-O films as a function of $R(O_2)$.

$R(O_2)$ (sccm)	Peak of FTIR (cm^{-1})	Phase	Assignments Vibration Mode	[Ref.]
1	485	CuO	Cu(II)-O	[5,6]
	630	Cu_2O	Cu(I)-O	[4,7,8]
	770	CuO	Cu(II)-O	[5]
	1046	Cu(OH)	Cu-OH	[6]
2	615	Cu_2O	Cu(I)-O	[9,10]
	765, 918	CuO		
4	610	Cu_2O		
	764, 905	CuO	Cu(II)-O	
	3438	Hydroxyls of absorbed water	Stretching vibration of OH^-	[4,5]
6	764, 904	CuO	Cu(II)-O	
	3407	Hydroxyls of absorbed water	Stretching vibration of OH^-	
8	764, 907	CuO		
	1641	Surface absorbed water	Bending vibration	[4,5]
	3495	Hydroxyls of absorbed water	Stretching vibration of OH^-	
10	764, 907	CuO	Cu(II)-O	
	3400	Hydroxyls of absorbed water	Stretching vibration of OH^-	

3.3 Electrical properties

Electrical properties of the films were evaluated by Hall effect and resistivity measurements in the van der Pauw configuration. We found that all studied films exhibited p-type conductivity. Variations of the resistivity, carrier concentration and mobility against $R(O_2)$ are shown in figure 7. Carrier concentration (n) was derived from the relation $n=1/eR_H$ where R_H is the Hall coefficient and e is the absolute value of the electron charge. The carrier mobility (μ) was determined using the relation $\mu=1/ne\rho$ where ρ is resistivity. Resistivity of the films increased from 2.01 attain maximum at 1.85×10^3 Ω.cm with $R(O_2)$ increased from 1 to 6 sccm and then decreased to 10.25 Ω.cm when $R(O_2)$ further increased to 10 sccm. The carrier concentration initially decreased with an increase $R(O_2)$ down to 6 sccm and then increased with an increase in $R(O_2)$. The variation of mobility value showed the similar behavior to resistivity. These results indicate that the resistivity, carrier concentration and mobility of the films are sensitive to $R(O_2)$. The films prepared at oxygen flow rate of 8 sccm revealed a low resistivity and high mobility values. So, they are identified as suitable candidates for fabrication as absorber layer in solar cell devices.

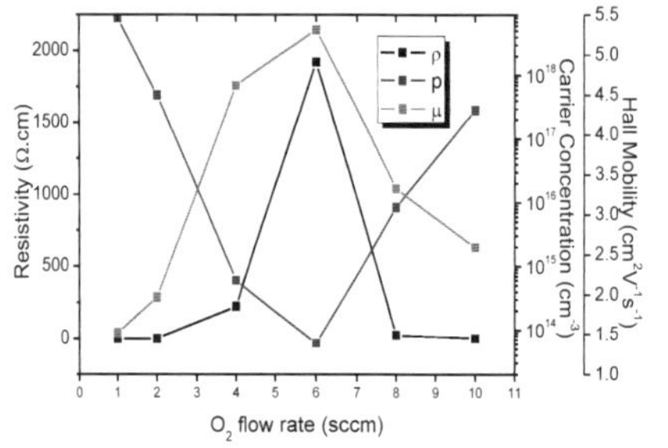

Figure 7. Variations of ρ, p and μ of Cu-O films as a function of R(O_2).

4. Conclusion

The influence of oxygen flow rate R(O_2) on the structural, optical and electrical properties of copper oxide thin films grown by reactive dc magnetron sputtering was investigated. The results revealed the structural properties of the obtained films strongly depended on R(O_2). The nanocrystallite copper oxide thin films with cubic structure were formed at R(O_2) values of 2 and 4 sccm. Then, transition to monoclinic structure at R(O_2) values of 6 and 8 sccm was observed. Out of the above R(O_2) ranges, amorphous phase was dominant. The average particle size of the films around 30 nm was observed by AFM. Influence of oxygen flow rate on optical and electrical properties of the as-grown films was also investigated. These results showed a better understanding of the growth behavior and other relevant properties of the deposited copper oxide thin films as a function of R(O_2). The effect of other growth parameters such as deposition temperature, deposition time and thickness on the optoelectronic properties of the films will be planned to investigate in the near future.

References

[1] P K Ooi, S S Ng, M J Abdullah, H Abu Hassan and Z Hassan 2013 *Mater. Chem. Phys* **140** p 243
[2] S Cho 2013 *Met. Mater. Int* **191** p 327
[3] B X Zhao, J C Zhou and L Y Rong 2010 *Trans. Nonferrous Met. Soc. China* **20** p 1429
[4] D GuoD, L Wang, Y Du, Z M Ma and L Shen 2015 *Mater. Lett* **160** p 541
[5] J S K Arockiasamy and J Irudayaraj 2015 *Ceramic International* **42** p 6198
[6] M Fterich, F B Nasr, R Lefi, M Toumi and S Guermazi 2016 *Mater. Sci. Semicon. Processing* **43** p 114
[7] M A Bhosale and B M Bhanage 2016 *Adv. Powder Tech* **27** p 238
[8] K Kaviyarasan, S Anandan, R V Mangalaraja, T Sivasankar and M Ashokkumar 2016 *Ultrasonics Sonochemistry* **29** p 388
[9] A Chen, H Long, X Li, Y Li, G Yang and P Lu 2009 *Vacuum* **83** p 927
[10] K Kardarian, D Nunes, P M Sberna, A Ginsburg, D A Keller, J V Pinto, J Deuermeier, A Y Anderson, A Zaban, R Martins and E Fortunato 2016 *Sol. Energy & Sol. Cells* **147** p27

Acknowledgment

This work is supported by King Mongkut's Institute of Technology Ladkrabang under grant number A118-59-038.

IOP Conf. Series: Materials Science and Engineering **211** (2017) 012026 doi:10.1088/1757-899X/211/1/012026

Tin sulphide prepared by sulphurisation process using metallic tin film precursor obtained from dc magnetron sputtering method

Ngamnit Wongcharoen* and Thitinai Gaewdang

Physics Department, King Mongkut's Institute of Technology Ladkrabang, Bangkok 10520, Thailand

*Email:ngamnit.wo@kmitl.ac.th

Abstract. Thin films of tin sulphide were prepared by sulphurisation process of metallic Sn precursor at temperature of 300-500°C in excess ambient of sulphur using a graphite box in N_2 atmosphere. Sulphurisation temperature at 300°C, the films were formed in SnS phase with orthorhombic structure. In contrast, the films were formed in SnS_2 phase with hexagonal structure when sulphurisation temperature higher than 300°C. From absorption spectra, direct band gap increased from 1.70 to 2.45 eV with increasing sulphurisation temperature. From the transient photoconductivity measurements, persistent photoconductivity behavior was observed in the films sulphurised at 300 and 350°C. The decay current data are better fitted with multiple exponential function resulting in the several slow decay times. Density of trap states corresponding to its decay time was also evaluated from the decay current data.

1. Introduction

In the past years thin films of SnS as a semiconductor of IV-VI group have attracted much attention because of their potential applications in the fabrications of photovoltaics, solar cells and optoelectronic devices [1]. Different values of band gap have been obtained for SnS, ranging from 1 to 2.33 eV [2], depending on the resulting structure obtained by different techniques and the type of electron transitions. In addition, both Sn and S are abundant in nature and non-toxic. The Sn-S system has three stoichiometric stable compounds, with variable Sn to S ratio such as SnS, SnS_2 and Sn_2S_3. Particularly, SnS shows the direct optical transition with p-type electrical conductivity. While SnS_2 found in polycrystalline nature with n-type conductivity and the compound Sn_2S_3 shows mixed valence compound with semiconductor nature. SnS_2 has CdI_2-like crystal structure that consists of densely packed atomic layers of tin located between two layers of S. This compound can exist in the form of several polytypes. The band gap of SnS_2 changes in the range of 2.12-2.44 eV depending on the method of desired materials and polytype. SnS_2 has n-type conductivity, a high optical absorption coefficient > 10^4cm^{-1} and relatively high charge carrier mobility 18.3-230 cm^{-2}V^{-1} s^{-1}. These properties make it promising for use in thin film solar cells. Tin sulphide films can be prepared by a variety of methods [2] with the purpose of manufacturing thin films suitable for use as a solar absorber in optoelectronic devices and photovoltaic applications. Among these methods, sulphurisation of Sn layers process is proposed low consumed materials, high growth rate, ease to control stoichiometric composition and ease to extend to large scale production. In this study, we prepared tin sulphide thin films on glass slide substrates by sulphurisation process of metallic Sn precursor at temperature of 300-500°C in excess ambient of S using a graphite box in N_2 atmosphere. Influence of sulphurisation temperature on structural, optical and electrical properties of tin sulphide films was investigated.

2. Materials and methods

The Sn precursor layers were deposited on glass substrates using dc magnetron sputtering from a home-made circular planar magnetron sputtering system. A circular planar magnetron of 50 mm diameter was used as the magnetron cathode. The magnetron target assembly was mounted on the top plate of the sputter chamber such that the sputtering could be done by sputtering down configuration. A continuously variable dc power supply of 630 V and 2.6 A was used as a power source for sputtering. A 50 mm diameter and 3 mm thick pure Sn (99.99%) used as sputter target. Pure argon was used as sputter gas. The flow rate of argon was controlled by AALBORG mass flow controller. Before deposition of each batch, the target was sputtered in pure argon atmosphere for 5 min to remove oxide layer contaminated on the surface of target. The sputtered Sn layers were sulphurised in a graphite box kept in tubular furnace together with 99.98 % purity of S flakes (Sigma Aldrich) in N_2 flow. Sulphurisation temperature (T_s) was varied from 300 to 500°C for a fixed time period of 60 min and then the films were allowed to cool down naturally to room temperature. The crystal structure of these films was checked by X-ray diffraction technique with a Brucker D 8 diffractometer using CuK_α radiation. The crystallite size (L) was calculated using Scherrer's formula, neglecting peak broadening due to residual stresses in the films, $L= 0.9\lambda/(\beta\cos\theta)$ where β is the broadening of diffraction line measured at half its maximum intensity in radian and λ is the wavelength of X-rays (0.1540 nm). Surface morphology was examined by LEO1455VP scanning electron microscope. Optical absorption measurements were performed with UV-VIS double beam spectrophotometer in the wavelength range of 300- 800 nm. Electrical properties of the films were evaluated by Hall effect and resistivity measurements in the van der Pauw configuration at room temperature.

3. Results and discussion

3.1 *Structural properties*

The XRD patterns of tin sulphide films prepared at different sulphurisation temperatures are shown in Figure 1. The diffraction patterns with 2θ in the range of 20-70° showed that the films sulphurised at 300°C are polycrystalline in nature, exhibited several peaks corresponding to orthorhombic SnS phase with file JCPDS 39-354. The evidence of other phase such as Sn_2S_3 cannot be observed probably due to its peaks is weak intensity. In contrast, XRD spectrum of the films sulphurised at 350°C showed the well-defined peaks at 2θ equal to 28.20, 32.08, 49.90 and 52.42° well matched the standard patterns JCPDS 40-1467 with the hexagonal SnS_2 phase and corresponded to crystallographic planes (100), (101), (110) and (111), respectively. The variation of lattice parameter "a" and "c" of hexagonal SnS_2 phase as a function of sulphurisation temperature are shown in Figure 2. The calculated values of crystallite size using Scherrer's formula are 30.12, 42.36, 48.28, 49.77 and 59.01 nm for the films sulphurised at 300, 350, 400, 450 and 500°C, respectively. SEM images of tin sulphide films obtained from different sulphurisation temperatures are shown in Figure 3. It can be observed that the film surface sulphurised at 300°C was uniform with tightly packed particles. On further increase of sulphurisation temperature to 350°C, SEM image revealed many hexagonal thin plates of one side about 0.8 μm. These films had the thin plate shape with random orientation because SnS_2 was a layered compound semiconductor with CdI_2 structure [3]. When sulphurisation temperature increased from 350 to 500°C, the side of hexagonal plates rapidly increased from 0.8 to 2 μm whereas the thickness slightly increased. It is worthwhile to note that the crystallite size calculated by Scherrer's formula well corresponds to the thickness of the SnS_2 films observed by SEM images.

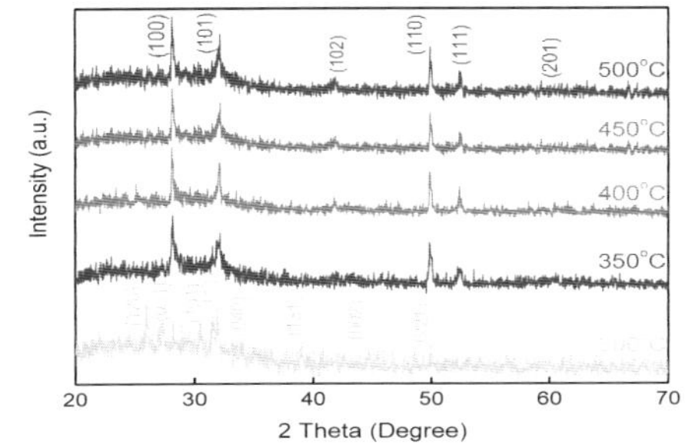

Figure 1. XRD patterns of Sn-S films as a function of sulphurisation temperature.

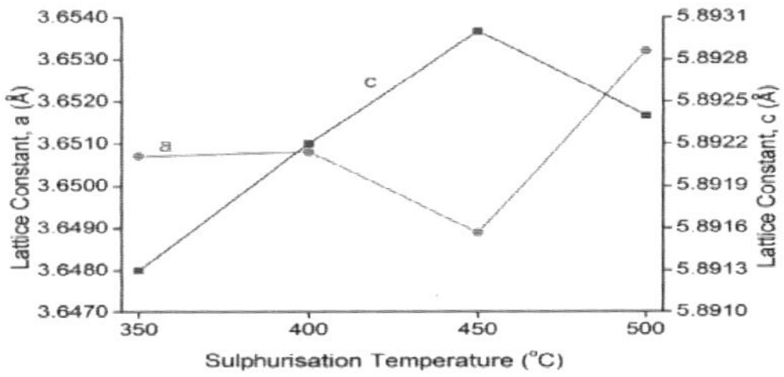

Figure 2. Lattice constants of SnS_2 films sulphurised between 350 and 500 °C.

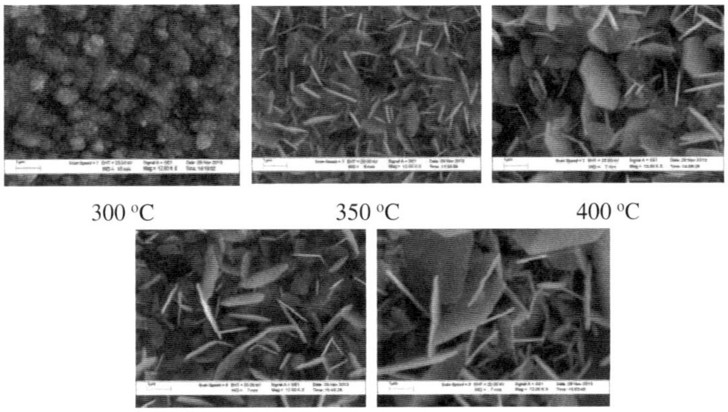

300 °C 350 °C 400 °C

450 °C 500 °C

Figure 3. SEM images of Sn-S films sulphurised between 300 and 500°C.

3.2 Optical properties

Figure 4 shows absorbance spectra of tin sulphide films obtained from five sulphurisation temperatures during 300-500°C. From the absorbance spectra, the absorption coefficient (α) was calculated using the equation $\alpha = 2.303 A/d$ where A is absorbance and d is film thickness. The absorption coefficient (α) and the incident photon energy (hν) related by $(\alpha h\nu)^2 = B(h\nu - E_g)$ where B and E_g are constant and band gap values, respectively. The E_g can be determined by extrapolation of the linear portion of the curve to the photon energy axis. Figure 5 shows the curve of $(\alpha h\nu)^2$ vs. photon energy (hν) of the films obtained from five different sulphurisation temperatures. The direct band gap type was clearly observed for all films sulphurised with different temperatures. The variation of E_g as a function of sulphurisation temperature is shown in Figure 6. The E_g increased from 1.70 to 2.45 eV with the increase of sulphurisation temperature from 300 to 500°C, respectively. The band gap (E_g) values of 1.38 eV for SnS phase and of 2.21 eV for SnS_2 phase were recently reported [4]. The difference in E_g values was mainly due to the difference in the deposition techniques employed and process parameters maintained during the growth of the films. The results indicated that E_g is sensitive to the sulphurisation temperature.

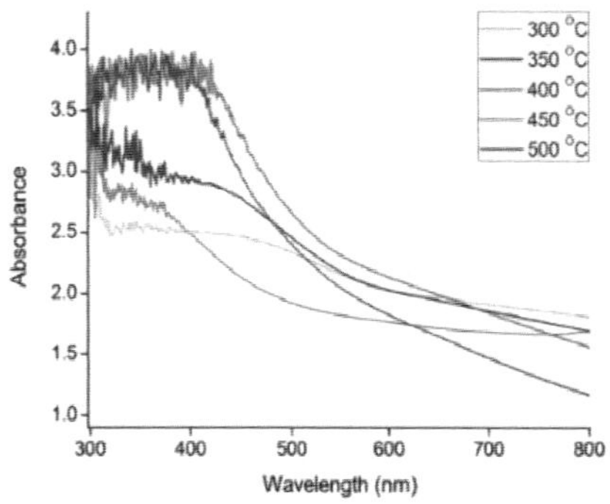

Figure 4. Absorbance of Sn-S films sulphurised between 300 and 500°C.

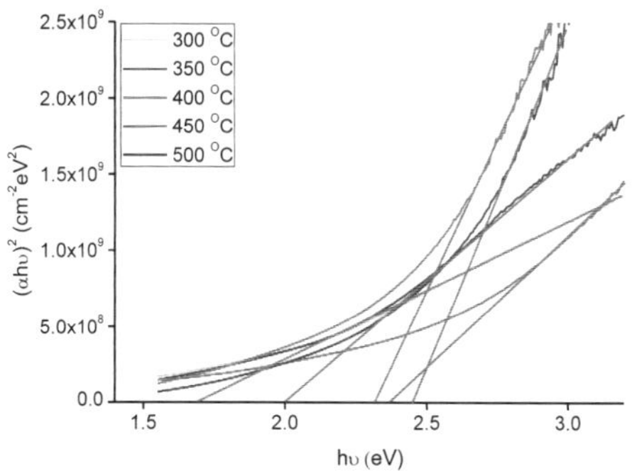

Figure 5. $(\alpha h\nu)^2$ vs. hν plot of Sn-S films sulphurised between 300 and 500°C.

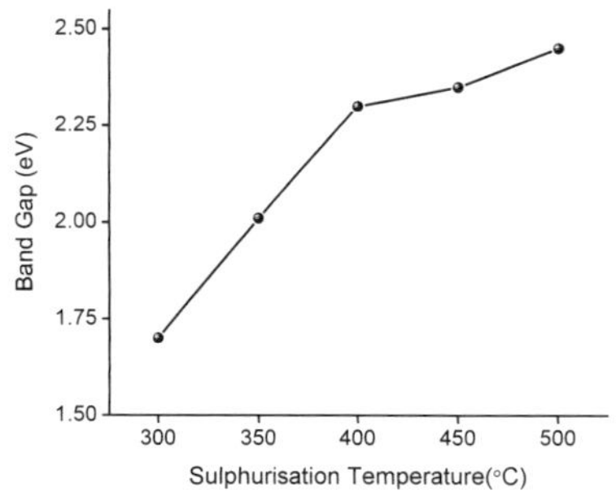

Figure 6. Variation of E_g of Sn-S films sulphurised between 300 and 500°C.

3.3 Electrical properties

Electrical properties of tin sulphide films were evaluated by Hall effect and resistivity measurements in the van der Pauw configuration. We found that the films sulphurised at 300°C exhibited p-type. In contrast, the films sulphurised at higher temperature (> 300°C) converted to n-type conductivity. Carrier concentration (n) was derived from the relation $n=1/qR_H$ where R_H is the Hall coefficient and q is the absolute value of the electron charge. The carrier mobility (μ) was determined using the relation $\mu = 1/qn\rho$ where ρ is resistivity. Resistivity of the films increased from 6.42×10^2 to $1.24\times10^3\Omega$.cm with an increase of sulphurisation temperature from 300 to 350°C. Further increase of sulphurisation temperature, resistivity sharply increased more than 2 orders ($>10^9$ Ω.cm). It is consistent to the sudden increase of E_g from 2.01 to 2.30 eV when the sulphurisation temperature increased from 350 to 400°C. The carrier concentration decreased from 1.02×10^{16} to 2.72×10^{14} cm^{-3} when the sulphurisation temperature increased from 300 to 350°C. However, large increase in mobility value from 0.96 to18.24 $cm^2V^{-1}s^{-1}$ when the sulphurisation temperature increased from 300 to 350°C was observed. Further increase of sulphurisation temperature, carrier concentration and mobility values cannot be obtained due to the films have very high resistivity values. The results indicate that the resistivity, carrier concentration and mobility of the films are sensitive to sulphurisation temperature.

3.4 Persistent Photoconductivity

The photocurrent response of the films sulphurised at 300 and 350°C was measured by two probe method using halogen lamp as a light source. Two silver electrodes were made on the sample surface. The photocurrent response of the films sulphurised at 300 and 350°C are shown in Figure 7. The sample was connected to the experimental setup and maintained in darkness under a constant applied bias of 30 V to stabilise the current. After the stabilisation period, the current was recorded in the following sequences: 30 s in darkness, 60 s under illumination and, finally, 90 s in the darkness. Under illumination, the current increased immediately because a lot of carriers were excited by light illumination. After the illumination period, the persistent photoconductivity (PPC) effect of the films sulphurised at 300 and 350°C occurred in darkness. The decay of PPC in Figure 7 can be described by multiple exponential functions [5] as equation (1).

$$I = I_{dark} + (I_0 - I_{dark})\sum_i \exp(-\frac{t}{\tau_i}) \qquad (1)$$

where I_{dark} is the current measured in darkness, I_0 is the PPC buildup level near the moment of light excitation being removed, τ is the PPC decay time constant. The normalised decay current in darkness as shown in Figure 8, which follows the illumination period, indicates the persistent of charge carrier traps in the films [5]. In Figure 8 the dash line is the measured data and the solid line is the least squares fit data by the multiple exponential functions as described above. Method for calculating the trap densities corresponding to the PPC decay constant was reported in the literature [6]. The decay time and density of trap states of the films sulphurised at 300 and 350°C were tabulated in Table 1. Our first decay time value in SnS phase is comparable to the previous reports [4,5] in which photoresponse times of about 20 to 40 s. The results suggested that the SnS films displayed shorter decay time values but slightly bigger density of trap states than SnS_2 films.

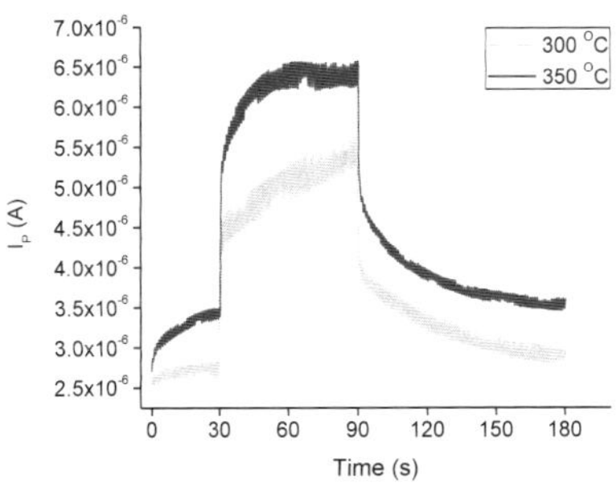

Figure 7. Photocurrent response of Sn-S films sulphurised at 300 and 350°C.

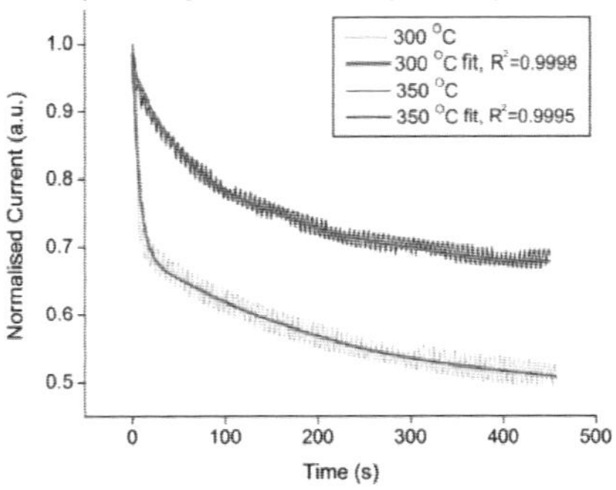

Figure 8. Decay time of Sn-S films sulphurised at 300 and 350°C.

Table 1. Decay time and density of trap states of Sn-S films sulphurised at 300 and 350°C.

Sulphurisation temperature (°C)	**Phase**	**Decay time (s)**	**Density of trap states ($\times 10^{10}$ cm^{-2})**
300	SnS	22.01, 534.75, 1076.35, 3211.02	2.29, 1.76, 1.66, 1.37
350	SnS_2	114.02, 448.43, 632.91, 1276.77, 1352.64, 4856.33, 10162.65	2.17, 2.03, 1.97, 1.82 1.71, 1.61, 1.54

4. Conclusion

Tin sulphide thin films were obtained through sulphurisation process of metallic Sn layers deposited on glass substrates by dc magnetron sputtering method. It is shown that the SnS single phase of orthorhombic structure was formed at sulphurisation temperature of 300°C whereas SnS_2 single phase was crystallised at sulphurisation temperature beyond 300°C. SnS films exhibited p-type but SnS_2 films showed n-type semiconductor. As the increase of sulphurisation temperature, band gap increased from 1.70 to 2.45 eV. From photocurrent response data, SnS films displayed shorter decay times but slightly bigger density of trap states than SnS_2 films.

References

[1] El- Nahass M M, Zeyada H M, Aziz M S and El-Ghamaz N A 2002 *Mater. Chem. Phys* **140** p 243
[2] Henry J, Mohanraj K, Kannan S, Barathan S and Sivakumar G 2015 *J. Exp. Nanosci* **10** p 78
[3] Shi C, Yang P, Yao M, Dai X and Chen Z 2013 *Thin Solid Films* **534** p 28
[4] Yang Y B, Dash J K, Xiang Y, Wang Y, Shi J, Dinolfo P H, Lu T M and Wang G C 2016 *J.* Phys.*Chem C* **120** 13199
[5] Jaing F, Shen H, Gao C, Li B, Lin L and Shen Z 2011 *Appl. Surf. Sci* **257** 4901
[6] Studenikin S A, Gelego N and Locivera M 1998 *J. Appl. Phys* **83** 2104

Acknowledgment

This work is supported by King Mongkut's Institute of Technology Ladkrabang under grant number 2560-0105023.

IOP Conf. Series: Materials Science and Engineering **211** (2017) 012027 doi:10.1088/1757-899X/211/1/012027

Two-stage collaborative global optimization design model of the CHPG microgrid

Qingfen Liao, Yeyan Xu , Fei Tang*, Sicheng Peng and Zheng Yang

School of Electrical Engineering, Wuhan University, Wuhan, China

*Email: tangfei@whu.edu.cn

Abstract. With the continuous developing of technology and reducing of investment costs, renewable energy proportion in the power grid is becoming higher and higher because of the clean and environmental characteristics, which may need more larger-capacity energy storage devices, increasing the cost. A two-stage collaborative global optimization design model of the combined-heat-power-and-gas (abbreviated as CHPG) microgrid is proposed in this paper, to minimize the cost by using virtual storage without extending the existing storage system. P2G technology is used as virtual multi-energy storage in CHPG, which can coordinate the operation of electric energy network and natural gas network at the same time. Demand response is also one kind of good virtual storage, including economic guide for the DGs and heat pumps in demand side and priority scheduling of controllable loads. Two kinds of storage will coordinate to smooth the high-frequency fluctuations and low-frequency fluctuations of renewable energy respectively, and achieve a lower-cost operation scheme simultaneously. Finally, the feasibility and superiority of proposed design model is proved in a simulation of a CHPG microgrid.

1. Introduction

CHPG system including energy storage is one of the most economical multi-energy systems to make full use of renewable energy sources. Energy storage is essential in the systems concluding renewable generation, due to the intermittent nature of renewable energy.

By multi-mode scheduling of energy generation, storage and consumption devices, a mcirogrid can operation autonomously, such as a group of houses[1]. Thermal storage is considered to dividing the use of electricity and heat apart on terms of time, in order to make full use of renewable energy and minimize the costs of purchasing energy[2-3]. However with the proportion of renewable generation becoming higher and higher, one of the bottleneck for full use of renewable energy is the limited power storage. Virtual energy storage has been proposed to make up for the deficiencies of power storage. Demand response is considered to be one kind of virtual power storage in demand side, to reduce the size of traditional power storage system by cooperating with battery. But some scheduling of demand response, such as smoothing the tie-line power[4-5], has forgotten to take users' comfort into account.

In addition to demand response, the emerging P2G (Power-to-Gas) technology can be another type of virtual storage. By generating gas through the use of electricity, P2G can utilize the excess power of renewable generation and the generated gas can be used in gas-to-power technologies, which can improve the proportion of renewable generation without increasing the capacity of power storage, and minimize the gas cost [6-7].

In this paper, a new multi-energy system-CHPG micro grid and its two-stage collaborative global optimization design model are proposed. P2G and demand response are used as virtual storage to reduce the operating cost and keep security of CHPG without affecting users' comfort, by a method of frequency division to smooth fluctuations caused by the intermittent nature of renewable energy sources.

2. Modeling framework of CHPG microgrid

The structure of CHPG is shown as Figure 1. The electricity demand is satisfied by WTs ,PVs and gas turbines; The heat demand is satisfied by heat pumps in demand side and gas turbines in supply side. Power, heat and gas systems are combined by conversion devices , which will offer diverse solutions to fully consume renewable energy. For example, P2G is used to produce gas by low-price power, in order to reduce the energy cost.

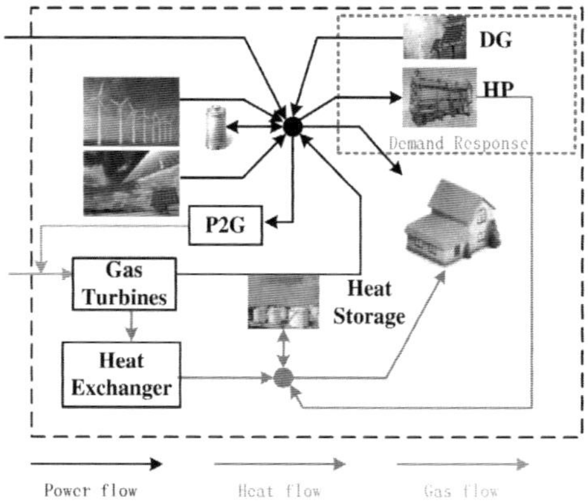

Figure 1. Structure of the CHPG microgrid.

2.1. Modeling of devices in the CHPG microgrid

The model of wind turbine, PV, gas turbine and battery are the same as[1-3].

2.1.1. Heat Pump. The model of a heat pump is shown as equation (1).

$$Q_{HP,i}(t)=\eta_{HP,i}P_{HP,i}(t) \tag{1}$$

where $\eta_{HP,i}$, $P_{HP,i}(t)$, and $Q_{HP,i}(t)$ are the efficiency, consumed power, and generated thermal power of the HP_i.

2.1.2. P2G. A typical P2G plant consists of two processes, the electrolysis and the methanation. The model of the two processes are proposed in [8] and [9] perspectively. Therefore, the model of the whole P2G process can be described as equation (2).

$$V_{CH_4}(t)=\eta_{P2G}P_{P2G}(t) \tag{2}$$

where η_{P2G}, $P_{P2G}(t)$, and $V_{CH_4}(t)$ are the efficiency, consumed power and flow rate of generated gas of P2G plant.

2.2. Modeling of the power characteristics of the CHPG microgrid

The power balance and voltage constraint are described as equation (3) and equation (4) respectively.

$$\sum P_{PV,i}(t)+\sum P_{WT,i}(t)+\sum P_{GT,i}(t)+P_{grid}(t)+P_{DG}(t)+P_{ES}^{D}(t)=P_{load}(t)+P_{P2G}(t)+\sum P_{HP,i}^{con}(t)+P_{ES}^{C}(t) \tag{3}$$

The variables on the left of equation (3) is the power of all the PV devices, all the WT devices ,all the GT devices, the tie-line power, produced power of demand response, and the discharge power of battery respectively. $P_{load}(t)$ and $P_{ES}^{C}(t)$ is the consumed power of uncontrollable load and the charging power of battery respectively.

2.3. *Modeling of demand reponse in the CHPG microgrid*

2.3.1. Economic guide for demand response. Demand response will change the characteristic of the load in the CHPG microgrid, making the flow of power bi-directional. By guiding demand response to smooth the peak-valley difference, the cost of purchasing power or gas will be saved. A peak-valley price mechanism for demand response is proposed, during 8:00 a.m to 20:00 p.m, the price is at peak.

2.3.2. Priority scheduling of controllable load—heat pumps(HPs). The fundamental requirement for any priority scheduling is not to reduce the user's comfort. Users' heat pumps will be forced to shut down when the room temperature T_t^i reaches $T_{\max}^i$ (the maximum temperature that users can bear), and will be forced to increase power when T_t^i reaches $T_{\min}^i$ (the minimum temperature that users can bear). As long as the room temperature is maintained within ($T_{\min}^i \sim T_{\max}^i$), adjustment of HPs' power will not affect the users' comfort.

T_S^i is proposed to judge the adjustability of the heat pump:

$$T_S^i = \frac{T_t^i - T_{\min}^i}{T_{\max}^i - T_{\min}^i} \tag{4}$$

T_S^i close to 1 or 0 means that the room temperature is close to $T_{\max}^i$ or $T_{\min}^i$, and the adjustment of HP may be forced to stop during the scheduling because of the reduce of users' comfort.

In order to compose a group of HPs with strong adjustability quickly, an efficient method for selection is proposed:

(1) When $T_S^{low} \le T_S^i \le T_S^{high}$, the power of the HP can vary within a larger range without reducing users' comfort,so the HP is good to control; (2) When $T_{S,t}^{high} \le T_{S,t}^i$ or $T_{S,t}^i \le T_{S,t}^{low}$, the operation status is prone to change; (3) When $T_S^i = T_S^{high}$, the HP can maintain the maximum operating status during the next Δt without reducing users' comfort. When $T_S^i = T_S^{low}$, the HP can shut down during the next Δt without reducing users' comfort. The adjustability of the HP group can be described as equation (5):

$$C_{HP}^t = \frac{I\left(T_S^{high} \ge T_{S,t}^i \ge T_S^{low}\right)}{N} \tag{5}$$

where $I(T_S^{high} \ge T_{S,t}^i \ge T_S^{low})$ and N is the number of strong adjustable HPs and the total number of controllable HPs respectively.

3. Two-stage collaborative global optimization design model

In order to make full use of renewable energy, WTs and PVs are under the MPPT control strategy, which will affect the stability of CHPG and larger-capacity storage is demanded. The proposed two-stage collaborative global optimization can make full use of P2G an demand response, to relieve the pressure of battery to smooth fluctuations of renewable energy and minimize the operation cost. The proposed optimization method is shown as Figure 2.

3.1. *Modeling of stability optimization for the CHPG microgrid*

The intermittent nature of renewable energy is the key factor affecting the stability of CHPG, which leads to the overcharge/excessive discharge of battery and large fluctuations of tie-line power, and will reduce the service life of battery and bring the security risks to other grids. Therefore, an intelligent energy storage/ utilizing scheme is needed to take advantage of demand response devices and P2G.

Demand response and P2G both have a quick reaction speed, and the impact of control on the other energy systems is minimal. Therefore, a method of frequency division to smooth fluctuation is proposed, as shown in Figure 3.

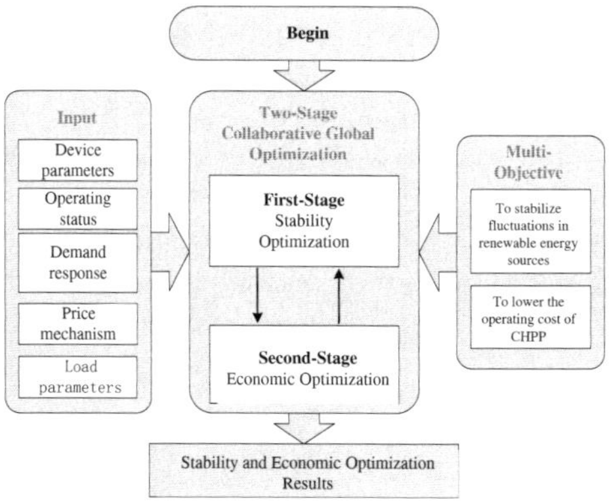

Figure 2. Diagram of the propsed method for CHPG microgrid.

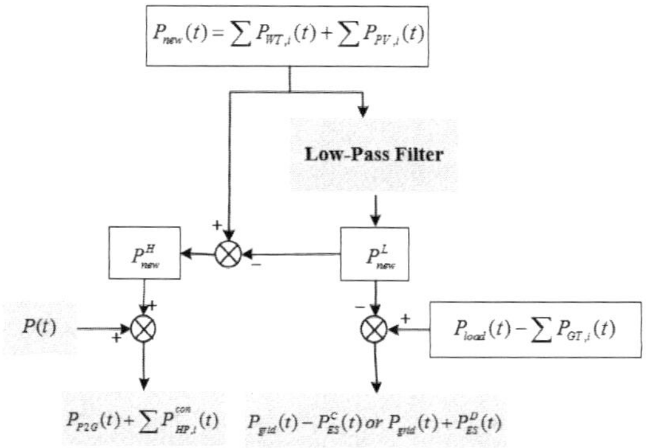

Figure 3. Method of frequency division to smooth fluctuation.

The high-frequency fluctuations are intermittent and fast-changing, but the low-frequency fluctuations are smooth and slow-changing. Demand response and P2G are used to smooth the high frequency fluctuations, in order to avoid overcharge/excessive discharge of battery and large fluctuations of tie-line power.

Given the energy balance constraint as equation(3), the model of Figure 4 can be described as equation (6) and equation (7).

$$P_{new}^{H}(t) = P_{P2G}(t) + \sum P_{HP,i}^{con}(t) - P(t) \tag{6}$$

$$P_{new}^{L}(t) = P_{load}(t) - \sum P_{GT,i}(t) - P_{grid}(t) + P_{ES}^{C}(t) - P_{ES}^{D}(t) \tag{7}$$

where $\sum P_{HP,i}^{con}(t)$ is the total power of all the controlled HP with strong adjustability. $P_{new}^{H}(t)$ and $P_{new}^{L}(t)$ are the high-frequency fluctuations and low-frequency fluctuations of all the renewable energy generators respectively.

Butterworth low-pass filter is used to divide fluctuations in this paper, modeled by equation (8):

$$[n, W_n] = buttord(W_p, W_s, R_p, Rs) \tag{8}$$

where n, W_n, W_p, W_s, R_p, and Rs is the lowest order, cutoff frequency, passband corner frequency, stopband corner frequency, passband ripple, and stopband attenuation respectively. $W_p < W_s$.

W_n is changing with the status of demand response during the scheduling, to make full use of virtual storage to smooth as large range as possible of the fluctuations.

W_p can be determined by equation (9).

$$\begin{cases} W_p = W_p^{nom}, C_{HP}^L \le C_{HP}^t \le C_{HP}^H \\ W_p = W_p^{nom} - \Delta W_p^H, C_{HP}^t > C_{HP}^H \\ W_p = W_p^{nom} + \Delta W_p^L, C_{HP}^t < C_{HP}^L \end{cases} \tag{9}$$

where W_p^{nom} is the rated passband corner frequency.

When $C_{HP}^t > C_{HP}^H$, the adjustability of HPs is stronger to smooth a larger range of fluctuations, so the passband corner frequency is reduced by ΔW_p^H to gain a larger range of high-frequency fluctuations.

When $C_{HP}^t < C_{HP}^L$, the adjustability of HPs is too weaker to smooth the rated range of fluctuation, so the passband corner frequency is increased by ΔW_p^L to reduce the range of high-frequency fluctuations.

The opteration of tie line and any battery is restricted by equation (10) and equation (11), respectively.

$$0 \le |P_{grid}(t)| \le P_{grid}^{m}(t) \tag{10}$$

$$\begin{cases} -\gamma_C Cap \le P_{ES}(t) \le \gamma_D Cap, SOC^L \le SOC(t) \le SOC^H \\ -\gamma_C Cap \le P_{ES}(t) \le 0, SOC(t) \le SOC^L \\ 0 \le P_{ES}(t) \le \gamma_D Cap, SOC(t) \ge SOC^H \end{cases} \tag{11}$$

where $P_{grid}^{m}(t)$ is the maximum permitted exchanged power on tie line; SOC^L, SOC^H, and $SOC(t)$ is the minimum permitted storage state, the maximum permitted storage state, and the real storage state, respectively. When the state is lower than minimum permitted state, batteries are forced to charge electricity to restore to the safe state.

3.2. Modeling of economic optimization for the CHPG microgrid

The economic optimization is executed under the stability optimization, as stability is the fundamental support for the operation of CHPG. The result of economic optimization is also the final result of stability optimization.

In order to minimize the operating cost, which consists of gas cost, tie-line cost, and demand-response cost, the objective function is described as equation (12).

$$\begin{aligned} \min F = \sum_{t=1}^{T} & [f_e(t)P + f_g(t)P_{grid}(t) \\ & + f_{gas}(t)(\sum F_{GT,i}(t) - V_{P2G}(t)/29.31)]\Delta t \\ s.t \quad & g_i(t) = 0 \\ & h_j(t) \le 0 \end{aligned} \tag{12}$$

where $f_{gas}(t)$ and $F_{GT,i}(t)$ is the price of purchasing gas and the purchase gas power of GT, respectively.

P2G can smooth the fluctuation in CHPG, and reduce the gas cost, shown in equation (12).

$g_i(t) = 0$ represents the equality constraints in CHPG, consisting of the energy balances of multi-energy systems and the frequency division model.

$h_j(t) \le 0$ represents the inequality constraints in CHPG, consisting of the constraints of all the devices, the constraint of tie-line power, and the constraints of node voltage.

4. Simulation

In order to verify the optimization design model proposed above, the simulation verification is done under a CHPG, whose paramaters is shown in Table 1. the scheudling period is 24 hours and the gap is 1h. ΔW_{HP}^{N} is 0.1, ΔW_{HP}^{H} is 0.02, ΔW_{HP}^{L} is 0.01.

$$f(t) = \begin{cases} 0.075, 8 \le t \le 20 \\ 0.05, other \end{cases} \tag{13}$$

$$f_e(t) = \begin{cases} 0.07, 8 \le t \le 22 \\ 0.05, other \end{cases} \tag{14}$$

$f_g(t)$ is the same as $f(t)$, as shown in equation(13). $f_e(t)$ is shown as equation(14).

where $f(t)$, $f_e(t)$,and $f_g(t)$ is the price for uncontrollable load, for DG & controallable load, and for tie-line power at the time t, respectively. The unit is \$/(kWh).

And the power prediction of WT, PV, and load is shown in Figure 4.

Simulation results of devices power prediction are shown in Table 2. At 13:00, when the power load is at peak, the total power of HPs is reduced to smooth the peak-valley difference. The proportion of DGs is increasing at the same time, to reduce the power purchasing from tie line.

The operating cost is decreasing along with the increasing of the P2G power, shown in Figure 5, because P2G is generating gas by lower-price electric power. Simultaneously, the operating cost is decreasing along with the decreasing of HP power, shown in Figre 6,because the total power of HP is reduced at the time when load power is at peak to reducing the purchasing cost from tie-line power and demand response.

Table 1. Device parameters and values of the CHPG microgrid.

Name	Parameter	Value	Name	Parameter	Value
Wind Turbine	Quantity	2	PV	Quantity	2
	$P_{WT,i}^{nom}$ (kW)	150		$P_{PV,i}^{nom}$ (kW)	75
Gas Turbine	Quantity	2	P2G	Quantity	1
	$P_{GT,i}^{nom}$ (kW)	200		P_{P2G}^{nom} (kW)	400
	$\eta_{HX,i}$	0.9		η_{P2G}	0.6
Battery & thermal storage	Quantity	1	Heat Pump	Quantity	450
	Cap/kWh	400		P_{HP}^{nom} (kW)	1.1
	W_{min}, W_{max}(kWh)	40, 180		$\eta_{HP,i}$	3
	δ_C , δ_D , δ	0.95, 0.95, 0.04		$T_{\min}^{i}$, $T_{\max}^{i}$	20, 30
	γ_C , γ_D	0.2, 0.4		$T_{S,t}^{low}$, $T_{S,t}^{high}$	0.4, 0.8
Tie Line	P_{grid}^{m} (kW)	600		C_{HP}^{L}, C_{HP}^{H}	0.4, 0.8

The relationship between cost and P2G is shown in Figure 5 and the relationship between cost and demand response is shown in Figure 6. During the whole peirod, the storage state of battery and the fluctuation of tie-line power are within the security range,. shown in Figure 7 and Table 2 respectively.

Table 2. Simulation results of devices power.

t/h	P_{GT} (kW)	P_{grid} (kW)	P_{ES} (kW)	P_{P2G} (kW)	P_{HP} (kW)	t/h	P_{GT} (kW)	P_{grid} (kW)	P_{ES} (kW)	P_{P2G} (kW)	P_{HP} (kW)
1	100.78	166.92	90.03	46.44	220.12	13	300.24	97.00	29.27	186.46	117.60
2	263.11	71.23	28.32	35.65	203.96	14	307.21	-11.01	-30.89	106.50	248.48
3	152.33	121.78	-33.00	113.81	224.69	15	238.05	89.87	2.30	101.22	228.73
4	277.37	144.19	17.54	144.35	231.62	16	357.01	-13.71	-51.56	169.48	222.07
5	359.77	69.21	6.22	45.99	238.11	17	159.86	109.49	45.52	132.75	264.01
6	242.37	169.83	10.03	32.80	239.84	18	168.63	156.94	17.57	73.34	274.79
7	197.80	200.28	-38.85	32.67	220.93	19	319.25	109.12	-35.34	129.96	244.21
8	166.05	280.29	22.63	208.02	205.19	20	197.11	150.63	-14.76	77.77	208.14
9	195.94	244.12	-16.68	64.32	263.37	21	245.45	139.88	-72.63	131.32	192.80
10	351.71	156.91	-44.72	219.42	226.95	22	159.77	149.97	61.39	85.82	193.42
11	343.77	197.84	4.44	292.79	341.15	23	317.16	95.17	8.34	300	144.11
12	354.47	173.00	40.85	178.92	332.81	24	142.10	193.30	17.28	237.52	331.24

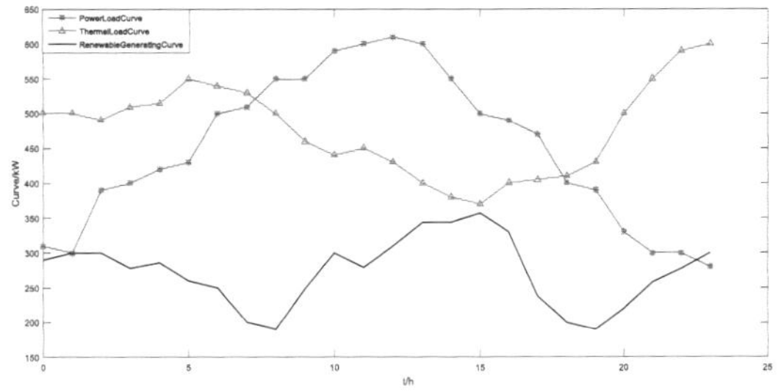

Figure 4. Power prediction curves.

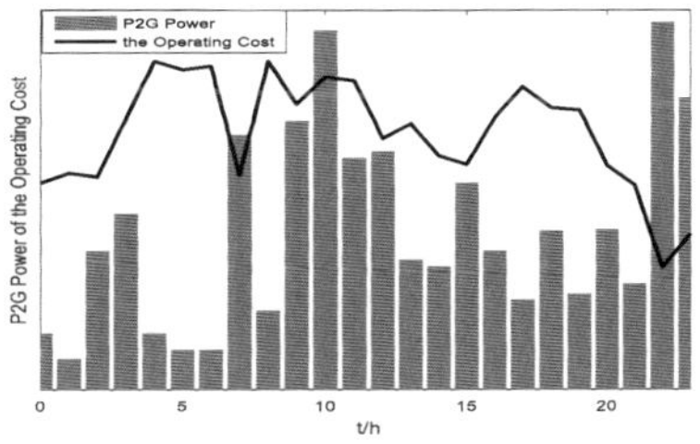

Figure 5. Relationship between cost and P2G.

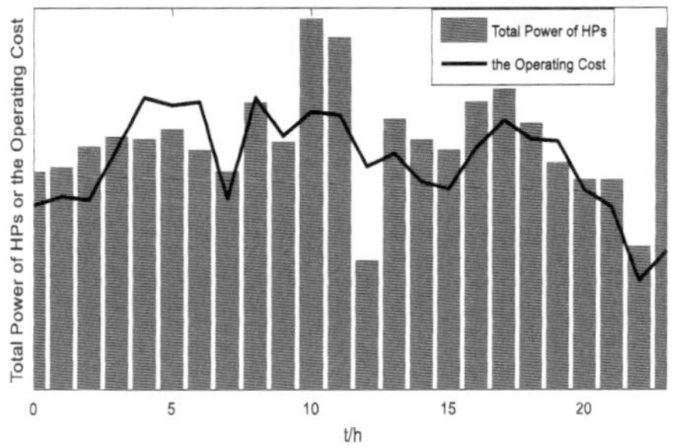

Figure 6. the Relationship between cost and demand response.

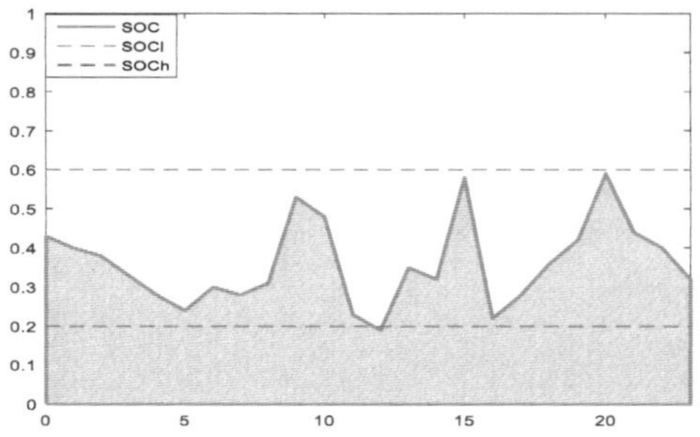

Figure 7. the storage state during scheudling period.

5. Conclusions

1. A new multi-energy mcirogrid, CHPG, is proposed, which consists of renwable generation, demand response, energy storage and P2G. Demand response in this paper, consists of economic guide for DGs, and priority scheudling of controllable HPs without affecting users' comfort. Demand response coorperated with P2G is also used as virtual storage, which can relieve the pressure for battery, and improve the economic and stable state of CHPG.

2. A frequency division method to smooth fluctuation is proposed ,considering the difference of reaction ability between battery and virtual storage. Virtual storage is allocated to smooth the high-frequency fluctuations to save the smoothing time and keep the security of battery and micro grid.

3. A two-stage collaborative global optimization design model is proposed. Encomic optimization is based on stability optimization, and the enconomic optimization result is the final optimal scheme of stability optimization. The simulation has verified that the proposed optimiaztion design model can minimize the operating cost by scheduling virtual storage and demand response, and keep the security.

References

[1] K X Perez, M Baldea and T F Edgar 2016 Soft-islanding a group of houses through scheduling of CHP, PV and storage *IEEE International Energy Conference(ENERGYCON)* pp1-6

[2] Min Jeong Kim, Hyoung-Yong Song, Jong-Bae Park, et al 2016 Optimization of CHP and thermal storage under heat demand *12th IEEE International Conference on Control and Automation(ICCA)* pp277-81

[3] Mbayer Abunku and Wim J. C. Melis 2015 Modelling of a CHP system with electrical and thermal storage *Power Engineering Conference(UPEC)* pp1-5

[4] Wang Dan, Ge Shaoyun, Jia Hongjie, et al 2014 A demand response and battery storage coordination algorithm for providing microgrid tie-line smoothing services *IEEE Transactions on Sutainable Energy* **5** No 2 pp476-86

[5] Sun Jianwei, Tang Shengwei, Wang Dan, et al 2014 An communication and battery performance evaluation of the demand response and battery storage coordination system potential for proving microgrid tie-line smoothing services *Advanced Materials Research* pp860-3

[6] GAHLEITNER G 2013 Hydrogen from renewable electricity: an international review of power-to-gas for stationary applications *International Journal of Hydrogen Energy* **38** No 5 pp2039-61

[7] GUANDALINI G, CAMPANARI S and ROMANO M C 2015 Power-to-gas plants and gas turbines for improved wind energy dispatch ability: energy and economic assessment *Applied Energy* **147** No 1 pp117-30

[8] Kakac S, Pramuanjaroenkij A and Zhou X Y 2016 A review of numerical modeling of solid oxide fuel cells *International Journal of Hydrogen Energy* **32** No 7 pp761-86

[9] KRAUSE T, ANDERSSON G, FROOHLICH K, et al 2011 Multiple energy carriers: modeling of production, delivery, and consumption *Proceedings of the IEEE* **99** No 1 pp15-27

IOP Conf. Series: Materials Science and Engineering **211** (2017) 012028 doi:10.1088/1757-899X/211/1/012028

Optimal placement of fast cut back units based on the theory of cellular automata and agent

Jun Yan* **and Feng Yan**

School of Electrical and Electronic Engineering, North China Electric Power University, Baoding, Hebei 071003, China

*Email: 594764469@qq.com

Abstract. The thermal power generation units with the function of fast cut back could serve power for auxiliary system and keep island operation after a major blackout, so they are excellent substitute for the traditional black-start power sources. Different placement schemes for FCB units have different influence on the subsequent restoration process. Considering the locality of the emergency dispatching rules, the unpredictability of specific dispatching instructions and unexpected situations like failure of transmission line energization, a novel deduction model for network reconfiguration based on the theory of cellular automata and agent is established. Several indexes are then defined for evaluating the placement schemes for FCB units. The attribute weights determination method based on subjective and objective integration and grey relational analysis are combinatorically used to determine the optimal placement scheme for FCB unit. The effectiveness of the proposed method is validated by the test results on the New England 10-unit 39-bus power system.

1. Introduction

Power system has been developed as one of the world's largest artificial network. With the increasing complexity of network structure, partial faults caused by improper handling may lead to cascading failures in the system, and eventually result in major system blackout accidents and bring out huge economic losses. After blackouts, guided by black-start power sources such as traditional hydroelectric generating sets, units without the self-starting capability will start generally, and then the entire grid will be restored. However, the output of hydro-power plants is generally smaller and its construction is harsh to the environment. Therefore, in the lack of traditional black-start power supply area, after technological transformation in conventional thermal power units, they are having fast cut back (FCB) function which could restore power rapidly, and act as black-start power.This technological transformation is of great significance[1, 2]. By FCB technology transformation of the units in different locations in the power grid, the impact on the subsequent recovery of the system is also different. So the research has critical significance to optimization layout scheme FCB [3].

In recent years, many researchers have carried out extensive research on system recovery, literature [3] uses discrete particle swarm optimization algorithm to achieve the goal of determining the framework of the network; the research focus of the literature [4] and the literature [5] is the optimization of the start-up sequence of the unit; literature [6] and [7] focus on the optimization of the reconstruction path. The optimization of unit start-up sequence and the optimization of the reconstruction path are simultaneously carried out by the literature [8] and [9]. However, the actual recovery process in the power grid [10] frequently, unpredictable situation such as over-voltage in lines, makes recovery process cannot be carried out according to pre-optimized scheme. Meanwhile, in the actual restoration process, the urgency of the situation makes dispatchers tense. So the order tends to be conservative and with local characteristics. These factors are not well considered in the above studies. Therefore, in order to determine the optimal and global recovery FCB unit layout schemes, they always differ from the actual situation.

In order to overcome above shortcomings, this paper introduces the concept of cellular automata [11] and agent [12], and the model of network reconfiguration is established. Based on the model, the process of network reconstruction in placement schemes of FCB units is simulated many times, and a number of indicators to evaluate the effectiveness of network reconfiguration is extracted. Several indexes are then defined for evaluating the placement schemes of FCB units. The attribute weights determination method based on subjective and objective integration and grey relational analysis are combinatorially used to determine the optimal placement scheme for FCB unit.

2.Important parameters involved in the process of network reconfiguration

Important parameters in the process of network reconfiguration include the following three parts: line's weighted value, importance degrees of nodes, and circuit switching failure rate. The calculation of line weights is about charging reactive power of path, and importance of nodes refer to the calculation method in literature [13].

This paper considers the operating overvoltage and the circuit switching failure rate caused by environmental factors. And in literature [14], the operating overvoltage is considered.

A lot of equipment in the system is exposed, so the probability of equipment failures will be greatly increased. In essence, the environmental dependence of outage means only the dependence exist in system components to the environment, rather than the correlation between the components and components. Harsh environment usually refers to unsuitable weather conditions such as wind, rain, snow and so on. If the failure frequency of components in normal or bad weather conditions are known, the system risk under the two climatic conditions can be evaluated separately, and the final risk index is weighted by the probability of the two weather conditions.

$$f_{to} = f_{ad} P_{ad} + f_{no}(1 - P_{ad}) \tag{1}$$

Equation (1) the failure frequency of f_{ad} and f_{no} are in the form of severe and normal weather conditions respectively.

3.Network reconfiguration model based on cellular automata and intelligent agents

3.1 Application of cellular automata in network re-configuration

In the power grid, the basic definition of cellular automata model for network reconfiguration is as follows:

1) Definition and state of cells. The cell is defined as the actual node and numbered cellular orderly arrangement from small to large. The state of the cell is represented by the state of the corresponding node in the network, such as: 1 indicates that the node is recovered; 0 indicates that the node is in a fault state.

2) Definition of cellular spaces. According to the actual situation of the power grid, this paper directly determines the space of each unit cell, and accurately reflects the distribution of the grid components.

3) Definition of cellular neighbors. In this paper, the definition of cellular floor is proposed, the other cell directly connected to a cell is defined as the first floor neighbor of the cell, the cell connected to a cell with crossing another cell is defined as the second floor neighbor, and so on.

4) Cellular transformation rule. The rule of cellular transformation is combined with the concept of intelligent agents, which is analyzed in detail in the next section.

3.2 Application of agents in network reconfiguration

All the bus nodes are all configured as bus agents. In the whole network, a facilitator agent (FAG) is configured to simulate the scheduling control of the intervention system. Next, the BAG(a bus agent) and FAG models are described respectively.

The purpose of BAG is to restore the load in corresponding bus. Each BAG has the following simple rules:

1) Target selection rules: If there are many restorative-available points or boundary points between the energized and de-energized area, then the BAG has to restore its own bus through the restorative-available point with the maximum capacity.

2) Communication rule: If the BAG is restored, the BAG tries to make a communication with its neighboring BAGs.

3) Joint rules: Joint mode exists between BAGs. When a BAG can only be returned to normal power through another BAG, there is a joint pattern between the two BAGs, that is, when the first level BAG failed to recovery, the current BAG directly labeled as recovery failure.

FAG is a special agent that facilitates the process of the multi-agent system, and it is not reflected in the restoration, but above the all of BAGs. After a power outage occurs, the FAG will start to perform the recovery process.

Simple rules of FAG are as follows:

1) Before each step of the recovery, parameters are obtained from each BAG, and nodes need to recovery also selected from these BAGs.

2) Through a re-operation at least, BAGs selected by FAG still fail to recover or another BAG fail to recover during the recovery process, FAG will select the new BAG to be restored.

3) When the top BAG of the BAGs in the combined mode fail to recover, the FAG will send the failure message directly to the rest of the BAGs.

4) All the BAGs in the concern list of FAG recover successfully or cannot recover, that means the end of network reconfiguration

Rules that FAG select BAGs to be restored are as follows:

According to the data in the critical matrix， FAG looks for three floors of neighbors from the restored BAGs. If there are power plant BAGs without recovery in it, they will be added to the list to be restored. If all the plant BAGs are recovery, FAG will find the neighbors in the two floors. If there is the load BAG without recovery, it will be added to the list to be restored. If the load BAGs are more than one, the node in max importance degree is selected by FAG, and then the recovery of BAGs is ended. If there are the load BAGs without recovery in the two floor, FAG will find the neighbors in the one floor and the max importance degree of BAGs, and it will be added to the list to be restored.

4.Evaluation system

The evaluating indicator of optimal placement scheme for FCB unit is defined: V1 is the numbers of recovery failed units, V2 is the times of line charging, V3 is the numbers of recovery failed lines, and V4 is the numbers of related recovery lines. In the literature [15], a method of determining the attribute synthetic weights of the subjective and objective integration is proposed. In this paper, this method is adopted to determine the weight of V1, V2, V3 and V4. Combined with the grey relational analysis[16]to evaluate the optimal placement scheme for FCB unit, the greater the grey weighted correlation degree, the better the scheme is. Figure 1 is the flow chart of the optimal placement scheme for FCB unit.

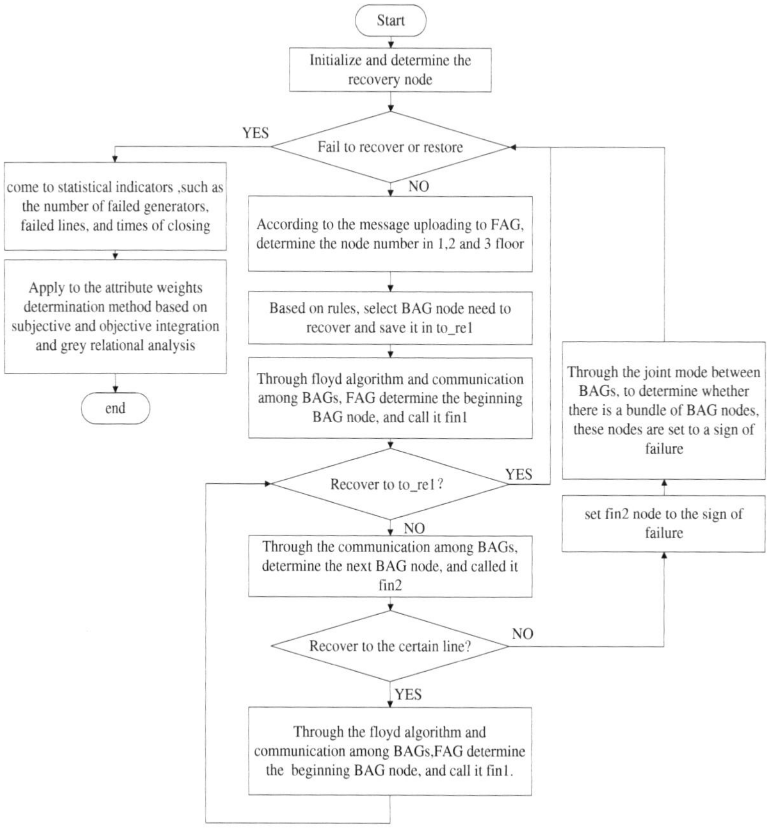

Figure 1. The flow chart of the network reconfiguration.

5.Simulation result

The effectiveness of the proposed method is validated by the test results on the New England 10-units 39-bus power system. Considering the generator node32 to the FCB unit node, Figure 2 illustrates the simulation results base on the network reconfiguration of the cellular automata. Red lines represent successful operation lines at the first time, yellow lines represent successful operation lines heavy after once re-operation, blue lines represent failed lines after once re-operation, and green lines represent not involved lines.

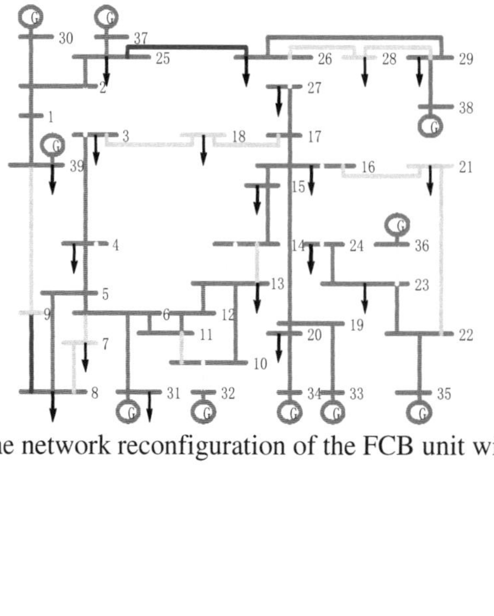

Figure 2. The network reconfiguration of the FCB unit with 39 nodes.

Particularly, on the way to recover node39, node9 failed to recover, and FCB send query command again. Node39 and node9 are not in joint state, so node39 cannot be considered fail to recover and node30 is selected as the new node to be recovered. The indexes involved in this reconstruction are shown in Table 1:

Table 1. Indicator values of the network reconfiguration.

V1	V2	V3	V4
0	43	2	35

The recovery process of the network reconstruction is shown in Table 2:

Table 2. Recovery target and recovery path at each time step.

Step k	recovery target node	Recovery path
1	13	32-10-13
2	31	13-12-11-6-31
3	4	6-5-4
4	39	5-8-9
		(node9 failed)
5	30	4-3-2-30
6	39	2-1-39
7	37	2-25-37
8	38	25-27
		(node27 failed)
9	15	4-14-15
10	33	15-16-19-33
11	34	19-20-24
12	36	16-24-23-26
13	35	23-22-35
14	27	16-17-27
15	38	27-26-29-38

According to the simulation method of network reconfiguration based on Cellular Automata, generator node30~39 are selected as the FCB nodes and execute the network reconfiguration 100 times. Indicators V1~V4 are obtained by statistics and averaged. Determine the weight of the four indexes according to the method of the subjective and objective integration. And then, grey relational analysis is combinatorially used to evaluate the optimal placement scheme for FCB unit.

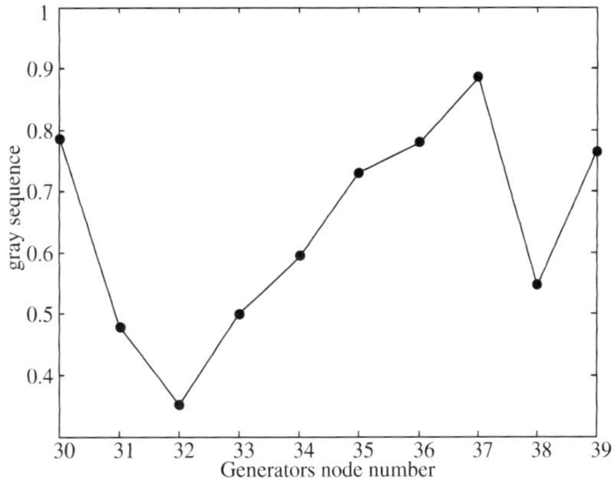

Figure 3. Weighted grey relational of the optimal placements scheme for FCB units.

From figure 3, transforming node37 into FCB node have largest correlation degree than node30 and node39, so node37 is the optimal transformation program of FCB unit. In addition, at the top of the diagram of the node30, node37 and node39 have the best effect than others under the diagram to transform to FCB units.

6.Conclusion

In this paper, we have presented a multi-agent consisting of several Bus Agents(BAGs) and a single Facilitator Agent (FAG). It is of great significance for the power restoration to transform generator units in proper position into FCB units with better black start performance.It also has important significance to guide the operator to make the corresponding FCB location plan. Future work should be directed towards expanding the proposed method to meet the requirements of the decentralized control environment, which has resulted from the world wide promotion of deregulation of electric power systems.

References

[1] SHEN Chong-qi, ZHOU Xin-ya, and YAO Jun 2007 *FCB function of coal-fired power unites and its application in power grid restoration* Shanghai Electric Power No 3 pp 251-4
[2] LI Jianhui, LU En, and WANG Ning 2013 *Optimal Placement of FCB Units for the Shortest Duration of a System Restoration* Guangdong Electric Power **26** pp 63-66 p 75
[3] LU En, NING Jian and LIU Haoming 2014 *An Ordinal Optimization Based Locating Scheme of Fast Cut Back Thermal Power Units* Power System Technology **38** pp 1216-22
[4] Pérez-Guerrero R, Heydt G T, Jack N J, et al 2008 *Optimal restoration of distribution systems using dynamic programming* IEEE Transactions on Power Delivery **23** pp 1589-96
[5] LIU Qiang, SHI Libao, ZHOU Ming, et al 2009 *Optimal start-up strategy of units during power system restoration* Electric Power Automation Equipment **29** pp 1-5
[6] LIN Zhenzhi and WEN Fushuan 1987 *A New Optimization Method for Determining Restoration Paths Based on Weighted Complex Network Model* Automation of Electric Power Systems **2** pp 740-1
[7] HAN Zhong-hui, GU Xue-ping and LIU Yan 2009 *Optimization of Restoration Paths Considering Unit Start-up Time Requirements at Early Stage of Power System Restoration* Proceeding of the CSEE **29** pp 21-26
[8] ZHU Dongxue, GU Xueping and ZHONG huirong 2013 *A Multi-Objective Optimization Method for Post-Blackout Unit Restoration* Power System Technology **37** pp 814-20
[9] ZHANG Can, LIN Zhen-zhi, WEN Fu-shuan, et al *Multi-Objective network reconfiguration strategy for power systems considering uncertainties* Journal of North China Electric Power University **39** pp 13-23
[10] TANG Si-qing, ZHANG Mi, LI Jian-she, et al 2006 *Review of Blackout in Hainan on September 26th ---Cause and Recommendations* Automation of Electric Power Systems **30** pp 1-7 p 16
[11] YU Qun, CAO Na and GUO Jianbo 2011 *Power System Self-Organized Criticality Simulation Model Based on Cellular Automata* Automation of Electric Power Systems **35** pp 1-5
[12] Nagata T and Sasaki H 2002 *A multi-agent approach to power system restoration* IEEE Transactions on Power Systems **12** pp 457-62
[13] LIU Yan and GU Xue-ping 2007 *Node Important Assessment Based Skeleton-network Reconfiguation* Proceedings of the CSEE **27** pp 20-27
[14] ZHANG Yu-qiong and GU Xue-ping 2005 *Computation and Statistical of Operating Over-Voltages in Black-Start of Power Systems* TRANSCATIONS OF CHINA ELECTROTECHNICAL SOCIETY **20** pp 92-97
[15] WU Jian, LIANG Chang-yong and LI Wen-nian 2007 *Method to determine attribute weights based on subjective and objective integrated* **29** pp 383-7
[16] SI Shou-kui and SUN Xi-qing 2011 *Mathematical Modeling* National Defence Industry Press

IOP Conf. Series: Materials Science and Engineering **211** (2017) 012029 doi:10.1088/1757-899X/211/1/012029

An energy internet-based distribution system morphology

W Yi[1]*, P Wang[2], X W Hu[1], X N Wang[1] and F Shi[1]

[1] School of Electrical Engineering, Shandong University, China
[2] Henan Electric Power Research Institute, China

*Email:yiwei2450@qq.com

Abstract. The future power distribution system will play an important role in the future energy system - energy Internet. With an eye to the diversification of energy and power fusion to the distribution system form, the prospect of the future integrated power distribution system for energy Internet is proposed, from the three aspects of structure, equipment form and industrial form: a multi-level AC-DC ring network as a framework, including the P2P interconnection and DC information satisfied power grid integrated network structure, is given; the concept of energy exchanger, the concept of energy server is put forward, and on this basis, the future distribution system of the equipment is prospected; the future distribution of the three types of industrial forms is analysed.

1. Introduction

In recent years, with the development of micro-grid, virtual power plant, active distribution network technology[1,2], the development of a variety of energy and electricity in traditional energy systems is generally carried out on the traditional power generation side. The main force of energy structure change - renewable energy to a large number of distributed access to the distribution network, the introduction of the exchange side of the difficult to count the interface; electric vehicle technology, small gas turbine, electric gas, hot and cold power supply, energy storage and other technologies[3-5], a variety of energy in the power system of the power distribution side of the interaction, resulting in such as random fluctuations, two-way trend and other new features[6], which makes the distribution network the most diverse area compared to traditional power grids and the distribution network will evolve into a key part of the future power system for the energy Internet. This question has led to many scholars thinking: [7] explores the structure and characteristics of future power distribution systems, pointing out that intelligent power distribution systems will be an important support for integrated energy systems; [8] suggest that future distribution networks will become power integration, multi-stream fusion of the energy Internet, and look forward to its key technical areas.

Therefore, it is of great forward-looking significance to research and look forward, combined with the concept of energy Internet, focusing on a variety of energy and power fusion complementary, to the distribution system morphology; however, such research is rare[9,10]. On the basis of the existing research results, this paper then draws lessons from the related technical characteristics of the energy system and information system, and looks forward to the future distribution system form from the three aspects of structure, equipment form and industrial form.

2. The morphology of future integrated distribution system oriented to the energy internet

The future integrated power distribution system for energy internet will be prospected from 3 aspects according to the correlation.

2.1. The structure of the future integrated distribution system

The essential difference between the future distribution system and the traditional distribution grid is the change in operation mode and equipment. The structure design of distribution grid and its macro topology are the most important research basis for the function and characteristics of the system.

In the network architecture, the future distribution grid will become a pattern with unified and standard interfaces in the high voltage and medium voltage, which is characterized by multi-layer AC / DC hybrid based on ring bus and flexible ad hoc network based on complex network theory. And the hierarchical structure is its main feature. The ring structure whose basic structural unit is regional DC and AC ring bus can be easily connected to various power, load and energy storage device, constituting the basic structural unit of one level and the device support for energy management unit as well. In normal operation, the internal bus of the unit works in a loop-close way with the ring and net structure. At the same level, different units are linked up to each other through the soft normally open points and other power electronic devices. What's more, power transfer or bidirectional power switch control can be realized by different ring buses. The level structure is set to 4 layers, called the regional distribution system, the local integrated power distribution system, the integrated micro grid and the DC information grid.

At the same time, in order to meet the needs of the energy Internet to achieve the goal of innovative development, the topology structure of the future integrated distribution system should draw lessons from the topology structure of the Internet which is typical for community, small world, fractal, and its realization method. Between the main bodies of the energy trade in the power distribution side, a more diversified connection mode should be established in order to serve for the power balance and industrial form under the energy Internet. The paper [11] designs a topology structure of energy Internet based on the simple hierarchy structure, and it introduces P2P connections within and across all levels. Besides, it is proved that the structure can enhance the robustness of the system. However, this method has some limitations from the point of the distribution grid, which only considers the structure of the simulation information system interconnection, and does not combine with the actual situation of the power system at the level of interconnection.

Combining with the characteristics of distribution grid and synthesizing and improving the existing research results, this paper designs a new topology structure of future distribution system, which takes the hierarchical AC / DC link network structure as frame and introduces the P2P connection and the DC information grid into the low voltage distribution system.

This structure has both advantages of the structure of the ring network and the P2P connection, which ensures that the high and the medium voltage distribution grid layers have a certain degree of interconnection and greatly enhances the interconnection and robustness of the low voltage distribution grid layer.

And it eventually forms the approximate full connected network in the whole distribution system level and constitutes the basic unit of a higher level of integrated power grid the energy Internet.

2.2. Equipment form for the future integrated distribution system

In this paper, the following 4 principles are followed in the prospect of the form of equipment for the future integrated distribution system.

- The equipment for the future integrated distribution system should be as far as possible with the characteristics of physical information fusion
- According to the control framework of the future distribution system which is typical for the overall coordination and the regional autonomy, the equipment should include not only the centralized controlling device

- According to the control framework of the future distribution system which is typical for the overall coordination and the regional autonomy, the equipment should include not only the centralized controlling device, but also the hierarchical autonomous control device.
- For the equipment which is similar in the function but different in levels, they are generally not given individual definition on the concept of the device but only the supplementary instructions to the level according to the way of definition in the electric power system.
- Consider both pure power distribution equipment and multi energy fusion equipment. As for the multi energy integration equipment, focus on those having interface, control and other interactive relations with power distribution system.

Therefore, a kind of equipment of future integrated distribution system which includes many new device concept is designed. For example, it perfects the concept of energy resources and puts forward the concept of energy converter. The device is composed as shown in Figure 1. In Figure 1, the rounded rectangles represent the outlook principle and the interlocking circle represents the characteristics of the device. Meanwhile, the smallest circles represent the specific device. The specific introduction is as follows.

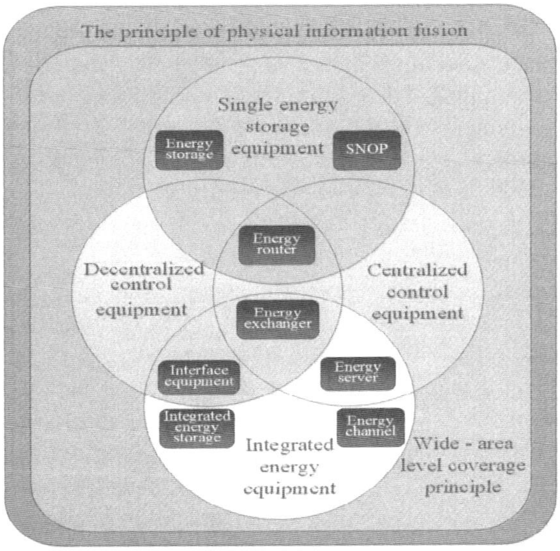

Figure 1. Equipment of future integrated distribution system.

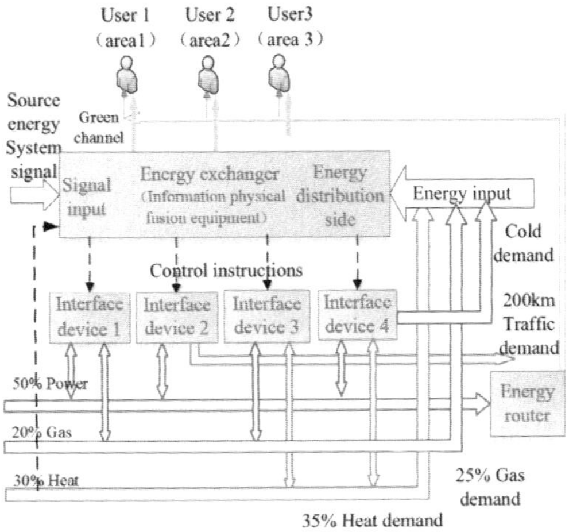

Figure 2. Conception of energy exchanger.

- **Electric energy storage and comprehensive energy storage.** In the future distribution system, the importance of energy storage device will be further strengthened. It will not only still play a buffer role in maintaining the balance of power, but also realize the sub packet transmission, integrated transmission and heterogeneous interconnection of the energy. And it can block the failure of the integrated energy system and realize the fine management of energy management unit among all levels.
- **Energy server.** The energy server in this paper is a concept of cyber-physical fusion device, whose information section is the centralized control center of the future distribution system, and the core of the system is analysis platform of big data and cloud computing. With rational use of the internet concept, its function can be designed as follows. Analyse the security posture of the system and dispose the energy allocation among the various autonomous regions through big data cloud computing in the whole distribution system level. And produce, store and distribute the energy in the high voltage distribution grid. In addition, it can deal with the response of complex demand side and complex business that is difficult to calculate in the bottom control equipment.
- **Energy exchanger**. Energy exchanger in this paper is a kind of typical cyber-physical fusion device. The information section of it is in charge of the comparison of energy demand information and energy input. At the same time, it's responsible for the control of all types of energy interface equipment to make various types of energy convert mutually to a certain proportion, ensuring the energy ratio meets the energy demand after conversion. The physical part is responsible for the distribution of non-electric energy to the energy channels of the user or the region on the next level while the electric energy is distributed by the energy router. Its conception is shown in Figure 2.
- **Energy router.** In this paper, the concept of energy router is similar to that in [12], which consists of solid-state transformers, integrated energy storage and regional coordination control system. And it is to be the core equipment of the physical layer connection and autonomy management of regional information in the future integrated distribution system.
- **The interface equipment between electric energy and multi-energy.** Interface is a device for the conversion between different energy forms. The future integrated distribution system will be mainly concerned with the interface equipment between power and other energy forms, such as inverter, electric vehicles and so on, which are the key to achieving the goal of energy integration.
- **Energy channel**. Energy channel is a kind of multi energy transmission channel. From the point of view of electric energy, electric energy will be the main energy source in the Internet. And the distribution network under the control of the energy router is responsible for distribution Therefore, it is designed to achieve chemical energy, heat and other energy packaging long distance flexible transmission, to achieve the energy form of isolation transmission, to improve transmission efficiency and other functions.
- **SNOP**.SNOP in this paper is a new type of controllable power electronic device, which is based on full control power electronic devices to achieve accurate control of the active power, reactive power and electrical decoupling in the future.

In the future, the equipment system and the traditional distribution system of power distribution system in the future constitute a form of the future power distribution system. It is in the concept of the energy Internet, and has a clear distinction with the smart grid era.

2.3. Industrial form of integrated distribution system in the future

In the context of the energy Internet, focusing on the distribution side of the user directly, the future of the integrated power distribution system may be in the energy information infrastructure industry, the

Internet + focus on business model and the Internet + 3 aspects of decentralized business model to expand its industrial form.

2.3.1. Energy information infrastructure industry. Future integrated distribution system is integrated with other energy physical interface device and information coordination and control equipment, formed the new structure and form of equipment, in the energy and information infrastructure industry brought great opportunities for the equipment and the reform of the system. The traditional distribution system and other source system networking transformation, the popularity of the interface devices between a variety of energy demand, new physical information fusion energy equipment R & D, production, alternative, maintenance, centralized control center and regional control system technology research and development and to be built, contains a huge market potential, gave birth to their new industrial structure.

2.3.2. Internet + centralized B2B/B2C business model .Internet + power grid can significantly improve the rate and efficiency of the exchange between power supply and demand for electricity supply and demand, the contradiction between supply and demand of electricity supply and demand has some mitigation effect; simultaneously, users can choose to make it possible. In the future, the market will gradually expand. Large users across the transmission system directly with the power plant B2C (business-to-customer), power plant and the general user for B2B (business-to-business) trading, power grid companies charge transmission and distribution services, and through scheduling to balance the distribution costs as its main additional profit model, but also through the sale of reactive power, standby and other ancillary services to expand business structure. B2C function, to guide users through the demand side response to efficient use of energy, reduce the cost of purchasing power to earn the difference as the main profit model. The characteristics of a certain number of users to enhance their market power to become a proxy user to participate in market competition and interaction.

2.3.3. Internet + P2P business model. The future distribution system has the characteristics of high level of interaction and real time, so it is more innovative in the industry form. In the future distributed generation, storage and electric vehicles can be carried out. In the future, users and operators can not only carry out a simple electric two-way transaction, but also can be put into other users. Networking business P2P transactions.In addition, the demand response technology should be developed to the future distribution system stage, where the system should be highly automated instead of dependent on manual operation. The response program is automatically triggered by the integrated microgrid control system. The future integrated power distribution system can provide users with personalized automatic manual combination of demand response choice.

However, in the future integrated power distribution system, there are still a lot of technical problems to be resolved in several fields, for example, the application of internet information technology field, the planning and reliability of future integrated power distribution system field, the device technical related to future integrated power distribution, the operation control technology field related to integrated power distribution, and so on.

3. Conclusion

Energy Internet is recognized as the trend of the future energy system development, the energy distribution network in the context of the concept of how to expand, how to transform the form, and become a problem. In this paper, the future integrated distribution system form is discussed, from the structure, equipment and industry 3 aspects: a multi-level AC / DC link network is proposed, which includes the P2P interconnection and direct current information. The energy converter is presented, and the energy server is improved, and the energy and information flow characteristics are analysed.

Reference

[1] Planas E, Andreu J, Gárate JI, Alegría IMD and Ibarra E 2015 *Renew. Sust. Energ. Rev* **43** pp 726-49
[2] Mnatsakanyan A and Kennedy SW 2015 *IEEE. T. Smart. Grid* **6(1)** pp 230-7
[3] Xu X, Jia H, Chiang HD, Yu DC and Wang D 2015 *IEEE.T.Power.Syst* **30(3)** pp 1212-21
[4] Varone A and Ferrari M 2015 *Renew. Sust. Energ. Rev* **45** pp 207-18
[5] Zakeri B and Syri S 2015 *Renew. Sust. Energ. Rev* **42(C)** pp 569-96
[6] Lopes JAP, Hatziargyriou N, Mutale J, Djapic P and Jenkins N 2007. *Electr.Pow.Syst.Res* **77(9)** pp 1189-203
[7] Wang C, Wang D and Zhou Y 2015 *Automat.Electron.Power. Sys* **39(9)** pp 1-9
[8] Zhao MA, Zhou X, Shang Y and Zhou L 2015 *Zhongguo Dianji Gongcheng Xuebao/proceedings of the Chinese Society of Electrical Engineering* **35(6)** pp 1289-98
[9] Tianjiao PU, Liu K, Chen N, Xianjun GE, Jiancheng YU and Dan W 2015 Zhongguo Dianji *Gongcheng Xuebao/proceedings of the Chinese Society of Electrical Engineering* **35(14)** pp 3511-21
[10] Zhang X, Li J and Fu H 2015 *Zhongguo Dianji Gongcheng Xuebao/proceedings of the Chinese Society of Electrical Engineering* **35(14)** pp 3532-40
[11] Zhao H, Cai W, Wang J and Jia S 2015 *Diangong Jishu Xuebao/transactions of China Electrotechnical Society* **30(11)** pp 30-6
[12] Huang AQ, Crow ML, Heydt GT and Zheng JP 2011. *P.IEEE* **99(1)(1)** pp 133-48

Acknowledgments

This work is supported by National Natural Science Foundation of China (No. 51407109).

IOP Conf. Series: Materials Science and Engineering **211** (2017) 012030 doi:10.1088/1757-899X/211/1/012030

Drying characteristics of whole Musa AA group 'Kluai Leb Mu Nang' using hot air and infrared vacuum

C Kulketwong[1]*, D Thungsotanon[1] and N Suwanpayak[2]

[1] Department of Engineering, King Mongkut's Institute of Technology Ladkrabang, Prince of Chumphon Campus,Chumphon, 86160, Thailand
[2] Department of General Science, King Mongkut's Institute of Technology Ladkrabang, Prince of Chumphon Campus, Chumphon, 86160, Thailand

*Email: chompoonud.ku@kmitl.ac.th

Abstract. Dried Musa AA group 'Kluai Leb Mu Nang' are the famous processing goods of Chumphon province, the south of Thailand. In this paper, we improved the qualities of whole Musa AA group 'Kluai leb Mu Nang' by using the hot air and infrared vacuum (HA and infrared vacuum) drying method which has two stages. The first stage of the method is the hot air (HA) and hot air-infrared (HAI) drying for rapidly reducing the moisture content and the drying times at atmospheric pressure, and the second stage, the moisture content, and color of the samples can be controlled by the HA and infrared vacuum drying. The experiment was evaluated by the terms of firmness, color change, moisture content, vacuum pressure and energy consumption at various temperatures. The results were found that the suitable temperature of the HAI and HA and infrared vacuum drying stages at 70°C and 55°C, respectively, while the suitable vacuum pressure in the second process was -0.4 bar. The samples were dried in a total of 28 hrs using 13.83 MJ/kg of specific energy consumption (stage 1 with 8.8 MJ/kg and stage 2 of 5.03 MJ/kg). The physical characteristics of the 21% (wb) of dried bananas can be measured the color change, $L^* = 38.56$, $a^* = 16.47$ and $b^* = 16.3$, was approximate the goods from the local market, whereas the firmness of them was more tender and shown a value of 849.56 kN/m^3.

1. Introduction

Banana is an important food and cereal after rice and corn for the world's human population. Thailand has been ranked as the third largest producer of bananas in Asia [1], has more than 30 varieties: Kluai Khai, Kluai Roy Wee, Kluai Nam Wah, Kluai Hom and Kluai Leb Mu Nang and M. acuminate. Kluai Leb Mu Nang is one important Thai banana cultivars. There is native and well known as local economic bananas at Chumphon province in southern Thailand. However, the productions of banana are needed an appropriate post-harvest technology and agricultural engineering to prolong the shelf life fruits in the storages and logistics process. Ripe banana is perishable and deteriorates rapidly after harvesting, especially [2]. Drying technology is wildly used in food industry preservation which has several techniques for the qualities of food dried production. Food drying is also a method of food preservation to remove moisture and water such as vegetables, fruits, and another agricultural production. Meanwhile, Drying is one of the technologies that can be used to preserve banana, there are many techniques to dry banana. Most common techniques for drying banana are HA drying [3] which is the oldest method of food preservation. In this process, heat transferred from the HA to the sample and removal of water to the air by convection. In addition, infrared (IR) drying is based on the operation of IR wavelength radiation from the heat source, and heat transported into the internal structure of the vegetables and fruits, thus increases its temperature [4]. Furthermore, the combination of HA and IR drying method is gaining popularity in food preservation because of uniform heating and inherent over only HA drying. However, drying methods have been developed for high qualities of food preservation product [5]. At present, HA and infrared vacuum drying are useful in food preservation for drying of thin slices [6] and the whole of fruits [7], [8] which are widely applied in agriculture and food technology. Therefore, the drying processes have been developed to produce the quality foods conservative productions for increasing efficiency and value.

In this study, the experiment was investigated to the terms of drying rate, color change, moisture content, vacuum pressure and energy consumption at various temperatures to improve the qualities of whole Musa AA group 'Kluai leb Mu Nang' by using the HA and infrared vacuum drying method.

2. Hot air and infrared vacuum drying setup

2.1. Structural dryer

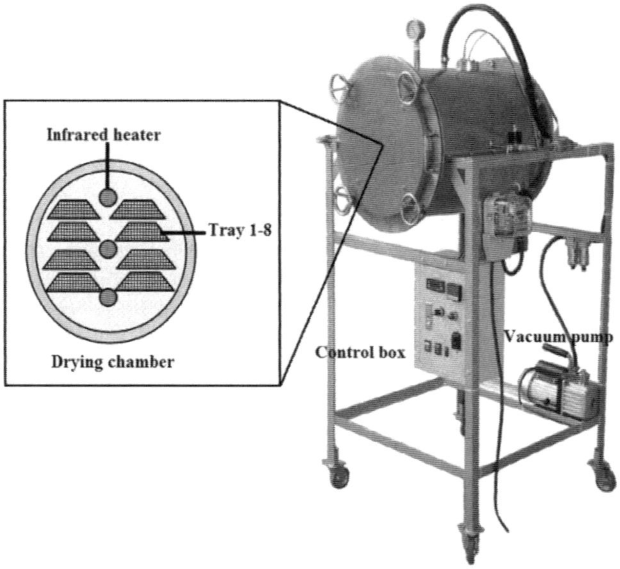

Figure 1. The combined of HA and infrared vacuum system.

Kluai Leb Mu Nang' with HA and infrared vacuum drying setup was developed at King Mongkut's Institute of Technology Ladkrabang Prince of Chumphon Campus (KMITL PCC) laboratory which has volume of stainless steel (food grade) chamber of 65.3 L, the chamber has 8 trays for the bananas drying in the uniform vertical dimension in which the sample can be received the external heat into the chamber thoroughly as shown in Figure 1. In addition, the apparatus have a bucket for water released from banana with 5% of the volume.

2.2. Infrared heater

An infrared heater is a high efficiency of heat transfer radiation which can be propagated into absorbed materials. In this experiment, we used an infrared heater 3 rods having a horizontal burning position which has diameter with 1.7 cm and length of 30 cm, the input power of 650 W, 220 volts. An infrared radiation is ranged from 3 to 10 mm wavelength by convection and conduction into the samples from the surface to the internal of bananas.

2.3. Vacuum pump

Vacuum is generated by a vacuum pump, depending on the requirements of suction power and pressure conditions. In this case, the vacuum pump (Single stage vacuum pump: RT, 185 L/min, 220V/50Hz) was connected to under the oven chamber as shown in Figure 1.

2.4. Materials

The bananas were produced by agriculturist in Pathiu district, Chumphon province, Thailand. We selected certainly ripe level (all yellow) and obtained from a local market, stored at room temperature of 27 °C. The samples were peeled by hand which have diameter of 2 to 2.5 cm and length with 11 to 12 cm. There were washed in 3% sodium chloride (NaCl) solution to control color, after that the samples were uniformly placed in a single layer on the trays for 10 minutes before drying process to measure initial moisture content of the fresh bananas, was determined using hot air oven (MEMMERT Model600) at 105 °C for 24 hours.

2.5. Drying process

In this experiment, the experimental conditions of drying (Figure 2) can be divided into two stages. The first stage, the drying characteristics of bananas (whole Musa AA group 'Kluai Leb Mu Nang') were combined the airflow and infrared radiation dryer system, temperature and moisture content can be controlled by using airflow 0.3 L/min of 60°C, 70°C and 80°C at atmospheric pressure to evaluate optimal temperature for HA and HAI drying. The second stage, the optimal temperature of the first stage can be used the standard to test the properly of vacuum pressure for HA and infrared vacuum drying of whole bananas which varied the temperature of 50°C, 55°C and 60°C with vacuum pressure range of -0.1 to -0.6 bar. Finally, the humidity, color, bananas texture, energy efficiency and physical properties were analysed to find the optimal conditions of HA and infrared vacuum drying. In addition, the study can be varied airflow at 0.0, 0.3 and 0.6 L/min.

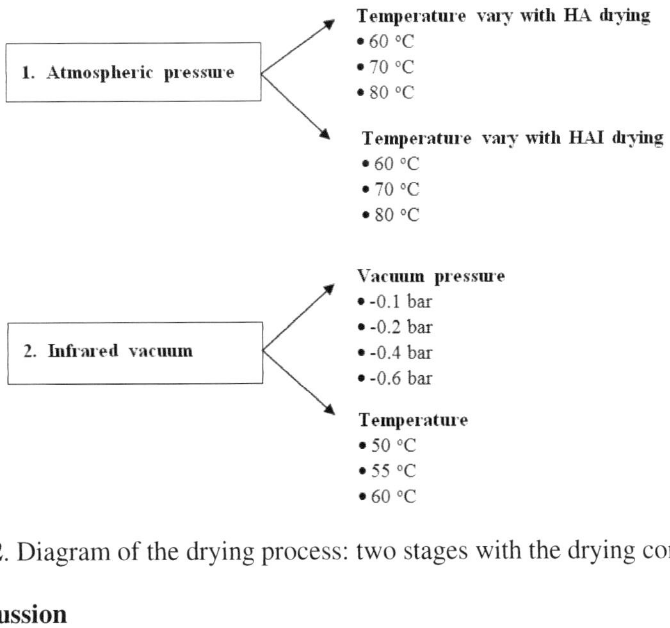

Figure 2. Diagram of the drying process: two stages with the drying conditions.

3. Results and discussion

A comparison between HA and HAI drying characteristics, we found that the HAI of 70°C can be preserved the qualities of whole bananas higher than HA drying because the drying rate with the slope in the graph between moisture content and time presented the least square linear regression as shown in Figure 3. Furthermore, the temperature drying of 60°C and 80°C in the chamber cannot be uniformly distributed temperature of the total trays (8 trays) effecting to color and texture of the samples and over dehydrated which made bananas dried to decrease quality, respectively. Therefore, the HA and HAI drying results showed the suitable temperature of 70°C at atmospheric pressure which is standard error estimate (SEE) of 1.085 less than another temperature of 4 hrs drying time was shown in Table 1.

Table 1. Drying rate of HA and HAI methods.

HA Drying								
Drying Time	4 hrs				6 hrs			
Temperature (°C)	y_0	a_0	R	SEE	y_0	a_0	R	SEE
60	73.529	-0.115	0.407	0.425	73.527	-0.114	0.481	0.466
70	73.568	-0.415	0.966	0.181	73.576	-0.422	0.980	0.187
80	73.393	-1.825	0.894	1.500	73.346	-1.790	0.890	2.061
Drying Time	8 hrs				10 hrs			
Temperature (°C)	y_0	a_0	R	SEE	y_0	a_0	R	SEE
60	73.524	-0.112	0.527	0.514	73.516	-0.109	0.550	0.570
70	73.572	-0.419	0.983	0.219	73.528	-0.403	0.934	0.528
80	73.358	-1.796	0.895	2.532	73.415	-1.817	0.904	2.936
HAI Drying								
Drying Time	4 hrs				6 hrs			
Temperature (°C)	y_0	a_0	R	SEE	y_0	a_0	R	SEE
60	73.519	-0.831	0.288	4.542	73.248	-0.628	0.331	4.021
70	73.579	-1.976	0.948	1.085	73.579	-1.976	0.948	1.483
80	73.937	-3.409	0.977	1.229	73.310	-2.939	0.978	1.409
Drying Time	8 hrs				10 hrs			
Temperature (°C)	y_0	a_0	R	SEE	y_0	a_0	R	SEE
60	73.169	-0.588	0.409	3.722	73.141	-0.578	0.485	3.564
70	73.573	-1.973	0.953	1.768	73.463	-1.932	0.964	1.813
80	72.701	-2.635	0.978	1.567	72.003	-2.373	0.976	1.810

Hence, the trays were alternated to above and below by producer for thoroughly qualities and physical properties of the bananas drying. Therefore, the suitable temperature of Klaui Leb Mu Nang drying with the atmospheric pressure was 70°C in which have dried of 48 hrs that accorded with the ordinary attribute which is corresponding to the Klaui Leb Mu Nang drying production of agriculturist at Chumphon province. The drying temperature and heating method of the first experiment at the atmospheric pressure were evaluated with the least square regression of drying rate in each period was shown in Figure 3 and the least square is shown by equation (1).

$$Moisture = y_0 + a_0 Time \quad (1)$$

where

Moisture is the wet basis moisture content (%wb).
y_0 is the initial moisture content (%wb).
a_0 is the slope of line (%wb/hr).

The a_0 was estimated by the drying rate for various temperature and technical heating. Table 1 showed the interpretations of drying condition and found that the drying rates increased, the temperature increased too. Figure 4 was shown the effect of HA and infrared vacuum drying of whole banana sample which receives the radiation from infrared heater and hot airflow which used to control the surface color and the suitable temperature for qualities of the bananas drying at vacuum pressure of -0.1 and -0.4 bar.

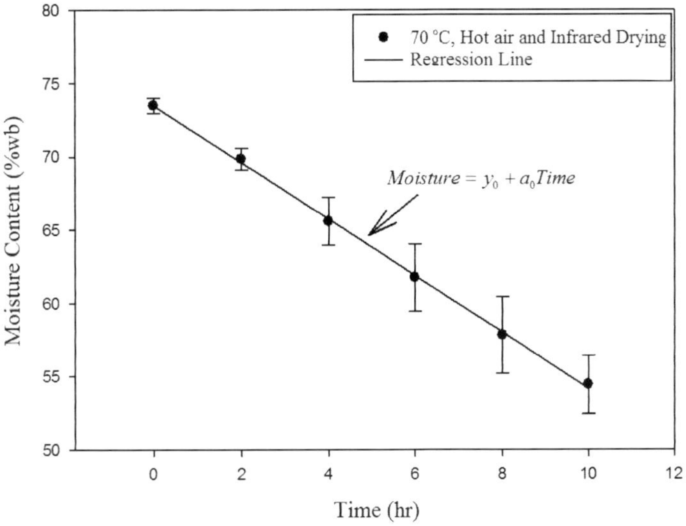

Figure 3. Graph of the drying rate of HAI drying.

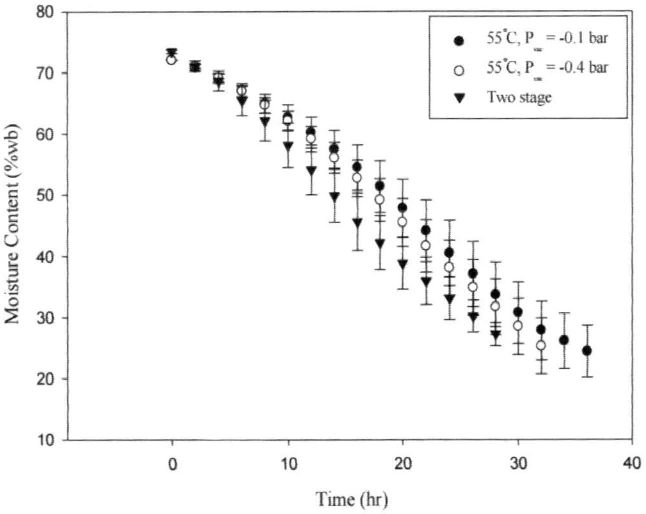

Figure 4. A comparison of the drying rate at -0.1 bar and -0.4 bar.

The experimental results of the HA and infrared vacuum drying have shown a suitable state for heating temperature of 55°C at -0.4 bar of vacuum pressure within 28 hrs (▾) drying time in Figure 4, which is the samples could rapidly dehydrated of moisture content by using SEC of 13.83 MJ/kg and the percentage of damaged sample was 18.2% less than the drying condition with -0.1 bar at the same temperature. The physical characteristics of the 21% (wb) of dried bananas can be measured the color change, $L^* = 38.56$, $a^* = 16.47$ and $b^* = 16.3$, was approximate the goods from the local market, where as the firmness of them was more tender and shown a value of 849.56 kN/m^3. The standard qualities banana are the color change, $L^* = 39.53$, $a^* = 11.20$, $b^* = 29.39$, and firmness of 990.91 kN/m^3 have shown in Table 2.

Table 2. The color change and firmness of the methods.

Drying Technique	L*	a*	b*	Firmness (kN/m^3)
HAI	26.83	9.76	26.22	943.73
HA and Infrared vacuum	38.56	16.47	16.30	849.56
Standard sample from the local market	39.53	11.20	29.39	990.91

However, HA and infrared vacuum drying technique are expensive for the system setup in which the temperature and vacuum pressure can be controlled by automatics the system, even if the drying time less than other the hot air dryer of the agriculturist, but the qualities of the dried samples can be preserved the same as dried banana production on the market by the dryer which do not alternate the trays during drying process.

4. Conclusion

The combined drying process of the whole Musa AA Group 'Kluai Leb Mu Nang' with the HA and infrared vacuum drying characteristics was determined in this work. The experimental results have shown the effect of temperature, infrared heating radiation and vacuum pressure on the whole bananas dried, the optimized drying temperature was found of 70°C at atmospheric pressure with HA and HAI dryer within 4 hrs period time. After that the samples have been dried by the infrared vacuum with the vacuum pressure of -0.4 bar using a suitable temperature was 55°C for 24 hrs. Therefore, The qualities of whole Musa AA Group 'Kluai Leb Mu Nang such as color, moisture content, temperature, texture and physical properties can be preserved by the combined HA and infrared vacuum technical drying for banana and food preservation. However, this experimental apparatus is a primary combined HA and infrared vacuum which has been developed for higher efficiency food preservative dryer in the future.

5. References

[1] Kanchanapoom K and Koarapatchaikul K 2012 In vitro induction of tetraploid plants from callus cultures of diploid bananas (Musa acuminata, AA group) 'Kluai Leb Mu Nang' and 'Kluai Sa. *Euphytica* 183 pp 111–7

[2] Swasdisevi T, Devahastin S, Ngamchum R and Soponronnarit S 2007 Optimization of a drying process using infrared vacuum drying of Cavendish banana slices *Songklanakarin J. Sci Technol* **29** pp 809-16

[3] Nimmol C, Devahastin S, Swasdisevi T and Soponronnarit S 2007 Drying and heat transfer behavior of banana undergoing combined low-pressure superheated steam and far-infrared radiation drying *Appl. Therm. Eng* **27** pp 2483–94

[4] Antal T 2015 Comparative study of three drying methods: freeze, hot air-assisted freeze and infrared-assisted freeze modes *Agron. Res* **13(4)** pp 863–78

[5] Praveen Kumar DG, Umesh Hebbar H, Sukumar D and Ramesh MN 2005 Infrared and hot-air drying of onion *J. Food Process Preserv* **29** pp 132–50

[6] Beigi M 2016 Energy efficiency and moisture diffusivity of apple slices during convective drying *Food Sci. Technol. (Campinas)* **36(1)** pp 145-50

[7] Michalska A, Wojdyło A, Lech K, Łysiak Grzegorz P and Figiel A 2016 Physicochemical properties of whole fruit plum powders obtained using different drying technologies *Food Chem* **207** pp 223–32

[8] Antal T and Kerekes B 2016 Investigation of Hot air- and infrared-assisted freeze-drying of apple *J. Food Process Preserv* **40** pp 257-69

Acknowledgments
The authors would like to thank King Mongkut's Institute of Technology Ladkrabang Prince of Chumphon Campus for providing the financial support and also thank Department of Engineering for support the research laboratory.